TOXIC METALS

TOXIC METALS

Pollution Control and Worker Protection

Marshall Sittig

NOYES DATA CORPORATION
Park Ridge, New Jersey, U.S.A.
1976

No liability is assumed for damages resulting from the use of the information contained in this book.

While attempting to give helpful advise, this book in no manner preempts what is rightfully the domain of a licensed physician specializing in toxicology and the ensuing therapeutics.

Any handler of toxic and hazardous substances should make himself familiar with the contents of one of the "First Aid in Poisoning" manuals available (usually free of cost) from federal and state governments. The phone number of the nearest poison control center, where information can be obtained on the effects of specific toxic substances, should be kept posted in a conspicuous place at all times.

Library of Congress Card Number: 76-24142
ISBN: 0-8155-0636-8
Printed in the United States

Published in the United States of America by
Noyes Data Corporation
Noyes Building, Park Ridge, New Jersey 07656

FOREWORD

This is the thirtieth volume of our Pollution Technology Review Series. It is a practical book about those metals and their compounds that are most likely to cause poisoning—in industry while being processed, and in the general environment while the disposal of their process-effluents is being implemented.

What is most important and effective in the prevention of toxic metals poisoning is thorough removal of the toxic agents in a manner that positively precludes ingestion or contact by all potential victims, and therein lies the emphasis and main endeavor of this book.

The book is based on government-sponsored reports (often elaborated by highly qualified personnel from industrial companies), U.S. patents, and on pertinent articles in authoritative journals. In the United States, we are fortunate in receiving direct help from the numerous surveys, together with active research and development programs that are being supported by the Federal Government to help industry control its wastes and harmful emissions.

In this book are condensed vital data that are scattered and difficult to pull together. Important techniques are interpreted and explained by actual case histories. This condensed information will enable you to establish a sound background for action.

Advanced composition and production methods developed by Noyes Data are employed to bring our new durably bound books to you in a minimum of time. Special techniques are used to close the gap between "manuscript" and "completed book." Industrial technology is progressing so rapidly that time-honored, conventional typesetting, binding and shipping methods are no longer suitable. We have bypassed the delays in the conventional book publishing cycle and provide the user with an effective and convenient means of reviewing up-to-date information in depth.

The Table of Contents is organized in such a way as to serve as a subject index and provides easy access to the information contained in this book. Bibliographic reference lists follow each chapter in order to provide the reader with easy access to further information on this timely topic.

CONTENTS AND SUBJECT INDEX

INTRODUCTION

The general question of toxic substances is now at the forefront of public attention. The public seems to go through a series of crises—first there was the general question of air and water pollution. Those problems are still with us but seem to have receded from crisis levels. Then there was the energy crisis when the Arab oil embargo threatened the pursuit of the American way of life—complete with air conditioning, high-speed automobiles and illuminated signs. That too seems to have receded with the adoption of the national 55-mile-an-hour speed limit and with the turning off of a few lights in office buildings.

The problem of toxic substances is, on the other hand—if anything—still in its ascendance. We have not yet properly defined all the toxic substances themselves, their levels of toxicity, their occurrence, how to detect them and how to remove them.

Among the toxic substances, the metals are particularly bad actors. Mercury has become infamous through the Minamata disease; cadmium through the Ouchi-Ouchi disease.

The problems with toxic substances in general, and with toxic metals in particular, is one of protecting people from poisoning. This in turn resolves itself into two problems—one of protecting the worker who may possess unique protective equipment but who may also be subjected to uniquely high concentrations—and the second of protecting the public from the results of factory emissions, transport spills, etc.

The purpose of this book is to survey the toxic metals and in each case to give leads to available information on:

- Toxicity
- Extent of exposure
- Detection
- Environmental standards
- Handling procedures
- Removal from air
- Removal from water
- Solid waste disposal
- The economic impact of controls

THE DEFINITION OF A TOXIC METAL

The question of defining a toxic metal is a complicated one and—perhaps surprisingly, but frankly—one that the author does not propose to discuss at exhausting length.

There are a variety of criteria for defining a list of toxic metals, none of which are in complete agreement with all the others. Table 1 gives some of the lists from various sources. The Council for Environmental Quality (1) listing in the first column shows "selected metals for which adverse human effects have been documented." As shown in the second column, the Office of Toxic Substances of the U.S. Environmental Protection Agency takes a slightly different view as indicated by reports they have commissioned on various metals. The third column lists those metals of immediate concern to EPA in 1975 (2) in their criteria for drinking water. The fourth column lists those metals of particular concern to OSHA as evidenced by their publications.

TABLE 1: LISTINGS OF TOXIC METALS

Metal	CEQ "Toxic Substances" (1971)	EPA Office of Toxic Substances	1975 EPA Drinking Water Criteria	OSHA Concerns
Antimony	–	x	–	–
Arsenic	x	–	x	x
Barium	x	–	x	–
Beryllium	x	–	–	x
Boron	–	x	–	–
Cadmium	x	–	x	–
Chromium	x	–	x	x
Copper	x	–	–	–
Indium	–	x	–	–
Lead	x	–	x	x
Manganese	x	–	–	–
Mercury	x	–	x	x
Molybdenum	–	–	–	–
Nickel	x	x	–	–
Selenium	x	x	x	–
Silver	x	–	x	–
Tin	–	x	–	–
Vanadium	x	x	–	–
Zinc	x	–	–	–

There are other metals that could be included. Thus, even iron has its toxic aspects as shown by the following quotation (7):

> "Inhalation of iron and iron oxides produces a benign siderosis (or pneumoconiosis). In addition to the benign condition, there may be very serious synergistic effects as well as other undesirable effects, such as chronic bronchitis. In the laboratory, iron oxide acts as a vehicle to transport the carcinogens in high local concentrations to the target tissue. Similarly, sulfur dioxide is transported in high local concentrations deep into the lung by iron oxide particles. The relationships between these conditions and dose and time are undetermined."

Then, there are the noble metals which may come into increasing use in catalytic exhaust purifiers for automobiles but which are not now the subject of wide exposure. The toxicology of platinum and palladium have been subjected to some initial scrutiny (8), but they will not be discussed at further length here. Some recent insight into the possible toxicity of automotive metallic emissions has also been published (9).

Finally there is that most toxic of all metals, plutonium, but it will not be discussed here because of its specialized use in the atomic energy field.

Thus, rather than worrying about whether or not a particular metal is or is not a toxic metal, this book will attempt to survey information on metals which have been the subject of some concern as to their toxicity—and on which data are currently available.

Another categorization of metals which it is necessary to discuss when defining the toxic metals is the category of what constitutes a "heavy metal." In many cases, authors and even reputable scientific authors have used the two categories almost interchangeably. In fact, the definition of heavy metals is perhaps one of the most ill-defined pieces of nomenclature in the entire chemical literature.

The classical definition of a heavy metal was as shown in the first column in Table 2, that of a metal precipitatable in acid solution by hydrogen sulfide. However, as shown in the second column in Table 2, the Environmental Engineers Handbook, (3) gives a very different listing which includes aluminum (of all things!) as well as iron. An EPA listing in a 1973 document relating to the development of criteria for the cement industry (4) gives yet another listing.

The point of the above discussion is that there is no acceptable current definition of heavy metals. In any event they are not synonymous with toxic metals and the term will be used as little as possible henceforth.

TABLE 2: LISTINGS OF HEAVY METALS

Metal	Classical Definition	Environmental Engineers Handbook (3)	EPA Definition (1973) (4)
Aluminum	-	x	-
Antimony	x	-	-
Arsenic	x	-	-
Bismuth	x	-	-
Cadmium	x	-	x
Chromium	-	x	x
Copper	x	x	x
Gold	x	-	-
Iron	-	x	-
Lead	x	-	x
Mercury	x	-	x
Nickel	-	-	x
Silver	x	-	-
Tin	x	-	-
Zinc	-	x	-

TABLE 3: SOME FACTS ABOUT THE TOXIC METALS DISCUSSED IN THIS BOOK

Metal	Use Level	Essential Micronutrient	Estimated Airborne Emissions	Recommended Limit in Air, mg/m^3	Special Toxic Effects
Antimony	45,000 tons/year	No	450 tons/year	0.5	Pulmonary diseases.
Arsenic	30,000 tons/year	No	9,500 tons/year	0.002	Causes dermatitis and possible skin cancer.
Barium	800,000 tons/year	No	11,000 tons/year	0.5	Soluble compounds are heart stimulants.
Beryllium	350 tons/year	No	150 tons/year	0.002	Most toxic; causes body-wide beryllium disease.
Boron	90,000 tons/year	Yes	10,000 tons/year	10 (B_2O_3) 3 (BF_3)	Boric acid poisoning, can even result in death.
Cadmium	6,000 tons/year	No	300-3,000 tons/year	0.05	Exposure can cause acute or chronic poisoning resulting in death.
Chromium	500,000 tons/year	Yes	12,000-18,000 tons/year	0.001 (Cr^{+6}) 0.1 (chromic acid) 0.5 (sol. salts)	Chromic acid can cause lung cancer.
Copper	3,000,000 tons/year	Yes	14,000 tons/year	0.2	Toxic to marine life and vegetation.
Indium	20,000 grams/year	No	NA	0.1	Injections have caused intoxication and degenerative changes in heart, liver and kidneys.
Lead	1,500,000 tons/year	No	9,000-18,000 tons/year	0.15	Cumulative poison; can affect blood, kidneys, nervous system, brain.
Manganese	1,200,000 tons/year	Yes	19,000 tons/year	5.0 0.1 (MMT)	Inhalation of oxides can cause chronic poisoning or manganese pneumonia.
Mercury	2,000 tons/year	No	500-800 tons/year	0.005 0.01 (alkyls)	Monovalent form harmless, divalent, organic and elemental mercury very toxic.
Molybdenum	28,000 tons/year	Yes	900 tons/year	5.0 (sol.) 10.0 (insol.)	Dusts may cause pulmonary disorders.
Nickel	190,000 tons/year	Yes	6,000 tons/year	0.007 (carbonyl) 1.0 (metal and insol. salts) 0.1 (soluble salts)	Can produce cancer of lung and sinus, esp. in carbonyl form.
Selenium	500 tons/year	Yes	900 tons/year	0.2 0.4 (SeF_6)	Respiratory and gastrointestinal irritations result.
Tin	82,000 tons/year	No	370 tons/year	2.0 (inorg.) 0.1 (organic) 10.0 (oxide)	Organic tin compounds are very toxic.
Vanadium	6,000 tons/year	Yes	18,000 tons/year	0.5 (dust) 0.005 (fume)	Causes respiratory irritation; may be carcinogenic.
Zinc	1,800,000 tons/year	Yes	151,000 tons/year	1.0 (chloride fumes) 5.0 (oxide fumes) 10.0 (stearate)	Inhalation of chloride fumes can be fatal.

Table 3 summarizes some information about the metals which will be discussed at greater length in this book.

It should be noted that the estimated airborne emissions in some cases exceed the use level. This is because uncontrolled emissions from the burning of fossil fuels may exceed the amount of metal produced and consumed in industry.

GENERAL TECHNIQUES FOR REMOVAL FROM AIR

The question of removal of toxic metals from air has generally taken second place to the question of their removal from aqueous effluents.

Where it is necessary to remove elemental metals or their compounds from air, it is frequently a question of particulate removal. The reader is referred for general information on this topic to an earlier Noyes publication, *Fine Dust and Particulates Removal* (5).

GENERAL TECHNIQUES FOR REMOVAL FROM WATER

Some common methods used for heavy metal removal used by industry are shown in Table 4 as taken from Sabadell (6). The reader will please excuse the use of the term heavy metal just after the author has criticized its use—but it is widely used in the literature.

TABLE 4: COMMON METHODS USED FOR HEAVY METAL REMOVAL BY INDUSTRY

1. Chemical precipitation
    a. Lime
    b. Alum
    c. Iron salts
2. Chemical treatment
    a. Oxidation
    b. Reduction
3. Ion exchange
4. Ultrafiltration
5. Electrochemical
6. Evaporative recovery

REFERENCES

(1) Council on Environmental Quality, *Toxic Substances,* Washington, D.C. (April 1971).

(2) Environmental Protection Agency, "National Interim Primary Drinking Water Regulations, *Federal Register,* 59566-59588 (December 24, 1975).

(3) Lipzak, B., Ed., *Environmental Engineers Handbook,* Philadelphia, Chilton (1974).

(4) Environmental Protection Agency, *Development Document for Proposed Effluent Limitations Guidelines and New Source Performance Standards for the Cement Manufacturing Point Source Category,* Report EPA 440/1-73/005, Washington, D.C. (August 1973).

(5) Jones, H.R., *Fine Dust and Particulates Removal,* Park Ridge, N.J., Noyes Data Corp. (1972).

(6) Sabadell, J.E., Ed., *Traces of Heavy Metals in Water: Removal Processes and Monitoring,* Washington, D.C., U.S. Environmental Protection Agency (1973).

(7) Booz-Allen Applied Research, Inc., *A Study of Hazardous Waste Materials, Hazardous Effects and Disposal Methods,* Report PB-221,466, Springfield, Va., National Technical Information Service (July 1973).

(8) Mayer, R.A., Prehn, W.L., Jr. and Johnson, D.E., *A Literature Search and Analysis of Information Regarding Sources, Uses, Production, Consumption, Reported Medical Cases and Toxicology of Platinum and Palladium,* San Antonio, Southwest Research Institute (April 1974).

(9) Holbrook, D.J., Jr., *Assessment of Toxicity of Automotive Metallic Emissions,* Report EPA-600/1-76-0102, Research Triangle Park, N.C., Environmental Protection Agency (January 1976).

ANTIMONY

Antimony, found primarily in mineral ores, is derived primarily from stibnite (antimony trisulfide), a mineral rich in antimony. Other natural sources of antimony are complex sulfide ores, which are ores of stibnite containing chiefly lead, copper, silver, and mercury. China has the world's largest antimony deposits, with known reserves amounting to almost 4 million tons. U.S. resources are very small, about 100,000 tons in known reserves, but the U.S. may have another 100,000 tons in undiscovered resources. A new find of antimony ore in the U.S. was reported in 1970, and the ore is currently being processed (1).

Although antimony and its compounds are industrially significant because of their unique contribution to the manufacture of many commonly-used products, such as alloys, paint, paper, plastics, and textiles, they are not ranked among the biggest volume chemicals in the U.S. Total production in 1974 of primary and secondary antimony was only about 41,000 tons, with the secondary production (i.e., antimony produced from scrap rather than from its ores or as a by-product of lead smelting operations) accounting for better than 50% of total production.

Antimony has been a cyclic marketplace performer, its price remaining quite stable over the years. Its price rose sharply in recent years because of tight supplies, but is now trending downward again. The price drop is attributed to the depressed economy, particularly in the automotive field (which cut down battery production) and to the use of stockpiled antimony. Federal legislation requiring certain textiles and other materials to be flameproofed is expected, however, to boost the demand for antimony. Consumption should rise at an annual rate of 4% through 1985.

Antimony has been manufactured in the U.S. for about 80 years. Although there have been reported incidences of pneumoconiosis among workers, and antimony compounds can be irritating to the skin and cause other disorders, no epidemiological studies have appeared. There is no evidence that antimony is a carcinogen, teratogen, or a mutagen.

While current knowledge does not permit ranking antimony and its compounds as major environmental contaminants, antimony has innate toxicity and inhalation of concentrations above its threshold limit value could be damaging. Careful control of antimony air emissions, careful monitoring of antimony-containing effluents, and proper disposal of toxic antimony compounds or those antimony compounds that degrade to form toxic substances should be encouraged.

TOXICITY

The toxicity of antimony and its compounds has been reviewed by T.D. Luckey et al (2). Antimony, according to Tucker (3), has some mysterious and potentially hazardous biological properties, however. It has been shown to shorten the life span of small mammals at extremely low concentrations, according to Schroeder et al (4).

It also has been shown to possess unpredictable and sometimes fatal effects in both the heart and the liver at dose levels well below those normally regarded as poisonous, according to Browning (5).

As Tucker points out (3), there have been no detailed studies of these phenomena or of the interactive effects of antimony with other metals, nor of its effect or fate in biological systems. Therefore, he goes on to say, antimony is in need of careful examination.

Analysis of sea birds killed in a massive and mysterious disaster in the Irish Sea in 1969 yielded antimony concentrations of 40 to over 400 ppm in livers and kidneys according to Holdgate (6). The tremendous variation in concentrations, with an upper range which might be close to the lethal threshold, points to possible industrial contaminations (perhaps from ocean dumping) and resultant marine hot spots whose existence had previously been unsuspected, according to Tucker (3). He goes on to point out that since marine species are not normally monitored for antimony, the meaning of these findings in terms of human nutrition is completely unknown.

According to Schroeder (7), heavy exposure to dusts can cause irritation of nasal passages, eyes and lungs. Ingestion is a much more serious matter; it is toxic to the heart and 100 mg is fatal, according to Schroeder (7). As little as 97 mg of antimony has been reported as lethal to humans, according to Arthur D. Little Inc. (1).

Antimony is found only infrequently in air, but following emission from an antimony (or copper or lead) smelter the antimony may be transported for some distance. There is evidence that it is persistent in surface soils following contamination. There is also abundant evidence of antimony uptake in human and animal tissues and organs with exposure to antimony, but some attempts to correlate tissue with age have been unsuccessful. Thus, it is uncertain whether antimony is actually accumulated. There is currently no evidence of biomagnification.

Occupational exposure to antimony ore, pure antimony, and antimony oxides has been implicated in increased incidence of pneumoconiosis, a disease of the lungs caused by the habitual inhalation of irritant minerals or metallic particles. In animal studies acute and subacute exposure to these forms of antimony resulted in a pulmonary phagocytic response (cell engulfing of foreign material) generally without any appreciable pneumonitis. With chronic exposure the macrophage (large phagocyte) response was more involved and there was an increase in fibrous tissue.

In contrast, occupational exposure to antimony trisulfide has been associated with cardiovascular changes. Experimental studies with this compound in rats, rabbits and dogs confirmed the cardiovascular effects. Exposure to gaseous forms of antimony produced more generalized systemic toxicity and, in the cases of antimony trichloride in men and antimony pentafluoride in rodents, the effects were probably attributable to both the antimony and the halide components.

EXTENT OF EXPOSURE

The flow of materials in the antimony industry is shown in Figure 1 (8).

According to Schroeder (7), potential exposures come from food and fluids in contact with ceramic enamels, pewter, and britannia metal; from clothing impregnated with antimony trioxide for flameproofing; and from air in dusts from the wear of rubber and fumes of type metal. Contact with certain kinds of red rubber is also a source. Exposures, however, are low, except in industry.

Antimony is used in metal and nonmetal products. As a metal, it finds a use in antimonial lead, solder type metals, ammunition (bullets) storage batteries, and in cable coverings. As a nonmetal, it is used in glass and clay products, in flameproofing chemicals and compounds, in ammunition (primers), in fireworks, and in plastics and rubbers.

Antimony oxide as well as antimony pentasulfide and antimony trisulfide, is used in various grades to impart flame resistance to many rubber, plastic, textile, paint and paper products. The oxide is used as a pigment, primarily in paints. As a decolorizer, antimony oxide is useful in glass, particularly optical glass. It is also used as an opacifying agent in porcelainized enamel. The trifluoride is used in the manufacture of pottery, while the tri- and pentasulfides are used to make fireworks and matches.

Antimony oxide and other antimony compounds are also used as catalysts in the manufacture of other chemicals, the oxide serving as a catalyst, for instance in the production of polyester resins.

The medicinal use of antimony compounds has decreased in recent years, primarily because of undesirable side effects of some of these compounds or their toxicity, but also, because of the advent of new type drugs. Antimony potassium tartrate is an effective emetic and has been used to treat certain tropical diseases, but other less toxic antimony compounds are replacing it. The barium, as well as the potassium tartrate, is used as a veterinary medicinal.

Antimony is a significantly useful commercial product but it is not an essential one. Although it has certain properties that promote its use (e.g., solidified from the molten state, it expands on cooling; thus it offers a unique property for type metal used in printing), antimony can be replaced by other materials.

FIGURE 1: SUPPLY-DEMAND RELATIONSHIPS FOR ANTIMONY

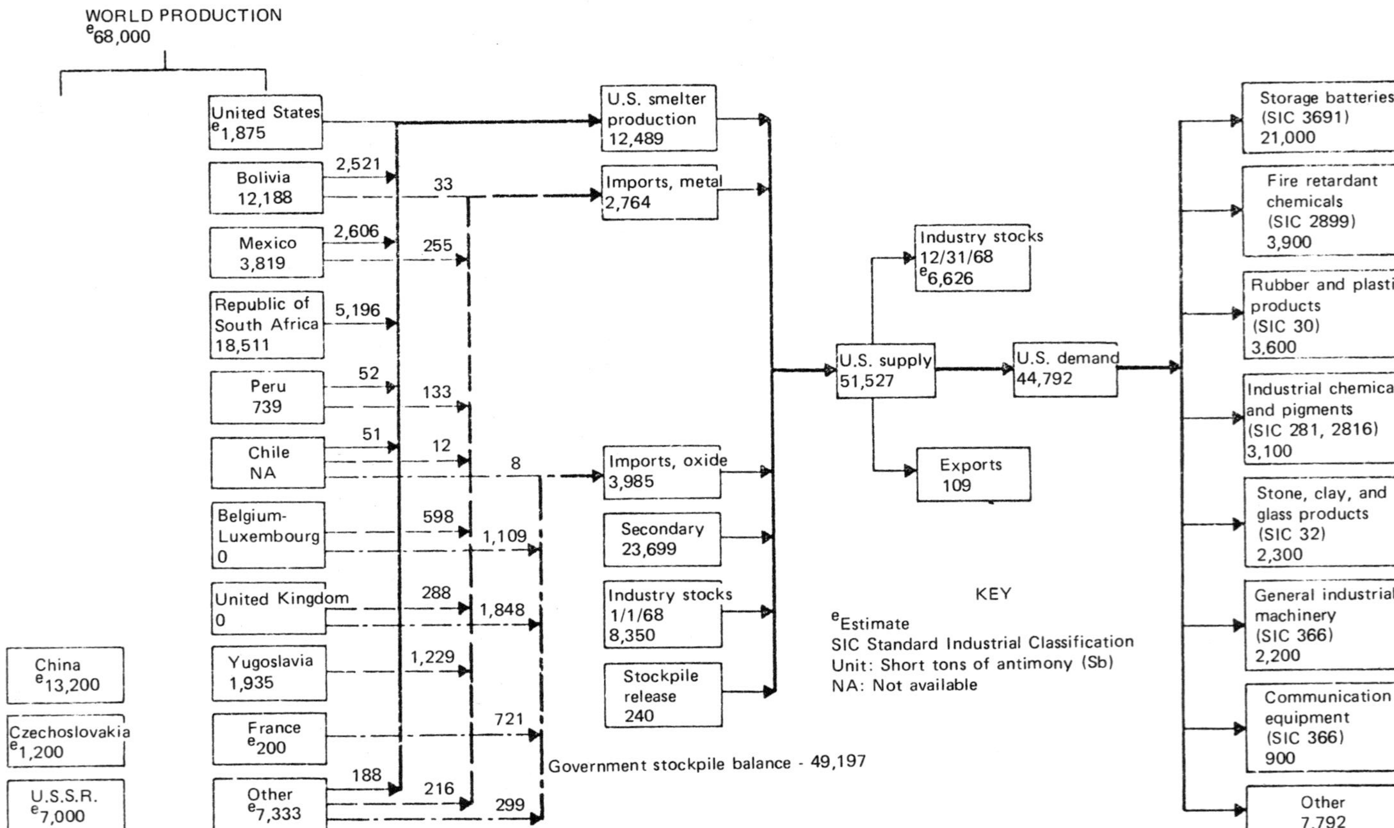

Source: Reference (8).

Alternative compounds have been used, when military needs required the major part of production or when less expensive materials have been required.

According to Flinn and Reimers (9), the emission of elemental antimony into the air is as follows (in tons per year):

	1969-71	Projected 1978	Projected 1983
From production	250	-	-
From consumption	100	-	-
Total	350	460	550

DETECTION

Traditional analytical methods are available for determining the antimony concentration in ores, minerals, and even rafter dusts, but the greater precision and sensitivity offered by modern instrumental methods is required for the trace analysis of this material and its compounds in air, water, and in biological specimens (1).

Colorimetric analysis of antimony has been employed widely for many years with the tetraethylrhodamine (Rhodamine B) method cited most frequently. The Rhodamine B test has a sensitivity of 6 γ. In more recent years, such analytical techniques as atomic absorption spectrophotometry (aas) and neutron activation analysis (naa) have come into greater use.

In its Official Methods of Analysis for 1975, the Association of Official Analytical Chemists (AOAC) describes the determination of pentavalent Sb as residue in foods by means of an American Conference of Governmental Industrial Hygienists (ACGIH)-AOAC method. In an aqueous hydrochloric acid solution, antimony reacts with tetraethylrhodamine (Rhodamine B) and forms a colored complex which can be extracted with organic solvents such as benzene or toluene. The color intensity is measured at 565 nm spectrophotometrically. This method resulted from the ACGIH-sponsored corrobarative tests on the Rhodamine B colorimetric method for antimony which the AOAC (then the Association of Official Agricultural Chemists) reported.

The AOAC also reports the development of an atomic absorption method for antimony in the food additive titanium dioxide. This method is based on fusion with potassium hydrogen sulfate and extraction with methyl isobutyl ketone.

Atomic absorption spectroscopy is widely used in determining the presence of antimony. Data gathered with the Perkin-Elmer Model 303 instrument at 2175 A wavelength with a 7A slit show the detection limit of 0.2 μg/ml with good sensitivity.

Spot tests are used by chemists to indicate the presence or absence of specific elements. Spot tests are not quantitative but they are frequently rated for the limits of concentration where they are effective. For spot tests a variety of analytical methods are offered for antimony.

1. Metallic antimony forms a sublimate of volatile antimony(III) chloride by means of pyrolysis with an excess of ammonium chloride.
2. Elemental antimony can be detected to 100 γ by a reduction-oxidation reaction with alkaline mercury cyanide solution. A strong alkaline solution of mercuric cyanide reacts with other antimony compounds, such as Sb_2O_3 or Sb(III) salts, to form black or gray, finely divided metallic mercury. Salts of Sb(V) must be reduced to Sb(III) in acid medium before this reaction can be obtained. A limit of identification of 10 γ antimony has been obtained with potassium antimonyl tartrate using this procedure.
3. Phosphomolybdic acid is used to determine Sb_2S_3.
4. 9-methyl-2,3,7-trihydroxy-6-fluorone is used to determine antimony(III).
5. Gallein gives a violet color in acid solution in the presence of antimony with a sensitivity of 0.005 γ.
6. Rhodamine B, in tests similar to the ACGIH-AOAC method mentioned above, can be used to identify pentavalent antimony.
7. Diaminotriphenylmethane dyes (Malachite Green and Brilliant Green) and triaminotriphenylmethane dyes (Crystal Violet) form colored dye complexes when used with antimony(V) compounds in hydrochloric acid. These complexes are soluble in toluene.
8. Antimony pentoxide and tetroxide remain following ignition of organic compounds containing antimony.
9. Diphenylamine or N,N-diphenylbenzidine dissolved in sulfuric acid imparts a blue color in the presence of antimony to the limit of 5 γ antimony.
10. Potassium antimony tartrate (tartar emetic of potassium antimonyl tartrate) may be detected using standard tests for trivalent antimony and tartrate. Limits of identification are 2.5 γ of the $K(SbO)C_4H_4O_6$ by forming antimony(III) sulfide and 5.0 γ $K(SbO)C_4H_4O_6$ by silver oxide reduction.

In the tentative method of analysis for antimony content of the atmosphere (1) the collected antimony must be oxidized to the pentavalent state before analysis with Rhodamine B is possible. The sensitivity of the antimony determination is 1.0 μg. In seven tests made, recovery of antimony ranged from 95 to 102.5% of the 2 to 10 μg sample. The Rhodamine B method is also recommended by the Analytical Chemistry Committee of the American Industrial Hygiene Association and published by the Intersociety Committee Methods for Ambient Air Sampling and Analysis. The determination by this method is sensitive to 1 μg Sb in 10 ml solvent, or 0.05 $\mu g/m^3$ in a 20 m^3 air sample.

Electroreducible cations of antimony can be determined by polarographic methods in samples collected from air. When there is little organic matter present and the sample is prepared carefully, polarography is a precise technique with sensitivity in the range that is adequate for industrial health purposes.

Analytical methods for determination of airborne particulate antimony are currently being developed at Arthur D. Little, Inc. (ADL) in conjunction with

Standards Completion Program, Analytical Methods Validation for NIOSH. Methods are also being developed for stibine in air samples collection medium and analytical methods are currently being tested.

Atomic absorption is recommended by the Methods Development and Quality Assurance Research Laboratory (MDQARL) in the Methods for Chemical Analysis of Water and Wastes under standard conditions. The precision and accuracy reported by the MDQARL for a mixed industrial-domestic waste effluent at concentrations of 5.0 and 15 mg Sb/l were ± 0.08 and ± 0.1 standard deviations respectively with recoveries of 96% and 97% respectively at these levels.

The test method for antimony listed in the Environmental Protection Agency Regulations on Test Procedures for the Analysis of Pollutants (10) is atomic absorption. A special notation refers to the MDQARL method.

There is no standard method described by the American Public Health Association (APHA), American Water Works Association (AWWA), and Water Pollution Control Federation (WPCF), for the examination of antimony in water and wastewater in their joint publication.

A method for direct determination of antimony in sea water by anodic stripping voltammetry at the mercury-coated graphite electrode has been developed especially for use on board ship (1). Typical results for the determination of the antimony content of sea water range from 0.18 to 0.48 $\mu g\ kg^{-1}$, consistent with the values found by others using neutron activation analysis of freeze-dried samples.

For monitoring purposes, the Canada Centre for Inland Waters developed a method (1) for automated atomic absorption determination of antimony in natural waters. When a tube furnace is used as a covalent hydride decomposition device instead of a conventional hydrogen argon-entrained air flame, sensitivity is improved considerably. Forty samples can be analyzed in an hour with a limit of detection of 0.5 μg/l at a wavelength of 217.6 nm.

In a discussion concerning automation of monitering equipment for marine pollution studies (1), a method for automated analysis of heavy metals and trace elements is suggested. When using atomic absorption to analyze for antimony, the detectability limit desired is 0.005 ppm. A fluctional concentration limit (FCL) of 2.2 ppm is recommended with 15.4 as the practical lower limit of range for analysis (FCL x 7).

ENVIRONMENTAL STANDARDS

The release into the environment of antimony and of such compounds of antimony as antimony trioxide, antimony pentasulfide, antimony sulfate, antimony trisulfide, antimony trifluoride, antimony pentafluoride and potassium antimony tartrate is provisionally limited to contamination levels of 0.005 ppm (mg/l) as Sb in air and 0.05 ppm (mg/l) as Sb in water (11).

These levels based on threshold limit values, were calculated by TRW in a study for EPA on recommended methods for disposal of hazardous waste.

They are offered as provisional limiting concentrations pending the development of ambient air quality standards that can be based on extensive evaluation of acute and chronic dose response criteria for exposure of all living and nonliving things.

The American Conference of Governmental Industrial Hygienists (ACGIH) has adopted 0.5 mg/m^3 as the Threshold Limit Value (TLV) for airborne concentrations of antimony and compounds (as Sb). This TLV represents conditions under which it is believed that nearly all workers may be repeatedly exposed day after day without adverse effect.

Under the Occupational Safety and Health Standards, announced in 1974 by OSHA, antimony and compounds (as Sb) are listed together as one entry in a table of air contaminants for which employee exposure limits are given as 8-hour time weighted averages. According to the OSHA standards, an employee's exposure to antimony and compounds (as Sb) shall not exceed the 8-hour time weighted average of 0.5 mg/m^3, for any 8-hour work shift of a 40-hour week (12).

The U.S. Environmental Protection agency has issued (13) its effluent limitation for Sb [recommended for best practicable control technology currently available (BPCTCA)—antimony mines] of 0.5 mg/l Sb for a 30-day average and 1.0 mg per liter for a 24-hour maximum. The EPA's National Interim Primary Drinking Water Regulations (14) are to be effective June 24, 1977. These regulations do not include antimony among the inorganic chemicals for which maximum contaminant levels are given.

The proportion of antimony in color additives is limited, according to the Food and Drug Administration (FDA) (15). This section states that the synthetically prepared color additive titanium dioxide (TiO_2) shall conform to the following:

Lead (as Pb), not more than 10 parts per million.
Arsenic (as As), not more than 1 part per million.
Antimony (as Sb), not more than 2 parts per million.
Mercury (as Hg), not more than 1 part per million.
Loss on ignition at 800°C (after drying for 3 hr at 105°C), not more than 0.5%.
Water-soluble substances, not more than 0.3%.
Acid-soluble substances, not more than 0.5%.
TiO_2 not less than 99.0% after drying for 3 hr at 105°C.

HANDLING PROCEDURES

Exposure to antimony dust and its salts can cause dermatitis. Antimony vapors, when inhaled, can cause respiratory and gastrointestinal problems. Antimony dust or vapor, under certain conditions, can also be a fire hazard. For these reasons, particular antimony compounds have to be transported according to prescribed procedures. Storage areas should be dry, well-ventilated, and protected from heat and sunlight. Fire extinguishers should be kept in the storage area.

The potential dangers from excessive dust and fumes that exist in the production of antimony also exist in those industries where antimony is used. For example, antimony dust may be generated during the machining and polishing of antimony-containing metal products. Antimony trioxide dust may form when the oxide is used as a pigment in the manufacture of paints and other products; antimony pentoxide dust may be generated in the making of glass and ceramics. Other hazards may exist from skin contact with antimony trichloride solutions during the dyeing of textiles or with antimony sulfide in rubber compounding operations.

The Coast Guard, Department of Transportion, includes only antimony pentachloride and antimony pentafluoride on the list of explosives and other dangerous articles and combustible liquids. Antimony pentachloride may be shipped dry or in solution but by cargo vessel only. It should be stored on deck, protected from the weather or under cover and carry a white label for corrosive liquid. Transport limits are given as follows:

for metal barrels or drums, 55 or 110 gallon capacity
for wooden barrels, 200 lb net weight
for wooden boxes, 200 lb gross weight
for portable tanks, 8,000 lb gross weight
for tank cars, must comply with DOT regulations (trailerships only)
for motor vehicle tank trucks, must comply with DOT regulations (trailerships and trainships only).

The regulations for antimony pentachloride solutions are the same as for $SbCl_5$ in dry form except for very detailed specifications for containers. Antimony pentafluoride is permitted to be transported by cargo vessel only in steel cylinders containing 1, 5, 25 and 200 lb SbF_5 and stored on deck protected from the weather or under cover. White labels are required signifying corrosive liquid. Electric wet batteries containing SbF_5 are listed by DOT.

REMOVAL FROM AIR

In general, antimony is not present in significant amounts in air but, when emitted from smeltering operations, it has been measured in concentrations higher than its permissible Threshold Value Limit. Antimony emitted from smelters can, evidently, be carried for some distance by the wind (1).

The mining and crushing of antimony ores offers potential for the generation of antimony dusts but, especially in recent years, carefully controlled operations and the use of dust collecting systems have minimized the dust hazards.

In the production of antimony metal from its ores, by-product dust and antimony oxide fume can be generated. Again, however, most large industries use local exhaust systems in the working areas and provide a direct exhaust for the smelting furnace, thus minimizing hazards to employees and controlling air emissions from the plant.

Air emissions at the antimony smelter in Laredo, Texas, were reported to be higher than permissible in 1972, but the company has installed additional baghouses, hoods and controls, which have helped to reduce these emissions considerably.

When the emissions of antimony and its compounds are of a particulate nature they can be controlled by standard dust collectors such as bag filters, scrubbers and electrostatic precipitators (16). In one study of the emission of trace elements from a steam power plant, an electrostatic precipitator removed about 96% of the potential antimony emissions from coal combustion (17). The same study indicated, however, that from mass balance considerations, some antimony compounds were escaping in the vapor phase. Hence a cooling method may be necessary to control all antimony emissions.

REMOVAL FROM WATER

There is very little information available on the removal of antimony from water. However, as pointed out by Luckey et al (2), antimony resembles bismuth in its chemical behavior more than it does arsenic or the other nonmetallic members of Group V of the Periodic Table. Thus, a technique such as that described by Sugano (18) which is applicable to bismuth should be applicable to antimony as well.

The technique developed by Sugano et al (18) involves extracting metal impurities from wastewater to which ferrous ions are added to provide at least two times the amount of ferrous ions on the mol basis to the amount of metal ions present, the resulting solution containing acid radicals. A base is added to the solution to raise the pH to form a suspension of metal hydroxides, following which an oxidizing gas is bubbled in solution to form crystals containing ferric ions and further containing the metal ions originally in solution, the precipitated crystals being thereafter separated to provide clean water.

In typical examples of the operation of this process, bismuth concentrations in wastewater were reduced from 300 ppm before treatment to 0.5 ppm after treatment and from 600 ppm to 1.2 ppm.

In Idaho, the leaching and electrolysis process used by one company to produce antimony generates no air emissions or solid waste. A recycling system installed a few years ago has reduced the antimony content of the liquid to a very low amount.

Currently, although antimony content is only 1.0 to 1.3 ppm, it is still too high to meet the suggested EPA proposed effluent limitations. The company is striving now to comply with the EPA requirements. The antimony effluent has not had any deleterious effect on vegetation in the area, but fish have not been able to survive in the local river. This effect, however, has not been attributed to the presence of antimony in the effluent. In general, there is no evidence that antimony oxide is a harmful pollutant. It is discarded into municipal systems.

Antimony is present in sea water at a normal concentration of 4 μg/l, but no excessive concentration of antimony in sea water (as the result of extracting minerals from the seabed or from waste disposal) has been reported. Because most salts of antimony are insoluble, there is little chance that large concentrations of antimony will be found in the ocean.

Marine animals, however, do concentrate antimony in their muscles, and several studies indicate that certain levels of antimony have been toxic to some fish.

Because of the possible consequences to human health that could result from eating seafood contaminated with high concentrations of antimony, it is advisable that any future large-scale extraction of antimony from the seabed be closely monitored.

SOLID WASTE DISPOSAL

As discussed by Arthur D. Little, Inc. (1), an antimony smelter at Laredo, Texas produces only an insignificant amount of liquid effluent, but does generate about 44,000 tons of solid waste (slag) each year. This slag, which contains 1% of antimony, is stored outdoors on company property.

Federal, state and local regulations govern the disposal of antimony wastes. Landfill and smelter are the principal disposal methods now used; ocean dumping has been practiced, but concern over the potential hazard to the ecosystem has practically ruled out this method. Some antimony compounds, such as antimony pentafluoride and antimony trifluoride, must be pretreated chemically before being buried. Encapsulation of nickel antimonide before disposing of it in a chemical waste landfill is recommended. Because the antimony halides, e.g., antimony trichloride can be decomposed by water to form toxic hydrogen chloride, care must be exercised in their disposal. Antimony pentafluoride and antimony trifluoride are listed as nonradioactive high hazard compounds by the EPA Office of Solid Waste Management Programs (19). The Office of Solid Waste Management Programs considers these two compounds of antimony unacceptable for normal dispoal methods.

Manufacturing process effluent streams containing the antimony fluorides can best be handled at the sites where they originated by the continuation of current, acceptable disposal processes (20). Small contaminated or surplus quantities of antimony trifluoride and antimony pentafluoride should either be returned to the manufacturers for reprocessing, or disposed of.

A proposed disposal process is to combine Methods 27d and 11 of the *Manufacturing Chemists Association Laboratory Waste Disposal Manual.* Briefly stated, the disposal concept is as follows.

Dissolve the antimony fluoride in the minimum quantity of dilute hydrochloric acid. Saturate with hydrogen sulfide. Filter, wash and dry the antimony sulfide precipitate; sell the precipitate to a primary antimony metal producer. Air strip the filtrate of dissolved H_2S, passing the effluent air into a controlled incineration device, equipped with a lime scrubber. React the stripped filtrate with an excess of lime; evaporate to dryness, and dispose of the lime-CaF_2-$CaCl_2$ mixture by land burial.

Antimony sulfates and sulfides are insoluble in water and therefore may be conveniently disposed of by landfill. However, antimony compounds are highly toxic and the disposal of dilute wastes containing these constituents by landfill should be in Class 1 sites located over nonwater-bearing sediments or with only unusable ground water underlying them. Particularly for antimony sulfide and sulfate wastes, recovery should be considered and preferred over disposal by landfill (20).

ECONOMIC IMPACT OF CONTROLS

No specific data are available regarding the economics involved with the control of antimony emissions. The general economics outlined (16) can be used to estimate the annualized cost of control for the particulate phase.

REFERENCES

(1) Arthur D. Little Inc., *Literature Study of Selected Potential Environmental Contaminants: Antimony and Its Compounds,* Washington, D.C., U.S. Environmental Protection Agency (Feb. 1976).
(2) Luckey, T.D., Venugopal, B. and Hutcheson, D., *Heavy Metal Toxicity, Safety and Hormonology,* New York, Academic Press (1975).
(3) Tucker, A., *The Toxic Metals,* London, Earth Island Ltd. (1972).
(4) Schroeder, H.A., et al, "Zirconium, Niobium, Antimony and Fluorine in Mice, etc." *Journal of Nutrition* 95, 95-101 (May 1968).
(5) Browning, E., *Toxicity of Industrial Metals,* 2nd Ed., London, Butterworth (1969).
(6) Holdgate, M.W., Ed., *The Seabird Wreck of 1969 in the Irish Sea,* London, Natural Environment Research Council (1971).
(7) Schroeder, H.A., *The Poisons Around Us,* Bloomington, Indiana University Press (1974).
(8) U.S. Bureau of Mines, *Mineral Facts and Problems,* Washington, D.C., U.S. Government Printing Office (1970).
(9) Flinn, J.E. and Reimers, R.S., *Development of Predictions of Future Pollution Problems,* Report EPA-600/5-74-005, Washington, D.C., U.S. Environmental Protection Agency (March 1974).
(10) U.S. Environmental Protection Agency, 40 CFR 136; 38 *FR* 28758 (October 16, 1973).
(11) TRW Inc., *Industrial and Municipal Disposal Candidate Waste Stream Constituents Profile Reports—Inorganic Compounds, Recommended Methods of Reduction, Neutralization, Recovery or Disposal of Hazardous Waste,* Report PB-224,591, Springfield, Va., Nat. Tech. Information Service (Aug. 1973).
(12) "Occupational Health and Safety Administration," *Fed. Reg.* 39(125),23540-41 (June 27, 1974).
(13) U.S. Environmental Protection Agency, *Development Document for Interim Final and Proposed Effluent Limitations Guidelines and New Source Performance Standards for the Ore Mining and Dressing Industry,* Report EPA 440/1-75-061, Washington, D.C., (October 1975).
(14) Environmental Protection Agency, "National Interim Primary Drinking Water Regulations," *Fed. Reg.,* 59566-59588 (December 24, 1975).
(15) Food and Drug Administration, 8 CFR Title 21, Chapter 1, Subpart D, Series 8.316.
(16) McFarland, A.R., Ed., *Control Technology for Toxic and Hazardous Air Pollutants,* Chicago, Illinois Institute for Environmental Quality (Dec. 1974).
(17) Bolten, N.E., et al., *Trace Element Measurements at the Coal Fired Allen Steam Plant—Progress Report,* Oak Ridge National Laboratory, June 1971-Jan. 1973, (March 1973).
(18) Sugano, I., Tsuji, T. and Kanamori, M.; U.S. Patent 3,931,007; assigned to Nippon Electric Co., Ltd. Japan (January 6, 1976).
(19) Environmental Protection Agency, *Report to the Congress on the Disposal of Hazardous Wastes,* Publication SW-115, Washington, D.C., (1974).
(20) Powers, P.W., *How to Dispose of Toxic Substances and Industrial Wastes,* Park Ridge, N.J., Noyes Data Corp. (1976).

ARSENIC

Arsenic is between germanium and selenium in the Periodic Table and as a member of Group V its physicochemical properties resemble those of phosphorus. Its principal valences are 3 and 5, and it is ubiquitous, being found in small amounts in soils and waters throughout the world, as well as in foods, particularly seafood. Arsenic is a constituent of a number of minerals. For industrial and commercial uses, it is obtained primarily from the ores of metals in which it is present as an impurity, removed as arsenic trioxide [arsenic(III) oxide, As_2O_3] during the smelting operation. This oxide is used in the manufacture of most other arsenic compounds, and is produced in the U.S. as a by-product in the smelting of copper ores (1).

TOXICITY

Arsenic is toxic, to some degree, in most chemical forms. Arsenical compounds may be ingested, inhaled, or absorbed through the skin. Industrial exposure to arsenic has shown that it can produce dermatitis, mild bronchitis, and other upper respiratory tract irritations including perforation of the nasal septum. However, because of the irritant qualities of arsenic, it is doubtful that one could inhale sufficient amounts to produce systemic poisoning.

Skin cancer can result from prolonged therapeutic administration of arsenic. Similar cancers have not been observed among industrial workers. Moreover, lung tumors which resulted from inhaling mixed industrial dusts were often thought to be the result of inhaling arsenic. Recently, this relationship has been questioned because animal experiments have failed to demonstrate that arsenic is a carcinogen. Therefore, the causal relationship between cancer and arsenic is disputed.

Arsenic is poisonous to both animals and plants, according to Booz Allen Applied Research, Inc. (2). The toxic properties of arsenic and its compounds have also been reviewed by Luckey et al (3), Schroeder (4) and Tucker (5). Tucker (5) points out that some kind of synergism exists between arsenic and lead. Although in general less toxic than lead, arsenic appears to increase its own toxicity or that of lead when both metals are present at the same biological site. According to NIOSH (1), no reports were found of occupational exposure to arsenic compounds resulting in nausea, vomiting, diarrhea, or peripheral neuropathy. Occupational exposures have been reported to cause hyperpigmentation, palmar and plantar hyperkeratoses, warts, contact dermatitis and sensitization, ulceration and perforation of the nasal septum, and conjunctivitis. Reversible ECG changes and severely reduced peripheral circulation resulting in gangrene of the fingers and toes have been reported. Cirrhosis of the liver has been observed, and one epidemiological study of a smelter population reported significantly increased mortality due both to cirrhosis of the liver and to cardiovascular disease. Another study of mortality among smelter workers found that cardiovascular mortality was not significantly increased in workers exposed to arsenic.

Mortality studies of orchardists exposed to lead arsenate spray and of copper smelter workers found no excessive cancer mortality. However, according to NIOSH (1), the conclusions in each of these reports are questionable. Other studies have reported cancer of the skin, lung, lymphatic system, and other organs. In general, attempts to produce cancer experimentally in animals have failed, but leukemia reportedly has been induced experimentally and teratogenic effects have been observed in animals.

One epidemiological study of a copper smelter reported that observed deaths due to cardiovascular disease exceeded the expected, but the difference was not statistically significant. Another study of a smelter population found that, compared to statistics for the state in which the smelter was located, mortality due to heart disease was significantly increased. In terms of length of employment, cardiovascular mortality was significantly increased in 4 of 5 cohorts, and the excess mortality was approximately the same in each of these 4 cohorts. In both smelter studies, exposures were to many compounds other than arsenic. However, the fact remains that arsenic apparently caused at least temporary ECG changes and may have caused increased cardiovascular mortality.

EXTENT OF EXPOSURE

Consumption of arsenic trioxide in the United States is estimated to range between 25,000 and 35,000 tons annually. Of this amount, 6,000 to 14,000 tons are produced in the United States. Various arsenic compounds are used as pesticides. Arsenic compounds are also used in pigment production, the manufacture of glass, textile printing, tanning, taxidermy, in antifouling paints, and to control sludge formation in lubricating oils. Metallic arsenic is used as an alloying agent to harden lead shot, and in lead-based materials. It is also alloyed with copper to improve its toughness and corrosion resistance.

Some occupations which have or in the past have had potential exposures to arsenic are listed by NIOSH (1), who estimate that 1,500,000 workers are potentially exposed to inorganic arsenic, including arsine and lead arsenate. Figure 2 shows primary arsenic emission sources and Table 5 presents arsenic emission data in detailed form from EPA data (6).

FIGURE 2: MATERIAL FLOW THROUGH THE ECONOMY SHOWING PRIMARY EMISSION SOURCES – ARSENIC

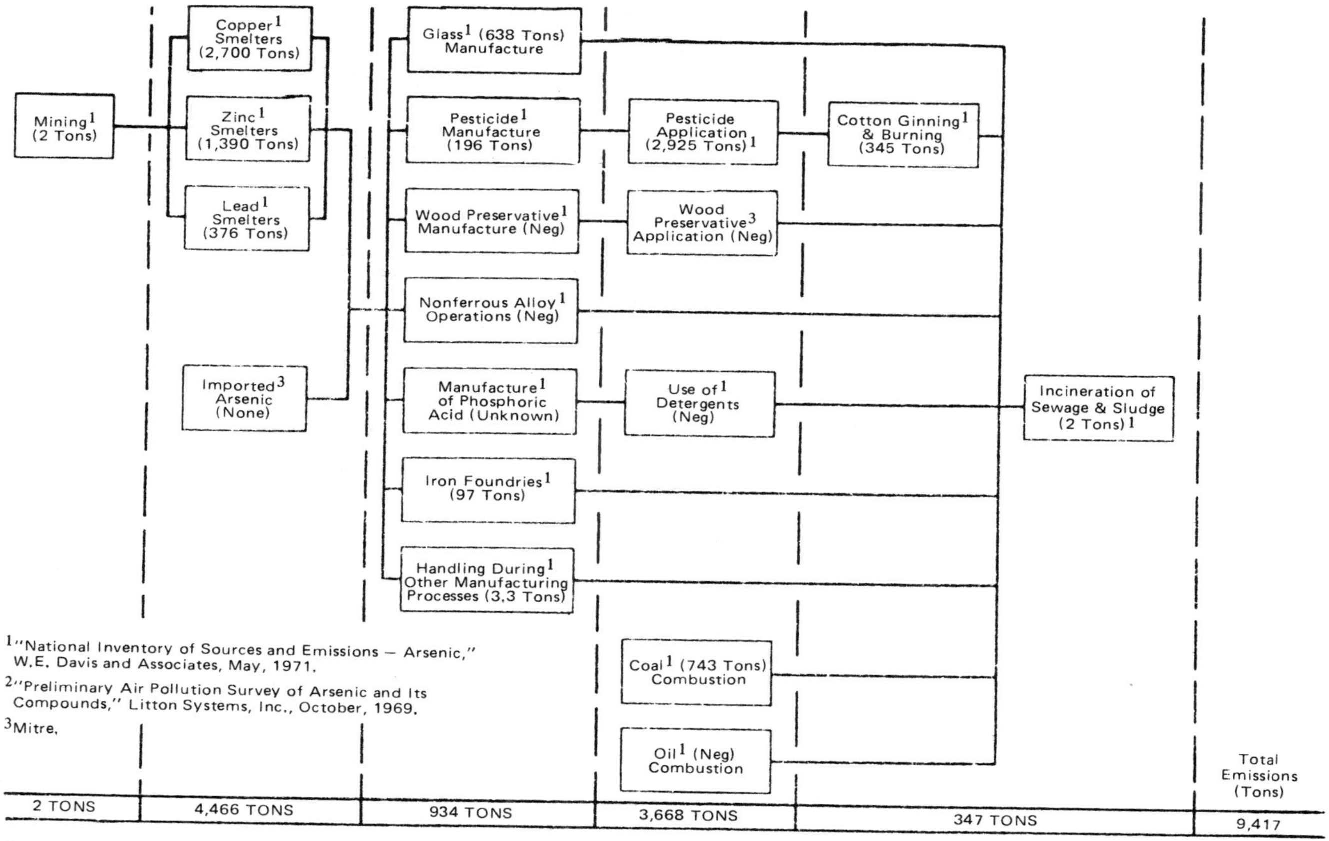

[1]"National Inventory of Sources and Emissions – Arsenic," W.E. Davis and Associates, May, 1971.

[2]"Preliminary Air Pollution Survey of Arsenic and Its Compounds," Litton Systems, Inc., October, 1969.

[3]Mitre.

Source: Reference (6)

TABLE 5: ARSENIC EMISSION

Source	Amount in Tons	% This Pollutant
Mining	<1	Neg.
Phosphate rock	Neg.	Neg.
Primary copper		
Roasting	900	10.07
Reverberatory furnaces	400	4.48
Converters	1,150	12.87
Material handling	250	2.80
Primary zinc		
Roasting	1,390	15.55
Primary lead		
Sintering	285	3.19
Blast furnace	80	0.90
Reverberatory furnace	11	0.12
Gray iron foundry	97	1.09
Cotton ginning	19	0.21
Nonferrous alloys	Neg.	Neg.
Phosphoric acid	Neg.	Neg.
Glass manufacture	638	7.14
Wood preservatives	Neg.	Neg.
Miscellaneous arsenic chemicals	3.3	0.04
Arsenic pesticide production	197	2.20
Pesticide, herbicide, fungicide use	2,925	32.72
Power plant boilers		
Pulverized coal	429	4.80
Stoker coal	49	0.55
Cyclone coal	15	0.17
Industrial boilers		
Pulverized coal	19	0.21
Stoker coal	67	0.75
Cyclone coal	9	0.10
All oil	Neg.	Neg.
Residential/commercial coal	6	0.07
TOTAL	8,939	100

Source Reference (6).

Two air pollution episodes in the US have shown that there is an arsenical air pollution potential at every smelter which refines arsenical ores. Arsenical compounds are used as insecticides and herbicides. Although the use of arsenical pesticides declined sharply after the appearance of DDT and 2,4-D, arsenical compounds are still used as desiccants, herbicides and sterilants, according to Booz Allen Applied Research, Inc. (2).

Some undetermined amounts of air pollution take place during spraying and dusting operations with arsenical pesticides. Pollution from cotton gins and cotton trash burning has been cited as an important source of agricultural pollution. While the emission rates from cotton trash burning have not been determined, as much as 1,258,000 $\mu g/m^3$ of exhaust air (580,000 $\mu g/min$) may be emitted during the ginning operation. This produced concentrations of only 0.14 $\mu g/m^3$ of arsenic in the air 150 feet from the gin.

Arsenic and arsenical compounds have been reported in aqueous waste products of the metallurgical industry, glassware and ceramic production, tannery operations, dye, and pesticide manufacture. The manufacture of Paris green and calcium meta-arsenate, both insecticides, was reported to produce wastewaters containing 362 mg/l of arsenious oxide (7). Arsenical wastes from these sources would be expected to assume less importance in the future, however, as noted on the previous page.

High arsenic levels have been encountered in raw municipal water supplies, necessitating arsenic removal by the water treatment industry. Groundwater supplies in Central and South America are frequently reported to contain excessive arsenic levels (8)(9). Shen and Chen (10) have described deep well waters in Taiwan which contain up to 1.1 mg/l arsenic, and are believed to be responsible for an endemic illness in the area called "blackfoot" disease.

DETECTION

No direct reading instruments are available for determining arsenic in the field. The dusts and fumes of inorganic arsenic compounds can be collected on a cellulose membrane filter with a pore size of 5.0 μm or less. Satisfactory methods are not available for collecting arsine and other arsenical gases, according to NIOSH (1).

The detection of lewisite (chlorovinyldichlorarsine) vapors in the atmosphere may be carried out by a technique developed by H.L. Stroterhoff (11). It involves the steps of placing a white silica gel paper spot and a frangible thin walled ampule containing a reactant solution consisting essentially of cupric carbonate, arsenious oxide, sodium hydroxide, piperidine in an aqueous solution of distilled water, in a recessed plastic film container means having an aperture means in both the upper and lower surfaces for exposing the silica gel paper spot to the atmosphere to be tested. The silica gel paper spot is exposed to the atmosphere suspected to contain chlorovinyldichlorarsine, simultaneously crushing the reactant-containing ampule to flow the reactant therein to the silica gel paper spot. The reactant solution and any chlorovinyldichlorarsine absorbed in the silica gel paper spot are reacted to produce a color change in the white silica gel paper to a color spectrum of brownish red to indicate the presence of chlorovinyldichlorarsine vapors in the tested atmosphere.

Several procedures have been developed for analysis of arsenic in air. Dubois and Monkman (12) compared 3 widely used methods on samples from a variety of sources. The methods tested were Gutzeit, silver diethyldithiocarbamate, and iodine microtitration. They concluded that the silver diethyldithiocarbamate method was superior to the others, and recommended it because of its sensitivity, accuracy, and suitability over a wide range of concentrations. The American Conference of Governmental Industrial Hygienists evaluated this method (13) by comparing test results obtained by 8 cooperating laboratories. It was found (13) (14) sensitive enough to detect, in a 10 m^3 (10,000 liter) air sample, 0.1 μg As/m^3 or a maximum of 1.5 μg As/m^3. However, at the recommended environmental limit of 2.0 μg As/m^3, only 0.06 μg (60.0 ng) of arsenic would be collected in the recommended 30-liter air sample.

In recent years, an analytical method involving arsine generation followed by analysis by atomic absorption spectrophotometry has shown increased sensitivity for arsenic (15)(17). An absolute sensitivity of 0.005 μg (5.0 ng) of arsenic has been obtained using sodium borohydride reduction, a balloon collection technique, and an electrodeless discharge arsenic lamp (16). Other methods with increased sensitivity are the heated graphite furnace (18) and anodic stripping voltammetry (89). Although these techniques are not in wide use, they show promise of attaining sensitivities equal to or greater than that achieved by atomic absorption. Atomic absorption spectrophotometry is the recommended analytical method because, in addition to possessing the required sensitivity, it is more widely known than alternative methods are at this time.

ENVIRONMENTAL STANDARDS

The National Institute for Occupational Safety and Health (NIOSH) recommends that worker exposure to arsenic and its inorganic compounds be controlled by requiring compliance with a recommended standard (1). The standard is designed to protect the health and safety of workers for a 40-hour week over a working lifetime. Compliance with all sections of the standard will prevent noncarcinogenic adverse effects of exposure to inorganic arsenic in the workplace air and by skin exposure, and should at the minimum materially reduce the risk of arsenic-induced cancer. The standard will be subject to review and will be revised as necessary.

"Arsenic" is defined as elemental arsenic and all of its inorganic compounds. "Exposure to arsenic" is defined as exposure at or above 0.002 mg (2.0 μg) As/m^3. Arsine and other arsenical gases should be controlled to the same concentration as other forms of inorganic arsenic.

These criteria for a standard for arsenic and its inorganic compounds are in a continuing series of criteria developed by NIOSH. The proposed standard applies only to the processing, manufacture, and use of inorganic arsenical products as applicable under the Occupational Safety and Health Act of 1970. When the inorganic arsenic criteria document was first published in January 1974, arsine and lead arsenate were excluded from the provisions of the recommended standard. They are included in these revised recommendations. The inclusion of arsine and other gaseous arsenicals poses some difficulties in that specific work practices are needed, and sampling methods need investigation. However, these difficulties should not be cited as cause for permitting continued exposure to arsine at concentrations above 0.002 mg (2.0 μg) As/m^3. NIOSH is working on these areas and will transmit recommendations directly to the Occupational Safety and Health Administration as soon as possible.

The standard was not designed for the population-at-large, and any extrapolation beyond general occupational exposures is not warranted. It is intended to (1) protect against the noncarcinogenic effects of inorganic arsenicals, (2) materially reduce the risk of arsenic-induced cancer, (3) be measurable by techniques that are valid, reproducible, and available to industry and official agencies, and (4) be attainable with existing technology.

It should be noted that the American Conference of Governmental Industrial Hygienists once recommended 0.5 mg/m^3 for arsenic, then recommended a change to 0.25 mg/m^3 in 1974, as contrasted to the NIOSH recommendation of 0.002 mg/m^3 in 1975 (1).

These recommended values can also be compared with a limit of 0.01 set by NIOSH in an earlier version of its recommended standard in 1973. Thus, as can be seen, standards for arsenic are clearly in transition.

HANDLING PROCEDURES

Inorganic arsenic shall be controlled so that no worker is exposed to a concentration of arsenic in excess of 0.002 mg (2.0 μg) per cubic meter of air as determined by a 15-minute sampling period, according to NIOSH (1).

Medical surveillance shall be made available and related records kept according to a proposed NIOSH standard (1) for all workers occupationally exposed to arsenic.

A. Preplacement and annual medical examinations shall include:
   1. Comprehensive preplacement or annual interim work history.
   2. Comprehensive preplacement or annual interim medical history.
   3. 14" x 17" posterior-anterior chest x-ray, giving particular attention to parenchymal and hilar changes.
   4. Careful examination of the skin to detect the presence of arsenic-induced hyperpigmentation, keratoses, or other chronic skin lesions. Skin examinations shall be repeated bimonthly if arsenic-induced skin lesions are detected. Care shall be taken to observe and record the location, condition, appearance, size, and any changes in all such lesions.
   5. Palpation of superficial lymph nodes to detect indications of neoplastic changes.
   6. Complete blood count to include differential.
   7. An evaluation of the advisability of the worker's using negative- or positive-pressure respirators.

B. A periodic sputum cytology examination is recommended for all workers occupationally exposed to inorganic arsenic. The frequency of this procedure should be determined by the responsible medical authority.

C. Proper medical management shall be provided for workers adversely affected by occupational exposure to inorganic arsenic compounds.

D. Initial annual examinations for presently employed workers shall be offered within 6 months of the promulgation of a standard incorporating these recommendations.

E. The medical representatives of the Secretary of Health, Education and Welfare, of the Secretary of Labor, and of the employer shall have access to all pertinent occupational medical records. Physicians designated and authorized by any employee or former employee shall have access to that worker's medical records.

F. Medical records shall be maintained by the employer or successors thereto for persons employed one or more years in work involving exposure to arsenic. Preplacement x-rays, x-rays for the 5 years preceding termination of employment, and all other medical records with pertinent supporting documents shall be maintained at least 30 years after the individual's employment is terminated. In the event that the employer ceases business without a successor, records shall be forwarded by registered mail to the Director, National Institute for Occupational Safety and Health.

As regards labelling, according to the proposed NIOSH standard (1):

A. Containers of arsenic compounds shall bear the following label in addition to or in combination with labels required by other statutes, regulations, or ordinances.

NAME OF COMPOUND
DANGER! CONTAINS ARSENIC
CANCER CAUSING AGENT
HARMFUL IF INHALED OR SWALLOWED, OR UPON SKIN CONTACT
AVOID CONTACT WITH SKIN, EYES AND CLOTHING
WASH THOROUGHLY AFTER HANDLING

Avoid breathing dust or spray mist
Keep container closed
Use only with adequate ventilation

B. The following warning sign shall be affixed in a readily visible location at or near entrances to areas in which there is occupational exposure to arsenic.

ARSENIC
DANGER! CANCER CAUSING AGENT

Dust or spray mist may be hazardous to health
Provide adequate ventilation

This sign shall be printed both in English and in the predominant language of non-English-speaking workers unless they are otherwise trained and informed of the hazardous area. All illiterate workers shall receive such training.

As regards personal protective equipment and work clothing, the proposed NIOSH standard specifies in the areas of:

A. Protective clothing —
   1. Where there is occupational exposure to inorganic arsenic compounds, protective clothing shall be provided by the employer. This may include underwear, gloves, coveralls, and a hood over the head and neck. When liquids are being processed in a manner that may result in splashes, impervious gloves, aprons, and splash goggles shall be used.
   2. Protective clothing shall be changed at least daily at the end of the shift.
   3. Work clothing shall not be taken home by employees. The employer shall provide for maintenance and laundering of protective clothing.
   4. The employer shall ensure that precautions necessary to protect laundry personnel are observed when soiled protective clothing is laundered.

B. Respiratory protection —
   1. Engineering controls shall be used wherever feasible to maintain arsenic concentrations below the prescribed limit. Compliance with the permissible exposure limit may not be achieved by the use of respirators except:
      a. During the time period necessary to install or test the required engineering controls.
      b. For nonroutine operations such as a brief exposure to concentrations in excess of the permissible exposure limit as a result of maintenance or repair activities.
      c. During emergencies when air concentrations of arsenic may exceed the permissible limit.
   2. When a respirator is permitted by paragraph B.1. of this section, it shall be selected and used pursuant to the following requirements:
      a. The employer shall ensure that no worker is being exposed to

arsenic in excess of the environmental limit because of improper respirator selection, fit, use, or maintenance.

b. A respiratory protection program meeting the requirements of 29 CFR 1910.134 as amended shall be established and enforced by the employer.

c. The employer shall provide respirators as specified below and shall ensure that the employee uses the respirator provided.

d. Respiratory protective devices described below shall be those approved under the provisions of 30 CFR 11 as amended.

e. The employer shall ensure that respirators are adequately cleaned, and that employees are instructed in the use of respirators assigned to them, and how to test for leakage.

f. Where an emergency may develop which could result in employee exposure to arsenic, the employer shall provide respiratory protection as listed below:
    1. Combination supplied air respirator, pressure-demand type, with auxiliary self-contained air supply.
    2. Self-contained breathing apparatus with positive pressure in the facepiece.

At the beginning of employment in an arsenic area, employees exposed to arsenic compounds shall be informed of the hazards, relevant symptoms of overexposure, appropriate emergency procedures, and proper conditions and precautions for safe use. Instruction shall include, as a minimum, all information which is applicable to the specific arsenic-containing product or material to which there is exposure. The information shall be posted in the work area and kept on file and readily accessible to the worker at all places of employment where arsenic is involved in unit processes and operations.

A continuing educational program shall be instituted to ensure that all workers have current knowledge of job hazards, proper maintenance procedures and cleanup methods, and that they know how to correctly use respiratory protective equipment and protective clothing. Information as required shall be recorded on the "Material Safety Data Sheet" or on a similar form approved by the Occupational Safety and Health Administration, U.S. Department of Labor.

As regards work practices using inorganic arsenic compounds, the NIOSH recommended standard (1) specifies that:

A. Readily accessible standby rooms under positive air pressure and in which the concentration of arsenic in the air is less than 2.0 μg As/m^3 shall be provided where there is occupational exposure to inorganic arsenic.

B. Arsenic shall be removed from work areas by vacuum cleaning or wet methods. Cleaning may be performed by washing down with a hose, provided that a fine spray of water has first been laid down. Sweeping or other methods which can stir the dust into the air shall not be used.

C. Waste material shall be disposed of in a manner which will prevent exposure of humans and animals as well as air and water pollution.

D. Arsenic trichloride shall be handled only in enclosed systems sufficient to prevent skin contact and to prevent worker exposure in excess of the environmental limit.

E. Where there is possibility of arsenic trichloride contact with the skin, emergency showers shall be provided in readily accessible locations. Eye-wash facilities shall also be conveniently located.

F. Procedures for emergencies, including fire fighting, shall be established to meet foreseeable events. Necessary emergency equipment, including appropriate respiratory protective devices, shall be kept in readily accessible locations. Only self-contained breathing apparatus with positive pressure in the face-piece shall be used for fire fighting. Appropriate respirators should also be available for use during evacuation.

G. Exhaust ventilation and enclosure of processes shall be used wherever practicable to control workplace concentrations.

H. Air from the exhaust ventilation system shall not be recirculated into work areas, and necessary measures shall be taken to ensure that discharge out-doors will not produce a health hazard to humans or animals.

I. Due to potential skin irritation associated with respirator use and arsenic dust exposure, workmen shall be permitted to leave the work area every 2 hours to wash their face and obtain a clean respirator.

Arsenic trichloride can cause irritation or ulceration on contact or may be absorbed through the skin with fatal results (19)(20). Since its vapor pressure at 25°C is sufficient to produce an air concentration of 14,000 ppm (104,000 milligram per cubic meter), (21) its handling requires complete enclosure.

As regards industrial sanitation practices when using inorganic arsenic, the NIOSH recommended standard (1) specifies that:

A. Employees exposed to arsenic shall be provided with separate lockers or other storage facilities for street clothes and for work clothes.

B. Employees exposed to arsenic shall not wear work clothing away from the plant.

C. Facilities for shower baths shall be provided for employees exposed to arsenic. Workers shall shower bathe before changing into street clothes. Shower baths shall be cleaned following use after each work shift.

D. Employees exposed to arsenic shall wash their hands and exposed skin before eating or smoking during the work shift.

E. No food shall be permitted in areas where arsenic is handled, processed, or stored.

F. Employees shall not smoke in areas where arsenic is handled, processed, or stored.

As regards monitoring and record keeping, the NIOSH recommended standard (1) specifies the following.

Workroom areas shall not be considered to have arsenic exposure if environmental levels, as determined on the basis of a professional industrial hygiene survey, are less than 2.0 μg As/m^3. Records of these surveys, including the basis for concluding that air levels are less than 2.0 μg As/m^3, shall be maintained. In workrooms where inorganic arsenic compounds are handled or processed, surveys shall be repeated at least annually and when any process change indicates a need for reevaluation. Requirements set forth below apply to areas in which 1 or more 15-minute breathing zone samples have indicated exposure at or above 2.0 μg As/m^3.

Employers shall maintain records of environmental exposures to arsenic based upon the following sampling and recording schedules.

A. In all monitoring, sufficient breathing zone samples shall be collected to characterize the potential exposure of workers at each operation or process.

B. The first environmental sampling shall be completed within 6 months of the promulgation of a standard incorporating these recommendations.

C. Environmental samples shall be taken within 30 days after first operation of a new process or process changes.

D. Samples shall be collected at least every 2 months for those work areas in which there is occupational exposure to inorganic arsenic.

E. Records of all sampling and of medical examinations shall be maintained by the employer or successors thereto for at least 30 years after the individual's employment is terminated. Records shall indicate the type of personal protective devices, if any, in use at the time of sampling. Records shall be maintained so that exposure information is available for individual employees, and each employee shall be able to obtain information on his own exposure. In the event that the employer ceases business without a successor, records shall be forwarded by registered mail to the Director, National Institute for Occupational Safety and Health.

Even outdoor agricultural uses of arsenic compounds may produce potentially hazardous exposures for nearby personnel. Engineering control methods used will depend on the equipment and techniques used to apply the chemicals. Protective clothing and respiratory protection may be needed as supplemental controls.

REMOVAL FROM AIR

Air pollution caused by arsenical compounds has been observed near gold and copper smelters as well as in the areas where arsenic is used for agricultural purposes. A small amount of arsenic can be measured in the air in most cities, according to Sullivan (22).

Arsenic is a common industrial nuisance wherever arsenical ores are smeltered. Before the advent of organic insecticides (e.g., DDT) the use of arsenicals as pesticides was increasing. However, since then, their use has leveled off and declined as the organic pesticides have taken their place. As a result, the supply of arsenic is greater than the demand, and the only economic incentive to remove arsenic from the exhaust fumes of smelters is the presence of other trace metals, such as tellurium, selenium, tin, zinc and antimony.

Control of arsenic emissions requires special attention to the temperature of exhaust gases since arsenic trioxide sublimes at 192°C. For this reason exhaust fumes must be cooled to approximately 100°C prior to removing them as particulates, as pointed out by Booz Allen Applied Research, Inc. (2).

Where fumes may be present, as in the sintering and roasting of arsenic-bearing ores, complete enclosure and exhaust ventilation of the operation are essential (21). Operations that agitate arsenic trioxide dust, e.g., grinding, screening, shoveling, sweeping and transferring, require control since the dust is very fine and disperses easily (23). When the operation has not been sufficiently enclosed and ventilated, supplemental protective clothing and respiratory protection may be needed until adequate engineering controls are installed, according to NIOSH (1).

In general, the removal of particulate material will control arsenic emissions if the control equipment operates at a temperature low enough (~100°C) to condense the arsenic fumes. An electrostatic precipitator has been reported to reduce the arsenic from 5 to 17 ppb to 0 to 4 ppb (24). Cooling flues, baghouses, and electrostatic precipitators have been used in the smelting industry. No data have been reported in the United States on their removal efficiency for arsenic.

However, at a chemical plant in the U.S.S.R., the efficiency for arsenic removal was greatly improved by using wet vacuum pumps instead of fabric filters. When the fabric filters were used, the arsenic content in the air frequently reached several hundred thousand micrograms per cubic meter. After the wet scrubbing vacuum pumps were installed, the removal is reportedly 100% effective (25).

In the cotton industry, removal of particulate material emitted from cotton gins should control the arsenic emissions (26). However, methods need to be developed to control the arsenical emissions produced by the burning of cotton trash.

REMOVAL FROM WATER

Only limited information was encountered on current levels of arsenic in industrial wastes, and on current treatment processes and removals obtained according to J.W. Patterson and R.A. Minear (27). Most of the literature describing industrial sources and treatment of arsenic is 30 years or more old. More recent industrial literature, while referring to arsenical wastes and to the severe pollution resulting from their discharge, presents no specific treatment processes or industrial waste values. More up-to-date information is available on the removal of arsenic from drinking water, and in fact the methods for arsenic treatment of both drinking water and industrial wastes are similar. The treatment methods and arsenic removal efficiencies are summarized in Table 6.

TABLE 6: SUMMARY OF ARSENIC TREATMENT METHODS AND REMOVALS ACHIEVED

Treatment	Initial Arsenic (mg/l)	Final Arsenic (mg/l)	% Removal
Lime softening	0.2	0.03	85
Charcoal filtration	0.2	0.06	70
Ferric sulfide filter bed	0.8	0.05	94
Coagulation with ferric sulfate	25.0	5 or less	80 or more
Coagulation with ferric chloride	3.0	0.5	98
Precipitation with ferric hydroxide	—	0.6	—
Precipitation with ferric hydroxide	362.0*	15-20*	94-96

*As arsenious oxide.

Source: Reference (27).

SOLID WASTE DISPOSAL

As regards arsenic trichloride, and as described by P.W. Powers (28), Rocky

Mountain Research, the only current producer of arsenic trichloride in the U.S., has indicated its willingness to accept excessive or waste arsenic trichloride for reprocessing, purification or disposal. It is fed into their reactor and purified in the same manner as a normal batch. Rocky Mountain Research was not willing to discuss the details of their process and hence an evaluation of the process and the purification technique for $AsCl_3$ were not available.

Rocky Mountain Research's alternate procedure would be to add water to the waste under ventilation control to hydrolyze the arsenic trichloride to arsenic trioxide which is much easier to handle. It is conceivable that any industry that handles $AsCl_3$ in significant amounts would have special facilities for hydrolyzing the $AsCl_3$ and scrubbing the evolved HCl.

Landfilling of arsenic trichloride is not considered as an acceptable disposal option because of: [1] the high degree of toxicity of the compound; [2] the nondegradable nature of the toxic arsenic component of arsenic trichloride, whereby the compound would not be converted to a less toxic or nontoxic form in the soil environment; and [3] the reactivity of arsenic trichloride with water to yield arsenic trioxide and corrosive hydrochloric acid. The corrosive potential of arsenic trichloride also indicates that the land burial of arsenic trichloride wastes containerized in metal drums is not an adequate method of disposal.

As described by R.S. Ottinger et al (29), various options exist for ultimate disposal of arsenical pesticide wastes.

Incineration: Compounds containing heavy metals including lead, mercury, and arsenic should be burned only if proper precautions are taken to control atmospheric emissions. Lead and mercury salts as well as arsenic can escape the combustion chamber and travel great distances, contaminating large areas. Low levels of these compounds may produce recognizable effects in plants or animals as well as people. Compounds containing chlorine produce hydrochloric acid, which can cause serious corrosion problems in the incinerator and in some cases significant downwind pollution problems.

Land Application: Some users, when faced with unwanted stocks of some of these pesticides, use land spreading as a means of disposal. This approach involves spreading the waste materials over large amounts of land in very light applications so as not to significantly affect any of the crops or plant life growing in the vicinity. The technique might be used for rare or infrequent disposal of small amounts of the material but in light of the fact that these compounds will build up in the soil by repeated application, this method cannot be considered as completely adequate and other methods should be considered first.

Landfills: The disposal of cacodylate wastes in sanitary landfills is generally not acceptable because of the potential danger of ground and surface water pollution, as well as possible occupational hazards resulting from on-site handling. There are, however, certain approved sites located over non-water-bearing sediments or with only unusable groundwater underlying them which are completely protected from flooding and surface runoff and drainage such as those designated as Class 1 sites in California. The disposal of small stocks of cacodylate wastes or empty containers contaminated with cacodylates in Class 1 sites is considered as adequate, provided special handling techniques are employed to protect site personnel.

The solid waste salt by-products from cacodylate manufacture contains sodium chloride, sodium sulfate, and 1 to 1.5% cacodylate contaminants. The volume of this waste material currently in storage amounts to 60,000,000 lb. There is a definite need for the development of an economical process capable of extracting the cacodylate constituents from the salt waste so that the latter can be readily and safely disposed of in municipal landfills (3).

The disposal of arsenate or arsenite compounds in sanitary landfills is generally not acceptable because of the potential danger of ground and surface water pollution as well as possible occupational hazards resulting from on-site handling. There are, however, certain approved sites located over non-water-bearing sediments or with only unusable groundwater underlying them which are completely protected from flooding and surface run-off or drainage such as those designated as Class 1 sites in California. The disposal of small stocks of arsenate or arsenite pesticides or empty containers used to ship arsenates or arsenites in Class 1 sites is considered as adequate, provided special handling techniques are employed to protect site personnel.

Long-Term Storage: Storage of cacodylates and cacodylate wastes, until they can be used or reprocessed, is a satisfactory waste management option. These materials are stable compounds and require the minimal storage precautions needed for any toxic material. Storage in the original containers is recommended but they should be periodically checked for corrosion or breakage of the containers. Bulk quantities of waste, such as the salt by-products from cacodylate manufacture, can be stored in concrete vaults or weatherproof bins.

Large, weatherproof and siftproof storage bins or silos are currently being used for the storage of arsenic compounds, especially arsenic trioxide (29). Waste arsenate or arsenite pesticides already packed in multiwall paper bags could also be safely stored in these large weatherproof bins. This approach is relatively expensive. Land and storage equipment are required for long, unknown periods of time before these materials can be moved by sale or disposal to other outlets. Considering the fact that these arsenic compounds are poisonous in all forms, long-term storage has to be considered as adequate and currently practical.

P.W. Powers (28) has tabulated hazardous waste constituents for which landfill disposal is considered an acceptable waste disposal alternative. The arsenic compounds included are as follows:

Material	Recommended Disposal Method
Cacodylic acid	1
Calcium arsenate	2
Calcium arsenite	2
Copper acetoarsenite	2
Copper arsenate	2
Lead arsenate	2
Lead arsenite	2
Magnesium arsenite	2
Manganese arsenate	2
Nickel arsenide	3
Potassium arsenite	2
Sodium arsenate	2
Sodium arsenite	2
Sodium cacodylate	1
Zinc arsenate	2
Zinc arsenite	2

The methods cited above are as follows:

1. Long-term storage in concrete vaults or weatherproof bins; small amounts may be disposed of in a chemical waste landfill.
2. Long-term storage in larger weatherproof and sift-proof storage bins or silos; small amounts may be disposed of in a chemical waste landfill.
3. Encapsulation followed by disposal in a chemical waste landfill.

ECONOMIC IMPACT OF CONTROLS

No specific economic data are available on the cost of arsenic removal from air (30). The general economics associated with particulate control technology which is appropriate for arsenic removal at exhaust stream temperatures of less than 100°C have been considered (30). Costs of cooling the gas stream in order to deal with the particulate phase alone requires additional costs. No treatment costs directly related to wastewaters were encountered in the literature search reported by Patterson and Minear (27). Very recently, however, OSHA has issued a comprehensive report (31) on the economic/inflationary impacts of a proposed new standard for inorganic arsenic.

REFERENCES

(1) Nat. Inst. for Occupational Safety and Health, *Criteria for a Recommended Standard . . . Occupational Exposure to Inorganic Arsenic,* Publ. No. NIOSH 75-149, Wash., D.C. (1975)

(2) Booz Allen Applied Research, Inc., *A Study of Hazardous Waste Materials, Hazardous Effects and Disposal Methods,* Vol. II, Report PB-221,466, Springfield, Va., Nat. Tech. Information Service (July 1973)

(3) Luckey, T.D., Venugopal, B. and Hutcheson, D., *Heavy Metal Toxicity, Safety and Hormonology,* New York, Academic Press (1975)

(4) Schroeder, H.A., *The Poisons Around Us,* Bloomington, Indiana University Press (1974)

(5) Tucker, A., *The Toxic Metals,* London, Earth Island Ltd. (1972)

(6) Goldberg, A.J., *A Survey of Emissions and Controls for Hazardous and Other Pollutants,* Wash., D.C., Environmental Protection Agency (Feb. 1973)

(7) Cherkinski, S.N. and Genzburg, F.I., "Purification of Arsenious Wastewaters," *Water Poll. Abstr.* 14, 315-16 (1941)

(8) Trelles, R.A. and Amato, F.D., "Treatment of Arsenical Waters With Lime," *Water Poll. Abstr.* 23, 125 (1950)

(9) Viniegra, G. and Marquez, R.E., "Chronic Arsenic Poisoning in the Lake Region, Section 4, Treatment of Drinking Water," *Water Poll. Abstr.* 38, 430-31 (1965)

(10) Shen, Y.S. and Chen, C.S., "Relation Between Blackfoot Diseases and the Pollution of Drinking Water by Arsenic in Taiwan," *Proc. Second Internat. Conf. Water Poll. Research, Tokyo* 1, 173-190, New York, Pergamon Press (1964)

(11) Stroterhoff, H.L., U.S. Patent 3,966,412; June 29, 1976; assigned to U.S. Secretary of the Army

(12) Dubois, L., and Monkman, J.L., "Determination of Arsenic in Air and Biological Materials," *Am. Ind. Hyg. Assoc. J.* 22, 292-95 (1961)

(13) ACGIH Committee on Recommended Analytical Methods, "Determination of Arsenic in Air," *Manual of Analytical Methods Recommended for Sampling and Analysis of Atmospheric Contaminants.* Cincinnati, American Conference of Governmental Industrial Hygienists, pp Arsenic 1-6 (1958)

(14) Intersociety Committee: "Tentative Method of Analysis for Arsenic Content of Atmospheric Particulate Matter," *Methods of Air Sampling and Analysis,* Washington, American Public Health Association, pp 289-92 (1972)

(15) Fernandez, F.J. and Manning, D.C., *Atomic Absorption Application Study No. 468-The Determination of Arsenic and Selenium at Sub-microgram Levels.* The Perkin-Elmer Corporation, Instrument Division, Norwalk, Conn., 10 pp (June 1971)

(16) Fernandez, E.J., "Atomic Absorption Determination of Gaseous Hydrides Utilizing Sodium Borohydride Reduction," *Atomic Absorption Newsletter* 12, 93-97 (1973)

(17) National Institute for Occupational Safety and Health, Division of Laboratories and Criteria Development: "Arsenic in Urine and Air - Physical and Chemical Analysis Branch Method 139, NIOSH *Manual of Analytical Methods,* HEW Publication No. (NIOSH) 75-121, Cincinnati, U.S. Dept. Health, Education and Welfare, Public Health Service, Center for Disease Control, NIOSH, pp 139-1 to 139-8 (1974)

(18) Barnett, W., *Atomic Absorption Application Study No. 523 - Determination of As and Se with the HGA,* The Perkin-Elmer Corporation (May 1973)

(19) Buchanan, W.D., "Toxicity of Arsenic Compounds," Browning E. (ed): *Elsevier Monographs on Toxic Agents.* New York, Elsevier Publishing Co. (1962)

(20) Delephine, S., "Observations upon the Effects of Exposure to Arsenic Trichloride upon Health." *J. Ind. Hyg.* 4, 346-64, 410-23 (1923)

(21) "Arsenic and Its Compounds (Except Arsine), Revised 1964." AIHA Hygienic Guide Series. *Am. Ind. Hyg. Assoc. J.* 25, 610-13 (1964)

(22) Sullivan, R.J., *Air Pollution Aspects of Arsenic and Its Compounds,* Report PB-188,071, Springfield, Va., Nat. Tech. Information Service (Sept. 1969)

(23) Patty, F.A., "Arsenic, Phosphorus, Selenium, Sulfur, and Tellurium," in Patty, F.A. (ed): *Industrial Hygiene and Toxicology,* ed 2 rev.; *Toxicology* (D.W. Fassett, D.D. Irish, ed). New York, Interscience Publishers, vol. 2, pp 871-80 (1962)

(24) Chambers, C.A., Foter, M.J. and Cholak, J., "A Comparison of Particulate Loadings in the Atmosphere of Certain American Cities," *Proc. Third Natl. Air Pollution Symposium,* pp 24-32, Pasadena, Calif. (1955)

(25) Matsak, V.G., "The Utilization of Air, Dust and Smoke Purification Equipment," *Survey of the U.S.S.R. Literature on Air Pollution and Related Occupational Diseases* 3, 141 (1960)

(26) *Control and Disposal of Cotton Ginning Wastes; A Symposium,* Public Health Service Publication 999-AP-31 (1966)

(27) Patterson, J.W. and Minear, R.A., *Wastewater Treatment Technology,* Report No. 73-1, Chicago, Illinois Institute for Environmental Quality (1973)

(28) Powers, P.W., *How to Dispose of Toxic Substances and Industrial Wastes,* Park Ridge, N.J., Noyes Data Corp. (1976)

(29) Ottinger, R.S., Blumenthal, J.L., DalPorto, D.F., Gruber, G.I., Santy, M.J. and Shih, C.C., *Recommended Methods of Reduction, Neutralization, Recovery or Disposal of Hazardous Wastes, Vol. V, Pesticides and Cyanides,* Report PB 224,584, Springfield, Va., Nat. Tech. Information Service (Aug. 1973)

(30) McFarland, A.R., Ed., *Control Technology for Toxic and Hazardous Air Pollutants,* Chicago, Illinois Institute for Environmental Quality (Dec. 1974)

(31) Occupational Safety and Health Administration, *Inflationary Impact Statement - Inorganic Arsenic,* Wash., D.C. (1976)

BARIUM

Chemical inertness, high density and low cost make barite especially suitable as a weighting agent for oil and gas well-drilling muds, which accounted for over 73% of barium consumption in 1968. Thus, the demand for barium as barite becomes a function of oil and gas drilling activity and fluctuates accordingly. The low cost of barite precludes the employment of materials that might otherwise be used as substitutes in the drilling-mud market. This overwhelming dependence on a single demand creates a problem, as the long-term stability of the market is somewhat speculative. Barium as barite and barium compounds also are important in the production of glass, paints, rubber and many chemicals making up the remaining 27% of demand.

The demand for barium in the year 2000 for the United States has been forecast to fall between 0.81 and 1.60 million short tons. The major use will continue to be for drilling mud and the high of the demand range is contingent on increased oil-well drilling activity. Demand will be influenced upward also by the production of various barium compounds having properties that suggest possible large scale markets. An example is the existing potential for an appreciable use of barium chloride for water treatment and water pollution abatement, provided lower manufacturing costs can be achieved.

Conversely, technological shifts in fuel supplies could reduce the oil and gas well drilling activities to such an extent that demand for barium by the year 2000 for this use would represent only 50% of total consumption, and thus cause the demand to gravitate toward the low end of the range.

TOXICITY

Schroeder (2) points out that too little is known of the subtle toxicity of many metals. He goes on to say that "when breathed or swallowed, barium shows little toxicity. It may accumulate in the lungs, giving rise to a mild lung disease. Added to diesel oil to control smoke in exhausts, it appears in high concentrations behind idling buses and trucks, and we probably inhale a lot of it when we are delayed in traffic behind an evil-smelling bus. Barium is also a natural contaminant of dusts, for there is plenty on the earth's surface. Huge amounts are swallowed for contrast in x-rays of the stomach and intestine."

Luckey et al (3) points out that barium closely resembles calcium in its metabolism, but differs from calcium in greater excretion in the feces. They go on to point out that the intestinal mucosa is highly permeable to soluble barium salts and is involved in rapid flow of barium salts in and out of the blood from which it disappears very rapidly.

Barium sulfate, which is nontoxic, is poorly absorbed in the digestive tract and hence is used as a clinical and nutritional marking substance. Barium deposits in the bones and also exerts a prolonged stimulative effect on all types of muscles.

Barium chloride and carbonate are toxic to rats orally at levels from 0.2 to 0.5 gram per kilogram as shown in Table 7 taken from Luckey et al (3).

TABLE 7: TOXICITY OF SOME BARIUM SALTS

Compound	Animal	Route	Toxic dose	Dosage expressed in mg of compound/Kg	Dosage expressed in mg of metal/Kg	Death in
Barium acetate	rabbit	oral	LD	236	127	1 day
	rabbit	oral	LD	815	438	1 hour
	rabbit	sc	LD	96	51.6	
	rabbit	iv	LD	12	6.4	
Barium carbonate	mouse	oral	LD	200	112.3	
	rat	oral	LD	800	450	
	rat	iv	MLD	20	13.9	
Barium chloride	rat	sc	LD	45-90	31.3-62.6	
	guinea pig	sc	LD	55	38.2	
	rabbit	sc	LD	40-75	27.8-52.1	
	rabbit	iv	LD	15	10.4	
	cat	sc	LD	18-60	12.5-41.7	
	cat	iv	LD	40-60	27.8-34.8	
	dog	sc	LD	10-20	6.9-13.8	
	mouse	oral	LD	70-140	48.7-97.4	
	rat	oral	LD	355-535	247-372	

ource: Reference (3)

etention of inhaled fine dusts of barium sulfate forms harmless nodular granution in the lungs, according to Luckey et al (3).

iner (4) states that soluble barium salts are very toxic, and if ingested, have a rong stimulating effect on all muscles, including the heart. Symptoms of barm poisoning are excessive salivation, vomiting, colic, diarrhea, convulsive trems, slow, hard pulse and increased blood pressure. There have been very few ported deaths due to barium poisoning, however.

soluble barium compounds, such as common barium sulfate, are generally nonxic. Inhaled barium compounds are known to cause a benign pneumoconiosis lled baritosis, which does not cause symptoms of emphysema, bronchitis or duced respiratory capacity according to Miner (4).

Tests on animals with exhaust solids from a diesel engine using fuels with barium-base smoke-suppressant additives indicated that the LD_{50} of these exhaust solids was in excess of 10 g/kg of body weight. Very limited animal studies with these exhaust solids do not appear to have demonstrated any acute health effects from the concentrations used; no information is available on chronic effects.

EXTENT OF EXPOSURE

According to the U.S. Bureau of Mines (1), there are no environmental problems connected with the mining and beneficiation of barite other than the normal land disturbance generally associated with mining. Flotation plants have the usual tailing waste disposal ponds. The wastes generated by flotation and grinding are relatively inert and cause no appreciable amount of air or water pollution.

Some chemical processing of barite to barium chemicals is subject to possible pollution problems. The production of barium chloride produces considerable quantities of hydrogen sulfide gas which cannot be released into the atmosphere. This toxic gas is successfully recovered and converted into by-product sulfides.

Waste ponds of oil-well drilling muds are usually abandoned after completion of the well. Such waste ponds could be a nuisance in some areas. However, the land should not be difficult to reclaim as the need arises.

Table 8 and Figure 3 show primary barium emission source data accumulated by EPA (5).

TABLE 8: BARIUM EMISSIONS

Source	Amount in Tons	Percent This Pollutant
Barium mining	30	0.28
Blast furnace	112	1.04
Open hearth	38	0.35
Basic oxygen furnace	20	0.19
Electric arc furnace	36	0.33
Gray iron foundry cupola	50	0.46
Glass manufacture	40	0.37
Barium milling and handling	2,700	24.94
Barium chemicals	4,400	40.64
Paint, varnish, etc. manufacture	30	0.28
Well drilling mud	70	0.65
Power plant boilers		
Pulverized coal	2,311	21.35
Stoker coal	266	2.46
Cyclone coal	80	0.74
All oil	29	0.27
Industrial boilers		
Pulverized coal	102	0.94
Stoker coal	358	3.31
Cyclone coal	51	0.47
All oil	22	0.20
Residential/commercial boilers		
Coal	32	0.30
Oil	49	0.45
Total	10,826	100.02

Source: Reference (5)

FIGURE 3: MATERIAL FLOW THROUGH THE ECONOMY SHOWING PRIMARY EMISSION SOURCES—BARIUM

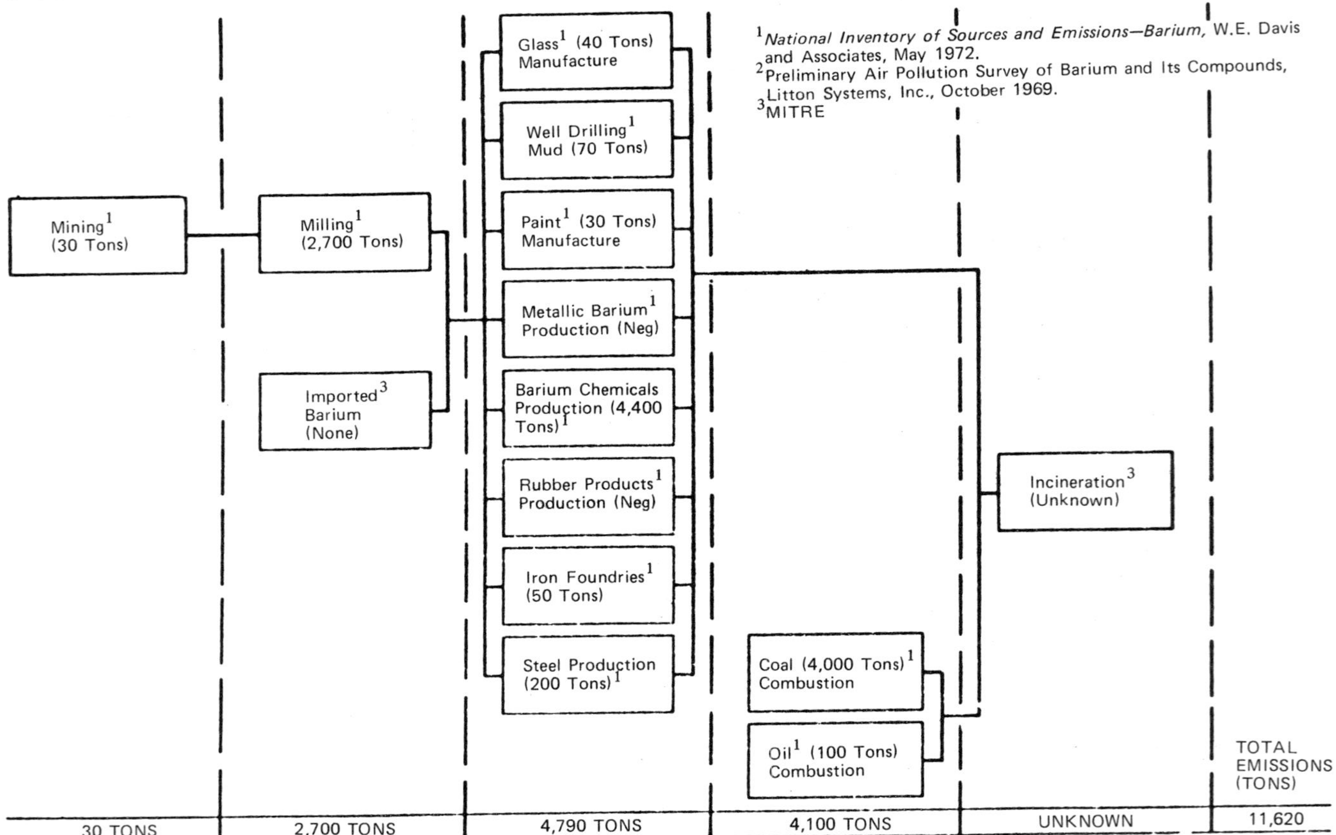

[1] *National Inventory of Sources and Emissions—Barium,* W.E. Davis and Associates, May 1972.
[2] Preliminary Air Pollution Survey of Barium and Its Compounds, Litton Systems, Inc., October 1969.
[3] MITRE

Source: Reference (5)

The emission factors presented in Table 9 are the best currently available for emissions of barium compounds to the atmosphere. They were determined through a combination of methods consisting of: [1] direct observation of emission data and other related plant processing and engineering data; [2] estimates based on information obtained from literature, plant operators and others knowledgeable in the field; [3] calculations based on experience and personal knowledge of metallurgical processing operations; and [4] specific analytical results where available.

TABLE 9: EMISSION FACTORS

Mining	100 lb/1,000 tons barite mined
Processing	5 lb/ton barite processed
Barium chemicals	50 lb/ton barite processed
End product uses of barium	
Oil and gas well drilling	110 lb/1,000 tons barite used
Glass manufacture	2 lb/ton barium processed
Paint manufacture	2 lb/ton barium processed
Rubber tire wear	0.3 lb/million miles
Other emission sources	
Coal	15 lb/1,000 tons of coal burned
Cast iron	5 lb/1,000 tons of process weight

Source: Reference (6)

Barium as barite (barium sulfate), is used as a white pigment for paint. Barium is employed in metallurgy, glass, ceramics and dyes manufacture, and in vulcanizing of rubber. Barium has been also reported present in explosives manufacturing wastewater. Very little information is available in the literature on levels of barium in industrial aqueous wastes, however.

DETECTION

Dusts and fumes of barium compounds may be collected by any of the usual methods for collection of particulate matter. For concentrations of barium encountered in ambient air, sampling has to be done with membrane filters (7). Barium solids from diesel engine exhaust samples have been collected on analytical-grade filter paper (8).

Any barium sample soluble in hydrochloric acid can be analyzed by spectrographic methods (9). Low concentrations of barium can be quantitatively determined by emission spectrography or atomic absorption (7). Thompson et al (10) reported that when barium is analyzed by atomic absorption, the minimum detectable limit is 0.02 $\mu g/m^3$ based on an air sample of 2,000 m^3.

To convert all the barium collected on a filter to a soluble form for spectrographic analysis, the filter is ashed and the ash is dissolved in a hydrochloric acid and nitric acid mixture (7). Golothan (8) separated the soluble and insoluble barium compounds by digesting the solids on the filter paper with hydrochloric acid. This acid solution was analyzed directly for the soluble barium compounds. The solid residue (remaining after the acid treatment) was ashed and treated with acid and subsequently analyzed to determine the amount of insoluble barium compounds in the original sample.

Industrially, barium is usually determined by precipitating it as insoluble barium sulfate, separating the precipitate, and weighing it (11)(12). When barium is associated with other alkaline earths such as calcium and strontium, the calcium is separated as calcium nitrate by dissolving it in a mixture of alcohol and absolute ether. The barium is then precipitated as barium chromate from the slightly acidified (acetic acid) mixture (12). Barium may be monitored in air by emission spectroscopy (13). Barium may be monitored in water down to a detection limit of 60 ppb by atomic absorption spectroscopy as noted by Sabadell (14).

ENVIRONMENTAL STANDARDS

The permissible concentration of barium in domestic water supplies is 1.0 ppm as noted by Mancy (15). Also, effluents to storm sewers or streams should contain a maximum of only 1.0 ppm of barium.

The American Conference of Governmental Industrial Hygienists has recommended an 8-hour threshold limit of 500 $\mu g/m^3$ for occupational exposure to the soluble compounds of barium in air. No ambient air quality standards for barium and its compounds are known to exist for either the United States or any other country.

HANDLING PROCEDURES

There are no special handling procedures published for barium compounds. Obviously in view of the toxic nature of the soluble salts of barium, care should be taken to avoid their ingestion, however.

REMOVAL FROM AIR

No information has been found on the abatement of barium or its compounds in the environmental atmosphere according to Miner (4). Since most of the barium processed by industry is in solid form, the conventional methods for removal of solids, such as bag filters, electrostatic precipitators and wet scrubbers, should prove effective in preventing their escape to the atmosphere. No information has been found on the control of emission of barium-containing exhaust solids from diesel engines operating with fuel containing barium-base smoke-suppressant additives.

REMOVAL FROM WATER

Very little information is available in the literature on levels of barium in industrial wastewaters, on existing methods of treatment of levels of barium removal achievable, or on costs associated with such methods of treatment according to Patterson and Minear (16).

Treatment of barium contained in an explosives manufacturing wastewater has been described in the literature (17). Barium was removed by precipitation as barium sulfate, upon addition of sodium sulfate. Barium sulfate is extremely insoluble, having a maximum theoretical solubility of approximately 1.4 mg/l as barium, and the treatment process is presumably quite effective. Coagulation of suspended solids in the wastewater (including the finer nonsettleable barium sulfate particles) with ferric sulfate provided solids removal. It has been reported that the above process of precipitation of barium sulfate by addition of sodium sulfate is used by barite producers to recover barium from solution (20).

The second barium removal treatment process discussed in the literature is ion exchange. This is discussed in specific relation to barium removal from nuclear wastes (18)(19). Ayres (18) has reported extremely high removals of 99.999% for radioactive barium-140 by ion exchange, while Pressman, Lindsten and Schmitt (19) report 99% removal of nonradioactive barium in a U.S. Army mobile ion exchange unit. These authors also found 99.9% removal of barium by an Aquamite-30B electrodialysis demineralization unit. No costs were reported for sulfate precipitation, ion exchange or electrodialysis.

Based upon relative economics, however, and the very low solubility of barium sulfate, the precipitation method of barium removal would probably be the method of choice by industry. The effectiveness of this method could presumably be enhanced by use of coagulation and filtration to remove fine, nonsettleable particles of barium sulfate.

SOLID WASTE DISPOSAL

The land disposal of barium compounds may be handled by various techniques as described by P.W. Powers (21), and as follows:

Material	Recommended Disposal Method
Barium fluoride	1
Barium nitrate	2
Barium sulfide	2

where the applicable methods are:

1. Precipitation with soda or slaked lime. The resulting sludge should be sent to a chemical waste landfill.
2. Chemical reaction with water, caustic soda and slaked lime resulting in the precipitation of the metal sludge which may be landfilled.

ECONOMIC IMPACT OF CONTROLS

No data are currently available on the economic impact of controls for barium.

REFERENCES

(1) Lewis, R.W., in *Mineral Facts and Problems,* Wash., D.C., U.S. Bureau of Mines (1970).

(2) Schroeder, H.A., *The Poisons Around Us,* Bloomington, Indiana University Press (1974).

(3) Luckey, T.D., Venugopal, B. and Hutcheson, D., *Heavy Metal Toxicity, Safety and Hormonology,* New York, Academic Press (1975).

(4) Miner, S., *Air Pollution Aspects of Barium and Its Compounds,* Report PB-188,083, Springfield, Va., Nat Tech Information Service (Sept. 1969).

(5) Goldberg, A.J., *A Survey of Emissions and Controls for Hazardous and Other Pollutants,* Wash., D.C., Environmental Protection Agency (Feb. 1973).

(6) Davis, W.E. and Assoc., *National Inventory of Sources and Emissions, Section 1: Barium,* Report PB 210,676, Springfield, Va., Nat Tech Information Service (May 1972).

(7) Morgan, G., Director, Division of Air Quality and Emissions Data, National Air Pollution Control Administration, (May 1969) quoted in Reference (4).

(8) Golothan, D.W., *Diesel Engine Exhaust Smoke. The Influence of Fuel Properties and the Effects of Using Barium-Containing Fuel Additives, SAE-670092,* Society of Automotive Engineers, (1967).

(9) Abernethy, R.F. and F.H. Gibson, "Rare Elements in Coal," U.S. Bur. Mines, Inform. Circ. 8163, (1963).

(10) Thompson, R.J., G.B. Morgan and L.J. Purdue, "Analyses of Selected Elements in Atmospheric Particulate Matter by Atomic Absorption," Preprint, presented at the Instrument Society of America Symposium, New Orleans, La. (May 5-7, 1969).

(11) "Barium," in *Encyclopedia of Science and Technology,* Vol. 2 (New York: McGraw-Hill, p. 94, 1960).

(12) "Barium" and "Barium Compounds," in *Kirk-Othmer Encyclopedia of Chemical Technology,* Vol. 3 (New York: Wiley, pp. 77, 80, 1964).

(13) Lawrence Berkeley Laboratory, *Instrumentation for Environmental Monitoring—Air,* Publ. LBL-1, Vol. 1, Berkeley, Univ. of Calif. (Feb. 1, 1973).

(14) Sabadell, J.E., Ed., *Traces of Heavy Metals in Water: Removal Processes and Monitoring,* Report EPA-902/9-74-001, Wash., D.C., Environmental Protection Agency, (Nov. 1973).

(15) Mancy, K.H., in NBS Publication 361, *Analytical Chemistry—Key to Progress on National Problems,* Wash., D.C., U.S. Govt. Printing Office (1972).

(16) Patterson, J.W. and R.A. Minear, *Wastewater Treatment Technology,* 2nd Ed., Chicago, Illinois Inst. of Environmental Quality (Feb. 1973).

(17) "Taming Explosives," *Indust. Eng. Chem.* 46 (12), 20A (1954).

(18) Ayres, J.A., "Treatment of Radiocative Waste by Ion Exchange," *Indust. Eng. Chem.* 43, 1526-1531 (1951).

(19) Pressmen, M., Lindsten, D.C. and Schmitt, R.P., "Removal of Nuclear Bomb Debris; Strontium-90—Yttrium-90, and Caesium-137—Barium-137 From Water with Corps of Engineers Mobile Water Treatment Equipment," *Wat. Poll. Abst.,* 36, 42-43 (1963).

(20) *The Economics of Clean Water, Vol. III, Inorganic Chemicals Industry Profile,* U.S. Dept. Interior, Washington (1970).

(21) Powers, P.W., *How to Dispose of Toxic Substances and Industrial Wastes,* Park Ridge, N.J., Noyes Data Corp. (1976).

BERYLLIUM

According to Schroeder (1), "Beryllium is a bad actor, the most toxic of the environmental industrial metals. We should avoid it at all costs." However, as pointed out by the U.S. Bureau of Mines (2), beryllium and its compounds have certain unique properties which, regardless of relatively high cost, promise a growing and diversified application in special and critical end uses. Presently, beryllium-copper alloys and beryllium metal are the principal use forms. A small quantity is employed in the ceramic and chemical industry.

TOXICITY

Almost all the presently known beryllium compounds are acknowledged to be toxic in both the soluble and insoluble forms, depending on the amount of material inhaled and the length of exposure according to Durocher (3). Soluble beryllium compounds, such as beryllium sulfate and beryllium chloride, commonly produce acute pneumonitis; insoluble compounds, such as metallic beryllium and beryllium oxide, can produce chronic pulmonary disease (berylliosis). However, it should be noted that the toxic beryllium effect is not limited to berylliosis but instead is a body-wide systemic disease.

As pointed out by Tucker (4), in the case of beryllium there is a public health hazard as well as an industrial occupation risk. This came to light in 1949 when the Atomic Energy Commission discovered that one of its large beryllium plants was leading to airborne contamination at concentrations up to 0.1 microgram per cubic meter as much as three-quarters of a mile away (5). Further investigation uncovered 10 cases of beryllium poisoning, one fatal, among the residents of the area not employed at the factory.

Inhalation of beryllium or its compounds is highly toxic to humans and animals, producing body-wide systemic disease commonly known as beryllium disease (6). Both acute and chronic manifestations of the disease are known. The effects of beryllium intoxication can be mild, moderate, or severe, and can prove fatal, depending on the duration and intensity of exposure.

FIGURE 4: MATERIAL FLOW THROUGH THE ECONOMY SHOWING EMISSION SOURCES—BERYLLIUM

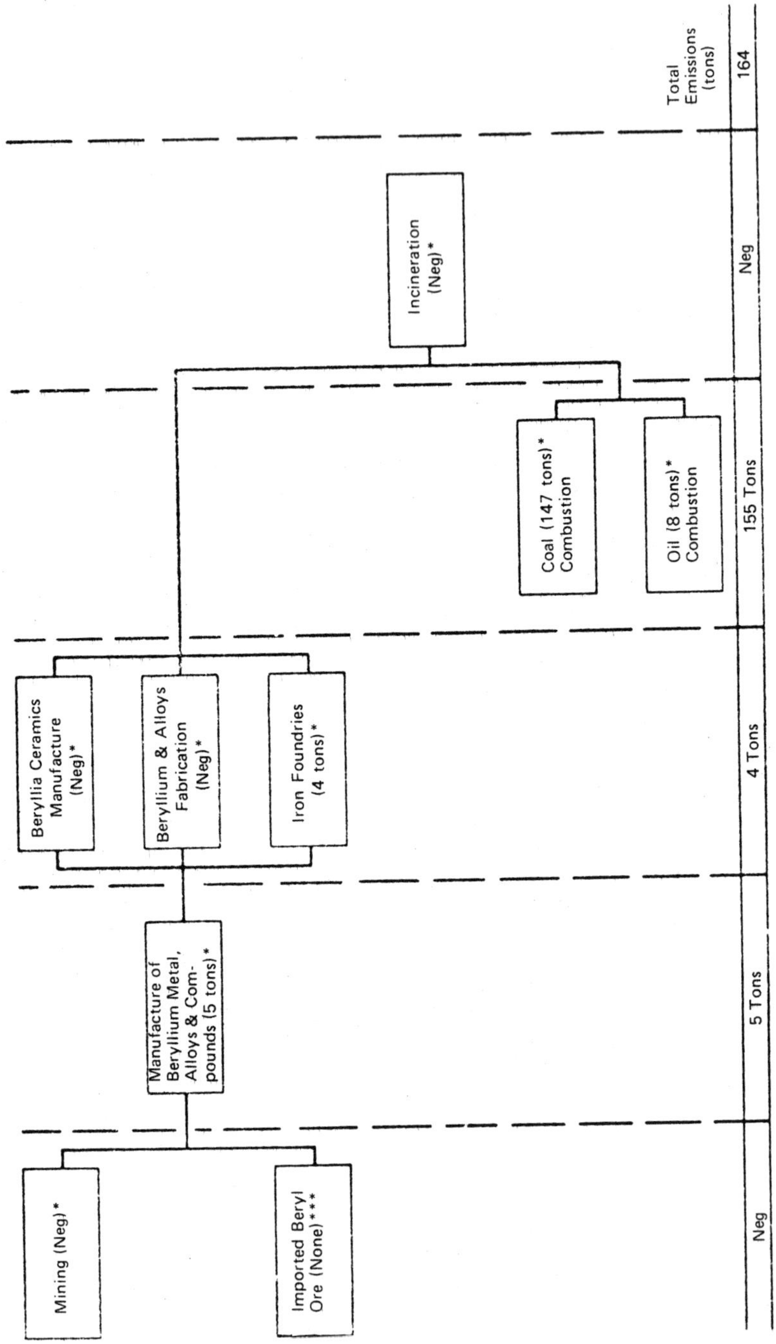

*National Inventory of Sources and Emissions–Beryllium, W.E. Davis and Associates, September 1971.
**Preliminary Air Pollution Survey of Beryllium and Its Compounds, Litton Systems, Inc., October 1969. ***MITRE.

Source: Reference (8)

Acute beryllium disease is manifested by a chemical penumonitis ranging from transient pharyngitis or tracheobronchitis to severe pulmonary reaction. Chronic beryllium disease generally occurs as lesions in the lung, producing serious respiratory damage and even death. However, every organ system may be involved in response to beryllium exposure, except for the organs in the pelvic area. The chronic form is characterized by a delay in onset of disease, which may occur weeks or even years after exposure.

Cancer has been produced experimentally in animals, and cases of cancer have been found in humans afflicted with beryllium disease. However, insufficient information exists at this time to causally relate beryllium poisoning to development of cancer in humans.

Beryllium and its compounds can produce dermatitis, conjunctivitis, and other contact effects; however, these manifestations are rare. There is some evidence that beryllium in soil is toxic to plant life. The toxicity of beryllium and its compounds has been reviewed by T.D. Luckey et al (7).

EXTENT OF EXPOSURE

The major potential sources of beryllium in the atmosphere are industrial (6). The processes of extraction, refining, machining, and alloying of the metal produce toxic quantities of beryllium, beryllium oxide, and beryllium chloride, which if allowed to escape into the atmosphere would cause serious contamination. Recognition of the serious hazards to health from these sources has led to adaptation of control procedures minimizing this potential. However, beryllium in limited quantity is emitted from these industrial processes, and danger also exists from accidental discharges. One major source of beryllium contamination (the use of beryllium in fluorescent light tubes) was discontinued in 1949.

Other sources could be the use of metallic beryllium in rocket fuels, and the combustion of coals. Rocket fuels could present a hazard in the handling and storage of the powdered metallic beryllium used as an additive in the fuels. Also, the exhaust fumes, which contain oxidized beryllium as well as other compounds of beryllium, would be of significance in local soil and air pollution if not contained. As beryllium is a normal constituent (above 2 ppm) of coals, the combustion of coal may add a significant quantity of beryllium to the atmosphere (6). Figure 4 and Table 10 detail some sources of airborne beryllium emissions (8).

TABLE 10: BERYLLIUM EMISSIONS

Source	Amount in Tons	% This Pollutant
Mica, feldspar mining	Neg	Neg
Gray iron foundry cupola	4	2.77
Ceramic coatings	Neg	Neg
Beryllium alloys and compounds	5	3.64
Beryllium fabrication	Neg	Neg
Power plant boilers		
Pulverized coal	86	59.62
Stoker coal	10	6.93

(continued)

TABLE 10: (continued)

Source	Amount in Tons	% This Pollutant
Cyclone coal	3	2.08
All oil	2	1.39
Industrial boilers		
Pulverized coal	8	5.55
Stoker coal	13	9.01
Cyclone coal	2	1.39
All oil	2	1.39
Residential/commercial boilers		
Coal	1	0.69
Oil	8	5.55
Total	144	100.01

Source: Reference (8)

DETECTION

The most common method of sampling beryllium concentrations in air employs a high-volume sampler which draws air to be analyzed through a filter for the appropriate sampling period (9). The high-volume sampler, which samples at a rate of 1.5 to 1.7 m^3/min when operated 24 hours, will provide an adequate sample for the measurement of beryllium in ambient air in concentrations as low as 0.002 $\mu g/m^3$. Cellulose-fiber filter paper with very low ash content is widely used; cellulose ester and fiber glass papers are also used, and are particularly suitable for use when collecting contaminants of very small size.

Sampling practices for beryllium have been spelled out by NIOSH (10) as regards both general room air sampling and breathing zone sampling. Atmospheric beryllium may be analyzed by colorimetric, fluorimetric, or spectrographic procedures (11). These methods are accurate and sensitive but do not discriminate between the various compounds of beryllium.

The morin method is suitable for measuring quantities of beryllium in the air surrounding beryllium plants; its lower limit of detection is approximately 0.01 μg of beryllium (11). In this procedure, the beryllium is removed from the filter, processed to remove interfering elements and to form an alkaline solution, and fluorescent morin solution ($C_{15}H_{10}O_7 \cdot H_2O$) is added. Beryllium acts to reduce the fluorescent characteristics of the morin solution, as a direct function of the quantity of beryllium present. Measurement of the fluorescence by fluorimeter against standard samples indicates the quantity of beryllium. [Note: Keenan (12) states that the procedure is sensitive to 0.00002 μg/ml of the solution.]

The colorimetric method commonly called the Zenia method is suitable for large concentrations from which at least 2 μg of beryllium can be obtained, and is used in monitoring the atmospheres of industrial plants. The procedure involves removal from the filter paper, removal of interfering elements, and addition of the Zenia (p-nitrobenzeneazoresorcinol) solution. Zenia solution added to a beryllium solution forms a reddish brown solution, whose absorption of light is a function of its beryllium content. Measurements can be made subjectively, to give a quick spot-check (13) or by means of a spectrophotometer. Sensitivity to 0.5 μg/ml of solution is claimed (12).

The spectrographic method is considered suitable for measuring concentrations in the general atmosphere, as it permits detection of trace quantities down to 0.003 μg. Also, automatic detectors are available that are capable of handling large numbers of samples rapidly and accurately. This method adds a thallium standard to the beryllium-Zenia solution produced in the colorimetric method, concentrates the solution, subjects it to an electric arc, and photographs its spectra. The ratio of the intensity of the beryllium line at 2348.6 A to that of the thallium line at 2379.7 A is determined, giving a measure of the concentration of beryllium. McCloskey (14) reported on a variation of this method using aluminon reagent in place of the Zenia solution that was more sensitive than the normal spectrographic procedure.

Another method for automatic monitoring of beryllium in air, under development by the U.S. Air Force, was reported on by Braman (15) of the ITT Research Institute. This device attempted to use the nuclear reaction of beryllium for the detection of beryllium. Design goals called for an alarm signal at 25, 2 and 0.01 $\mu g/m^3$ of beryllium over sampling periods of 30 seconds, 1 hour, and 24 hours respectively.

Emission spectroscopy has been used by the National Air Pollution Control Administration for beryllium analysis of samples from the National Air Sampling Network (16), The samples are ashed and extracted to eliminate interfering elements. The minimum detectable beryllium concentration by emission spectroscopy is 0.0008 $\mu g/m^3$ for urban samples and 0.00016 $\mu g/m^3$ for nonurban samples. The different sensitivities result from the different extraction procedures required for urban samples (17). The National Air Pollution Control Administration uses atomic absorption to supplement analyses obtained by emission spectroscopy. The method has a minimum detectable limit of 0.01 $\mu g/m^3$ based on a 2,000 m^3 air sample (16).

NIOSH (10) has specified the use of the atomic absorption spectrometer as perhaps the most economical method for the analysis of beryllium. This method is recommended by NIOSH for purposes of determining compliance with the recommended beryllium standard. Other methods of analysis for beryllium may be used provided their accuracy and sensitivity are determined to be equivalent.

The liquid sample containing the beryllium is aspirated into the flame of the burner assembly and the vaporized sample is irradiated by a lamp containing a cathode made of beryllium metal. The ground-state atoms of beryllium in the sample absorb the 2348.6 A resonance line of beryllium emitted by this lamp, thereby effecting a decrease in the intensity of light reaching a photodetector. The percent of the absorption so indicated is a measure of the concentration of beryllium in the sample.

For aqueous solutions, the working range for beryllium is linear from concentrations of 0.03 μg/ml to approximately 4 μg/ml. The sensitivity under the standard operating conditions is about 0.03 ug/ml beryllium for 1% absorption. This sensitivity can be increased with chemical concentration by solvent extraction.

High concentrations of aluminum (500 μg/ml) depress the sensitivity of the beryllium determination. High concentrations of silicon and magnesium also decrease the sensitivity. The interference can be controlled by adding oxine (8-hydroxyquinoline) to the sample and standards. Beryllium may be measured in wastewaters by atomic absorption spectroscopy down to a detection limit of 1.0 ppb. Beryllium may also be measured in water by spectrofluorometric methods as described by K.H. Mancy (18). Using 1-amino-4-hydroxyanthraquinone as the reagent, a sensitivity of 0.02 μg/ml can be attained with only chromate and lithium ions interfering. Using 8-hydroxyquinoline as the reagent, a sensitivity of 0.001 μg/ml can be attained but many more ions interfere. Using 2,3-hydroxynaphthoic acid, a sensitivity of 0.0002 μg/ml can be attained but an even greater number of other ions interfere as noted by Mancy (18).

ENVIRONMENTAL STANDARDS

NIOSH has proposed (10) a recommended standard for occupational exposure to beryllium as follows. No worker shall be exposed at his place of employment to a concentration of beryllium more than two micrograms of total airborne particulate beryllium per cubic meter of air (2 μg Be/m^3) (determined by the breathing-zone and general air methods), determined as a time-weighted average (TWA) exposure for an 8-hour workday, and no peak concentration of beryllium to which workers are exposed shall exceed 25 μg Be/m^3 as determined by a minimum sampling time of thirty minutes.

Prior to 1947 it was virtually impossible to correlate medical findings with exposure levels of workmen. One reason for the lack of such data was that, at the time, sensitive analytical methods for beryllium had not been developed, nor were air sampling instruments generally available that were suitable. Because of the steadily increasing accounts of respiratory illness and death resulting from beryllium exposure, and because of the similarities apparent in pneumoconiosis, a symposium was held at Saranac Lake, New York, in the fall of 1947 to review the entire beryllium problem (19). This was the 6th Saranac Symposium and information gained from the sessions, coupled with the research and recommendations of Eisenbud and coworkers (20)(21), provided the basis for the United States Atomic Energy Commission Control Requirements established in 1949.

Limited investigations by Eisenbud (20) suggested that maximum concentration at which the beryllium workmen investigated had been exposed did not exceed 15 μg Be/m^3 of air. For control of the acute disease, a value of 25 μg/m^3 was recommended by Eisenbud as the maximum permissible peak exposure. In addition, Eisenbud's studies of nonoccupational cases of beryllium disease (21) resulted in the conclusion that, in the human population around a beryllium production plant, the lowest concentration which produced disease was greater than 0.01 μg/m^3 and probably less than 0.10 μg/m^3.

There still remained, however, the problem of recommendations for chronic occupational exposure to beryllium. Eisenbud's comments in 1961 (22) are still appropriate in 1972. "There was not then, nor is there today, any substantial body of environmental information that could be correlated with clinical reports of occupational berylliosis, and such data as do exist are puzzling." Having no real empirical basis for the establishment of a limit for chronic occupational exposure, Eisenbud and Machle arrived at a figure of 2 μg/m^3 on the basis of information on animals and man and by analogy with industrial air limits for toxicity of heavy metals, such as lead, mercury, cadmium, and thallium.

As a result of the Saranac Symposium and Eisenbud's recommendations, the following limits of permissible concentrations of beryllium were adopted by the Atomic Energy Commission on the recommendation of an ad hoc committee:

1. The in-plant atmospheric concentration of beryllium should not exceed 2 micrograms per cubic meter as an average for an 8-hour day.
2. Even though the daily average may be less than 2 micrograms per cubic meter, no person should be exposed to concentrations greater than 25 micrograms per cubic meter at any time, however short.
3. In the neighborhood of the AEC plant handling beryllium compounds, the average monthly concentrations at the breathing zone levels should not exceed 0.01 microgram per cubic meter.

The AEC Division of Biology and Medicine established the control of beryllium hazards, and it was binding on all AEC installations handling beryllium. In addition, all AEC contracts involving the handling of beryllium carried a health and safety clause demanding adherence to these requirements. The first two involved in-plant controls, whereas the third represented the community air limit.

In 1955 the American Conference of Governmental Industrial Hygienists (ACGIH) adopted, on a tentative basis, an 8 hour, daily-weighted average exposure level of 2 μg Be/m^3. The American Industrial Hygiene Association, in 1956, also published a Hygienic Guide which recommended the AEC values.

In 1958 the AEC ad hoc review committee was dissolved in accordance with its own recommendation that henceforth proper governmental and industrial agencies be encouraged to adopt maximum allowable concentrations for general publication. The nationally recognized health and safety guides and standards for beryllium were then prescribed by the AEC as applicable to their activities. The contract clause was still maintained, however, to afford a means of standard enforcement to AEC contractors. These same standards were used as of 1972 for AEC industrial hygiene control.

In 1961, based on the Federal Walsh-Healey Act, the AEC was relied upon for surveillance of occupational exposures to beryllium for as long as the AEC contractual provisions remained in effect. By late 1962, the AEC had terminated all beryllium contracts and relied upon acquisition of their beryllium needs through direct purchase from the various commercial beryllium producers.

Other Beryllium Standards:

A. Foreign Standards (10)
   1. Exposure limits to beryllium, as specified by the AEC, have been adopted by the West German Government and serve as the basis for their industrial health practices for beryllium.
   2. The World Health Organization, in 1969, issued a joint statement with the International Labor Organization on permissible levels of occupational exposure to airborne toxic substances. The recommended safe concentration zone for beryllium and its compounds is 0.001 to 0.002 mg/m^3.

B. Standards for Beryllium Rocket-Motor Firing (10) — The National Academy of Sciences-National Research Council in 1966 developed criteria and standards for protection of off-site personnel from intermittent exposures to beryllium compounds arising from the firing of rocket motors. These limits are as follows.

1. For soluble beryllium compounds and beryllium oxide comparable to a product calcined at temperatures around 400°C, a standard of 75 μg-minutes/m^3 within the limits of 10 to 60 minutes, accumulated during any two consecutive weeks.
2. For beryllium oxide comparable to a product calcined at temperatures in excess of 1600°C, a standard of 1,500 μg-minutes/m^3 within the limits of 10 to 60 minutes, accumulated during any two consecutive weeks.

The above standards need adjustment for the overall concentration of soluble beryllium compounds in the effluent. The present occupational standards were established in 1949 out of a caution based upon a lack of definite data upon which to establish a standard and were considered at that time to be markedly low. There appears to be no more scientific basis for establishing these short-term exposure limits on a microgram-minute basis than there was available to develop the occupational standard. The standard was based upon national interest as well as protection of the worker and the general public. Present research indicates a lesser degree of toxicity to high-fired beryllium oxide than to the low-fired oxides.

The proposed national emission standard for beryllium (23) will be applicable to machine shops, ceramic plants, propellant plants, foundries, extraction plants, and incinerators designed or modified for disposal of toxic substances.

The standards are based upon information derived from many sources, including health effect levels, meteorology, technical analysis of control capability, and consideration of economic impact. The overriding considerations are health effects. These beryllium standards are limited, in general, to emissions to the ambient atmosphere.

The guideline used in the development of the standards is based on a maximum allowable concentration of beryllium for ambient air that has been in use by the Department of Defense and the Atomic Energy Commission for many years. The proposed standards offer the owner or operator the option of determining compliance either by emission testing or by measurement of ambient concentration levels in the vicinity of the plant. In addition, separate emission standards for beryllium-rocket motor firing are proposed.

National Emission Standards for Beryllium:

1. Total emissions to the atmosphere shall not exceed 10 grams of beryllium in a 24-hour day.
2. Total emissions to the atmosphere shall not exceed amounts which result in an out-plant concentration of 0.01 microgram of beryllium per cubic meter of air averaged over a 30-day period.

According to Tucker (4), Britain's Alkali Inspectorate has specified a maximum allowable beryllium concentration at the point of effluent discharges of 0.1 μg/m^3.

HANDLING PROCEDURES

Work practices specified by NIOSH (10) for occupational exposure to beryllium are as follows.

Emergency Procedures: Procedures, including fire-fighting procedures, shall be established and implemented to meet foreseeable emergency events, including those involving massive release of beryllium contaminants. (a) Where there is the possibility of massive beryllium contamination from accidents involving elevated temperature operations or large-scale spills, a general alarm, drench-type showers, and cleansing facilities shall be installed in such a manner as to provide prompt, immediate access by any employee; respirators shall be available for wearing during evacuation procedures if long distances need to be traversed; all full-facepiece or supplied-air respirators shall be available for employee use where equipment or operations cannot be abandoned. If large-scale spills occur, careful monitoring of beryllium air levels shall be instituted as soon as possible. Determinations shall be made as to when it is safe to re-enter a contaminated area with respiratory protection and to resume operations without respiratory protection. Thoroughly supervised cleanup operations shall be instituted as soon as air levels recede to the point where operations may be conducted.

(b) Full-facepiece or supplied-air respirators shall be kept ready for emergency use. They shall be inspected after each use and at least monthly to assure satisfactory working condition. Where used, air and oxygen cylinders shall be fully charged according to the manufacturer's instructions. Sand, soda ash, or commercial metal fire extinguishment powder shall be available for use as an extinguishing agent for beryllium; water and carbon dioxide shall not be used.

Exhaust Systems: Procedures shall be established to reduce exposure to airborne beryllium through implementation of adequate ventilation methods. Local exhaust and collection shall be designed and maintained to prevent the accumulation of beryllium dust and fumes.

General Housekeeping:

1. No dry sweeping shall be performed.
2. Emphasis shall be placed upon cleanup of spills, periodic repair of equipment and leaks, and proper storage of materials to prevent breakage.

Disposal: Beryllium waste and scrap shall be collected and disposed of in sealed bags or other sealed containers. Scrap shall be recycled or shall be disposed of by burial.

Nonworker Access to Beryllium Areas: Entry to any area where there is the possibility of exposure to beryllium shall be permitted only on the basis of need and all persons so entering shall be provided with the same protective clothing as required for employees regularly assigned to that area.

Educational Program: An educational program shall be instituted for all workers to include proper instruction in maintenance procedures, cleanup methods, and use of respiratory protective devices and protective clothing.

According to the NIOSH (10), medical surveillance as specified in this section is required for workers who are exposed to beryllium.

A. A comprehensive preplacement history and physical examination for all worker applicants shall be provided to include as a minimum a 14" x 17" chest roentgenogram, baseline pulmonary function [forced vital capacity (FVC) and forced expiratory volume at one second ($FEV_{1.0}$)], and a baseline weight.

B. Each worker exposed to beryllium shall receive an annual evaluation (except where a variance has been granted, based on minimal exposure) that includes:

1. Spirometry, including FVC and $FEV_{1.0}$.
2. A medical history questionnaire that includes presence and degree of respiratory symptoms, i.e., breathlessness, cough, sputum production, and wheezing.
3. A 14" by 17" chest x-ray.

C. Department of Health, Education, and Welfare (HEW) and Department of Labor (DOL) physicians and medical consultants and those physicians designated and authorized by the employees shall have access to medical records.

D. Medical records shall be maintained for at least 20 years.

As regards labeling, the NIOSH (10) specifies that:

A. The following precautionary label shall be applied to all shipping and storage containers or packages containing beryllium and beryllium compounds where exposures to dusts, fumes, powders, or liquids are likely to occur.

BERYLLIUM
(Name of Compound)
Dust, Fume, Powder, or Liquid
DANGER

Harmful if Inhaled and May Cause Immediate or
Delayed Injury

Use Only with Adequate Local Exhaust Ventilation and
Approved Respiratory and Personal Protective Devices

May Cause Rash or External Ulcers
Wash Thoroughly After Handling

B. The following warning sign shall be affixed in a readily visible location on processing and other equipment, on storage bins and tanks, and at or near entrances and areas where exposure to dusts or fumes of beryllium and its compounds are likely to occur.

DANGER

Beryllium Dust (or Fume) Areas

UNAUTHORIZED PERSONS

KEEP OUT

Breathing Dust (or Fumes) May Cause Immediate or
Delayed Injury

In the category of personal protective equipment, NIOSH (10) specifies the following regarding respiratory devices:

A. Respiratory protective devices shall be used any time the actual or projected level of beryllium will exceed an 8 hour time-weighted average concentration of 2 $\mu g/m^3$. These devices shall not be used as a substitute for proper engineering controls, but are appropriate for necessary periods where excessive atmospheric concentrations result from emergencies or from maintenance or repair.

B. If a variance order is granted under provisions of the Occupational Safety and Health Act of 1970 from the environmental limits, the provisions of this section shall still apply to safeguard workers.

C. Standard procedures shall be established for respirator use to include the following:

1. Procedures established for respirator usage shall insure that no worker is exposed to beryllium in excess of the standard because of improper respirator selection or facepiece fit.
2. Facepiece fit shall be checked by the wearer each time he puts on the respirator.
3. The respirator used shall be of the appropriate class as determined on the basis of the actual or projected atmospheric concentration of airborne beryllium at the worksite where the respirator is to be used as follows:
   a. Reusable half mask air-purifying respirators equipped with high efficiency filters shall be used in atmospheres containing not more than 25 μg Be/m^3 for any period of time.
   b. Reusable full-facepiece air-purifying respirators equipped with high-efficiency filter shall be used in atmospheres containing not more than 100 μg Be/m^3 for any period of time.
   c. Powered air-purifying respirators equipped with high-efficiency filters, operating with positive pressure during the inhalation phase of breathing shall be used in atmospheres containing not more than 1,000 μg Be/m^3. When equipped with a fume filter, respirators shall be used in atmospheres containing not more than 40 μg Be/m^3.
   d. Continuous flow air-line respirators or self-contained breathing units operating in the pressure-demand mode and providing positive pressure during the inhalation phase of breathing, shall be used in atmospheres that exceed 1,000 μg Be/m^3.
4. Only respirators approved by the U.S. Bureau of Mines and/or NIOSH under the provisions of 30 CFR Part II (37 F.R. 6244) shall be used.

D. An effective respirator program to include selection, use, maintenance and care, special problems, and effective respirator program evaluation shall be established. Compliance with the requirements of the latest revision of the American National Standard Practices for Respiratory Protection Z88.2 shall be considered prima facie evidence of compliance with this subsection.

In the area of protective clothing, NIOSH (10) specifies for both exposed workers and workers with minimum beryllium exposure as follows:

A. Beryllium-Exposed Workers

1. The employer shall provide each employee except those with minimal exposure with protective clothing, headgear, and shoes when the employee must work with and/or be exposed to beryllium, beryllium compounds, or beryllium containing products.
   a. Such clothing may be coveralls or similar fullbody protective clothing.
   b. Protective shoe covers or work shoes shall be worn during the working hours in areas where there is exposure to beryllium.
   c. Resin-impregnated paper or similar protective clothing can be substituted for fabric-type clothing.

2. Protective clothing shall be changed at least each day at the end of each shift.
3. The employer shall provide for maintenance and laundering of soiled protective clothing, which shall be stored, transported, and disposed of in sealed, nonreusable containers and labeled with easy-to-read letters.
4. Dust removal, by blowing or shaking of clothing, shall be prohibited; protective clothing shall be cleaned by power suction (vacuum) prior to removal; removal of contaminated clothing shall be allowed only in the change rooms.
5. The employer shall provide laundering controls and inform the employee and any third party which launders contaminated clothing of the potentially harmful effects of exposure to beryllium dust and of safe practices required in the laundering of beryllium-soiled work clothing.

B. Workers with Minimum Beryllium Exposure
1. The employer shall provide laboratory coats or equivalent protective clothing to each such employee who works with or is exposed to beryllium, beryllium compounds, or beryllium-containing products.
2. Clean protective clothing shall be supplied to each worker at least weekly.

NIOSH specifications (10) regarding the apprisal of employees of the hazards of beryllium are as follows:

A. Each employee exposed to beryllium shall be apprised of all hazards, adverse symptoms of overexposure, appropriate emergency procedures, and proper conditions and precautions for safe use or exposure, to include as a minimum, all information which is applicable to that specific product or material containing beryllium to which he is exposed.

B. The information shall be maintained on file and readily accessible to the worker at all places of employment where beryllium is involved in unit processes and operations.

C. Information shall be recorded on U.S. Department of Labor Form OSHA-20, Material Safety Data Sheet, or a similar form approved by the Occupational Safety and Health Administration, U.S. Department of Labor.

As regards sanitation practices to be followed in occupational exposure to beryllium, NIOSH specifies (10):

A. Food Facilities
1. Food preparation and eating should be prohibited in beryllium work areas.

B. Locker and Toilet Facilities (except where a variance has been granted, based on minimal exposure)
1. Separate locker facilities shall be provided for work clothes and street clothes.
2. Showers for exposed workers shall be required following a work shift and prior to putting on street clothes.
3. Locker-shower facilities shall be so arranged that the showers can serve to demarcate between potentially clean and contaminated areas.
4. Suitable provisions shall be made for the control of contaminated dust in work shoe storage and clothing hamper locations.
5. Handwashing and toilet facilities shall be arranged so that following use, workers need not re-enter a potentially contaminated area.

As regards monitoring and recordkeeping requirements for occupational exposure to beryllium, NIOSH specifies that:

A. Employers shall maintain records of environmental exposures to beryllium based upon the following sampling and recording schedule:
   1. Quarterly Requirements—Except where a variance has been granted, based on minimal exposure, as determined by selected sampling of representative or minimal exposed workers. It is recognized that sampling frequency is dependent upon the process and emissions; the quarterly sampling frequency is primarily for production operations.
      a. Breathing-zone samples shall be collected from employees at least quarterly for specific work operations. The first sampling period shall be completed within 180 days of the effective date of this standard.
      b. Samples shall be collected and evaluated as both time-weighted and peak concentration values.
   2. Thirty-Day Requirements—The sampling regime shall be every 30 days for work areas under the following conditions:
      a. The environmental time-weighted average or peak concentrations are in excess of the standard.
      b. Sampling, monitoring, and recordkeeping requirements of a 30-day schedule shall be required only until two consecutive 30-day periods have resulted in environmental levels which meet the standard.

B. Records shall be maintained for all sampling schedules to include the type of personal protective devices, if any, in use.

C. Workers shall be informed of the periodic results of samplings.

REMOVAL FROM AIR

Expanded industrial use of beryllium in the 1930s, particularly the large-scale production of fluorescent lamps using phosphors of beryllium oxide, produced a number of cases of pulmonary diseases which were initially attributed to causes other than exposure to beryllium. Recognition of the toxic effects of beryllium, however, beginning in 1943, resulted in establishment of the first Community Air Limit for metal substances, as well as in health safety procedures which were exceptionally effective in controlling further occurrences of the disease.

However, increased use of beryllium in the metallurgical industry, along with its proposed use as a high-energy fuel for rocket motors, suggests that further study should be made of the air pollution aspects of this highly toxic material. An annoted bibliography has been published dealing with beryllium and air pollution (24).

Abatement measures have been implemented industry-wide, with a very high degree of success (6). Conventional air-cleaning procedures have been employed, including the use of electrostatic precipitators, baghouses, scrubbers, etc. These procedures have enabled the beryllium industry to meet the industrial hygiene standards established for beryllium.

Efforts to reduce concentrations of beryllium in the atmosphere have generally centered on the problem of reducing industrial concentrations of beryllium, where exceptional success has been achieved. Control activity has mainly surrounded significant beryllium plants, where concentrations of beryllium dust, produced within the plant, are discharged to the outside atmosphere through

FIGURE 5: UNITIZED FABRIC TUBE FILTER, PREFILTER, AND HEPA FILTER FOR BERYLLIUM OR BERYLLIUM OXIDE MACHINING FACILITY

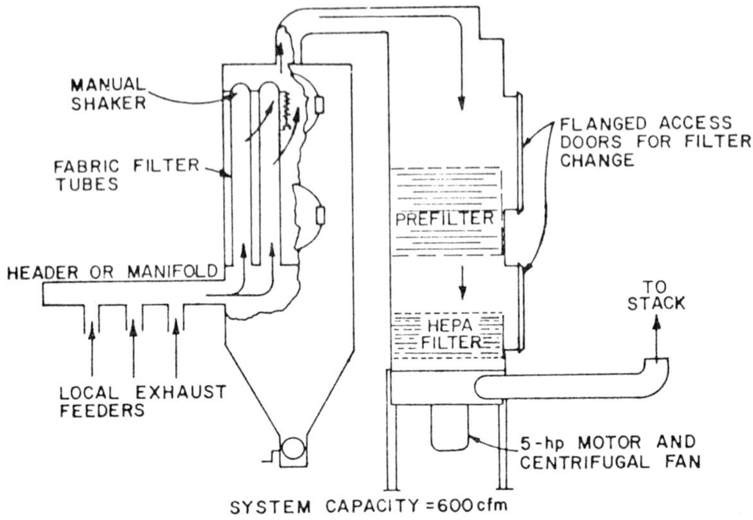

FIGURE 6: UNITIZED MULTIPLE DRY CYCLONE COLLECTOR, PREFILTER, AND HEPA FILTER FOR BERYLLIUM OR BERYLLIUM OXIDE MACHINING FACILITY

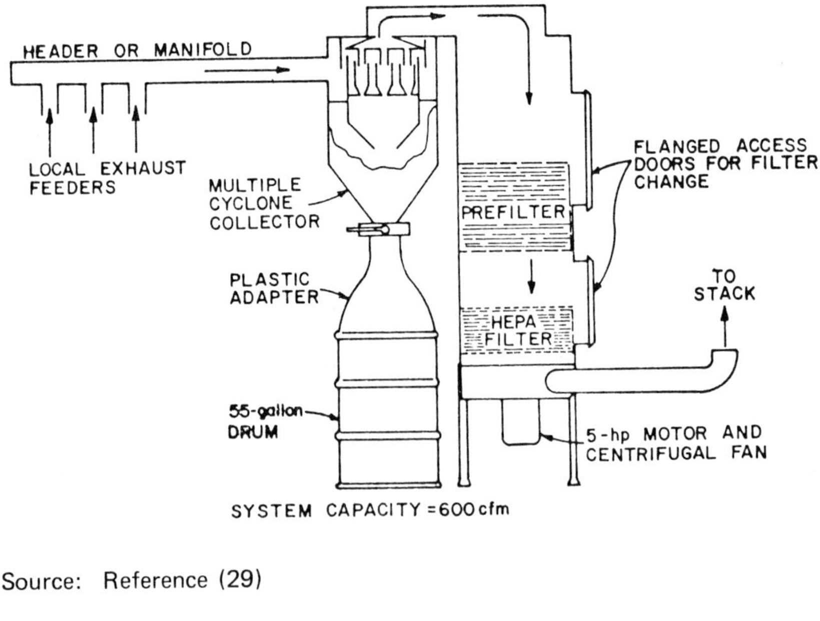

Source: Reference (29)

intent or by accident. Recognition of beryllium as a cause of industrial disease in the early 1940s has led to sweeping changes in manufacturing processes and controls designed to reduce the concentration of atmospheric beryllium both in-plant and out-of-plant (3).

Conventional air-cleaning devices have been used. For wet chemical processes, scrubbers, venturi scrubbers, packed towers, organic wet collectors, and wet cyclones have been found to be effective. For dry processes, conventional bag collectors, reverse-jet bag collectors, electrostatic precipitators, cyclones and unit filters have been used. Silverman (25) has published a listing of air-cleaning devices and their expected efficiency in handling effluents from various processes handling beryllium and its compounds.

The effectiveness of abatement activities can be appreciated by comparing present-day industrial exposure concentrations with concentrations existing prior to recognition of the significance of beryllium air pollution to industrial health. Breslin (26) states that current practices limit employees' exposure to beryllium to about 2 $\mu g/m^3$ or less, with out-of-plant neighborhood concentrations limited to 0.01 $\mu g/m^3$. By comparison, University of Rochester samples of beryllium concentration in a Cleveland factory as reported by Laski et al (27) revealed concentrations as high as 4,710 $\mu g/m^3$.

Another form of control is exemplified by discontinuance of the use of beryllium in fluorescent lamp tubes in 1949; the action eliminated an industrial exposure to beryllium which had been responsible for approximately one-third of all beryllium respiratory illnesses (28).

Emissions from beryllium-copper foundries have been controlled by using settling chambers or cyclones as precollectors followed by fabric filters as a final collector. Successful usage of fabric filters with dacron bags has been reported (29).

The emissions from machine shops handling beryllium metal, beryllium oxide and beryllium-copper alloys are primarily in the form of dusts, mists, and fumes. The nature and quantity of these emissions vary with the type of beryllium compound and the operation utilized. For example, the use of cutting fluids results in mist-type emissions. Also, secondary beryllium emissions in machine shops can occur during equipment cleaning.

Beryllium emissions can be efficiently controlled by passing the gas stream through gas-cleaning devices before venting to the stack (Figure 5). As an alternative to the use of fabric filters as precleaners, cyclones can be used (Figure 6). Silverman (25) indicates that small cyclone units when supplemented by a fabric filter together with an asbestos filter-aid can attain 99.9% by weight removal efficiency.

Beryllium emissions from the combustion of coal can exist as dust, mist or fumes. Control devices for beryllium such as venturi scrubbers, packed towers, cyclones, electrostatic precipitators, and bag collectors can be used. The choice of control equipment will depend on the nature of emissions and the size of operation.

REMOVAL FROM WATER

The tolerance for beryllium in irrigation waters has been given as 0.5 ppm for continuous water use and 1.0 ppm for short-term water use in fine-textured soils. However, very little information is available on removal of beryllium from water, most of the attention having been given to air pollution control.

A process for the recovery of beryllium from spent pickle acid has, however, been described by F.C. Senior and D. MacGregor (30).

SOLID WASTE DISPOSAL

As regards the options in disposing of beryllium and its compounds, recycling is the most desirable from the standpoints of users, producers and environmentalists (3). Beryllium and beryllium compounds are difficult to produce and the primary producers repurchase all available material at $10 to $20/lb contained beryllium. This situation is expected to continue indefinitely.

Liquid, solid or particulate beryllium-containing waste which is too dilute to recycle is buried on private property or in public landfills. Often waste is first burned to produce the insoluble, chemically inert oxides. This is easily and safely done, providing the exhaust gases are scrubbed to remove any particulates. These procedures were verified (31) with KBI Industries, two other large beryllium users, and the County of Los Angeles, California, which operates several landfills which receive liquid coolant wastes containing beryllium. All independently agreed with this analysis. Since there have been no new reported cases of berylliosis in 20 years and demand is expected to remain static for the indefinite future, it may be concluded that such practices are adequate at present and for the foreseeable future (3).

As regards landfill disposal of specific beryllium chemicals, Powers (31) recommends that in the case of beryllium carbonate, beryllium chloride, beryllium oxide, beryllium (powder) and beryllium selenate that "wastes should be converted into chemically inert oxides using incineration and particulate collection techniques. These oxides may be landfilled."

ECONOMIC IMPACT OF CONTROLS

Schulte and Hyatt (32) reported one particularly detailed study on the economics of industrial hygiene control in a beryllium machinery plant in the Los Alamos Scientific Laboratory (New Mexico) in 1958. According to Tepper (33), in many instances the recovery of valuable materials will significantly offset the costs of abatement of beryllium air pollution.

About 19% of the total capital cost of beryllium-copper foundry is needed for emission control (34). Economics of installing and operating a cyclone-fabric filter control series in a beryllium-copper alloy foundry is given in Table 11.

Typical capital cost for beryllium emission control from machine shops represents approximately 8% of the total investment (34). Details of the capital and operating costs are outlined in Table 12.

TABLE 11: EMISSION CONTROL COSTS FOR BERYLLIUM-COPPER ALLOY FOUNDRY

Equipment Class	Type	Gas Flow Rate cfm	Annual Mantenance Cost, $	Annual Power Cost, $	Installed Cost, $	Annual Operating Cost, $
Fabric Filter	High-efficiency	18,000	1,060	1,140	36,800	2,200
Dry Cyclone	Reverse	18,000	320	900	10,360	1,220

TABLE 12: EMISSION CONTROL COSTS FOR BERYLLIUM MACHINE SHOP

Item	Cost, $
Emission control equipment	50,000
Air conditioning with special filters	40,000
Installation	35,000
Annual maintenance	20,000
Annual power	12,000

Source: Reference (29)

REFERENCES

(1) H.A. Schroeder, *The Poisons Around Us,* Bloomington, Indiana University Press (1974).

(2) R.A. Heindl, in *Mineral Facts and Problems,* Washington, D.C., U.S. Bureau of Mines (1970).

(3) N.L. Durocher, *Air Pollution Aspects of Beryllium and Its Compounds,* Report PB 188,078, Springfield, Va., Nat. Tech. Information Service (Sept. 1969).

(4) A. Tucker, *The Toxic Metals,* London, Earth Island Ltd. (1972).

(5) M. Eisenbud, et al, *Journal of Industrial Hygiene* 31, 282 (1949).

(6) Booz-Allen Applied Research, Inc., *A Study of Hazardous Waste Materials, Hazardous Effects & Disposal Methods,* Report PB-221,466, Springfield, Va., Nat. Tech. Information Service (July 1973).

(7) T.D. Luckey, B. Venugopal, and D. Hutcheson, *Heavy Metal Toxicity, Safety and Hormonology,* New York, Academic Press (1975).

(8) A.J. Goldberg, *A Survey of Emissions and Controls for Hazardous and Other Pollutants,* Washington, D.C., Environmental Protection Agency, (February 1973).

(9) "Community Air Quality Guide for Beryllium," *Am. Ind. Hyg. Assoc. J.* 29, (1968).

(10) Nat. Inst. for Occupational Safety and Health, *Criteria for a Recommended Standard... Occupational Exposure to Beryllium,* Washington, D.C., (1972).

(11) J. Cholak, R.A. Kehoe, L.H. Miller, F. Princi, and L.J. Schafer, *Toxicity of Beryllium,* U.S. Air Force Systems Command Report ASD-TR-62-7-665, (April 1962).

(12) R.G. Keenan, "Analytical Determination of Beryllium," Chapter 5 in *Beryllium, Its Industrial Aspects,* H.E. Stokinger, Ed. (New York: Academic Press, 1966).

(13) H.M. Donaldson, R.A. Hiser, and C.W. Schwenzfeier, "A Rapid Method for Determination of Beryllium in Air Samples," *Am. Ind. Hyg. Assoc. J.* 22:280 (1961).

(14) J.P. McCloskey, "Spectrophotometric Determination of Beryllium in Airborne Dust Samples," *Microchem. J.* 12:401 (1967).

(15) R.S. Braman, *Research and Development of an Automatic Beryllium-Air Monitor, U.S. Air Force Systems Journal,* Report RTD-TDR-63-1112, (1963).

(16) R.J. Thompson, G.B. Morgan, and L.J. Purdue, "Analyses of Selected Elements in Atmospheric Particulate Matter by Atomic Absorption," Preprint, presented at the Instrument Society of American Symposium, New Orleans, La. (May 5-7, 1969).

(17) *Beryllium, Actual and Potential Resources, Toxicity, and Properties in Relation to its Use in Propellants and Explosives,* Naval Ordnance Laboratory, Bureau of Naval Weapons Report 7346, March (1961).

(18) K.H. Mancy, *Instrumental Analysis for Water Pollution Control,* Ann Arbor, Mich., Ann Arbor Science Publishers (1971).

(19) A.J.Vorwald, Ed., *Pneumoconiosis: Beryllium and Bauxite Fumes,* New York, Hoeber, Inc. (1950).

(20) M. Eisenbud, C.F. Berghout, and L.T. Steadman, "Environmental Studies in Plants and Laboratories Using Beryllium; The Acute Disease," *J. Ind. Hyg. Toxicol.* 30, 281-85 (1948).

(21) M. Eisenbud, R.C. Wanta, C. Dustan, L.T. Steadman, W.B. Harris and B.S. Wolf, "Non-Occupational Berylliosis," *J. Ind. Hyg. Toxicol.* 31, 282-94 (1949).

(22) M. Eisenbud, "Basis of the Presently Used Maximum Allowable Concentrations for Control of Beryllium Disease, " *Workshop on Beryllium,* Cincinnati, Ohio, The Kettering Laboratory, University of Cincinnati (1961).

(23) Environmental Protection Agency, *Fed. Register* 36, (234), 23243-23245 (December 7, 1971).

(24) *Beryllium & Air Pollution: An Annotated Bibliography,* Publ. No. AP-83, Research Triangle Park, North Carolina, Air Pollution Control Office, U.S. Environmental Protection Agency (Feb. 1971).

(25) L. Silverman, "Control of Neighborhood Contamination Near Beryllium-Using Plants," *A.M.A. Arch. Ind. Health* 19, 172 (1959).

(26) A.J. Breslin, "Exposures and Patterns of Disease in the Beryllium Industry," Chapter 3 in *Beryllium–Its Industrial Hygiene Aspects,* H.E. Stokinger, Ed., New York, Academic Press (1966).

(27) S. Laskin, R.A.N. Turner, and H.E. Stokinger, "An Analysis of Dust and Fume Hazards in a Beryllium Plant," 6th Saranac Symposium (1947) in *Pneumoconiosis: Beryllium, Beryllium Fumes, Compensation,* A.J. Vorwald, Ed., New York, P.B. Hoeber, Inc. (1950).

(28) "Fluorescent Lamp Makers Stop Use," *Beryllium Industry Hygiene Newsletter* 9 (1949).

(29) *Control Techniques for Beryllium Air Pollutants,* Environmental Protection Agency, (Feb. 1973), Pub. No. AP-116.

(30 F.C. Senior and D. MacGregor; U.S. Patent 3,787,306; January 22, 1974; assigned to Cationic Corp.

(31) P.W. Powers, *How to Dispose of Toxic Substances and Industrial Wastes,* Park Ridge, N.J., Noyes Data Corp. (1976).

(32) Schulte and Hyatt, *Workshop on Beryllium,* The Kettering Laboratory, University of Cincinnati (January 1961).

(33) L.B. Tepper, H.H. Hardy, and R.I. Chamberlin, *The Toxicity of Beryllium Compounds,* London, Elsevier (1961).

(34) *Background Information–Proposal National Emission Standards for Hazardous Air Pollutants: Asbestos, Beryllium, Mercury,* Environmental Protection Agency, Pub. No. APTD-0753, Wash., D.C. (Dec. 1970).

BORON

Boron is a versatile and useful element used mainly in the form of its many compounds, of which borax and boric acid are most well known. The largest single use of boron is in glassmaking where boron compounds add strength to the glass, especially above the temperatures at which ordinary glass softens. Significant quantities of boron compounds also are used in vitreous enamel, soaps, detergents, cleaners, metal fluxing and agricultural chemicals. However, these uses eventually may be overshadowed by the new uses for boron such as in atomic energy control rods, jet and rocket fuels, motor fuel additives, heat and radiation shields, steel hardening, and as the stiffening agent in extremely strong, lightweight, composite structural materials (1).

While boron is not an extremely rare element, few commercially attractive deposits of boron minerals are known. It is estimated that about half of the commercially attractive world boron resources, estimated at about 72 million tons of boron, are in southern California as bedded deposits of borax (sodium borate) and colemanite (calcium borate), or occur as solutions of boron minerals in Searles Lake brines. The boron deposits of Turkey are second only to those of the United States. Economically important boron mineral deposits also are known in the U.S.S.R., Argentina, Chile, and other countries.

Among the nations of the world the United States is not only the largest producer of boron, supplying 71 percent of the world demand in 1968, but also the largest consumer, requiring about 36 percent of the world output. U.S. demand for boron, which was 87,000 short tons in 1968, is expected to grow to the year 2000 at an annual rate ranging between 3.5 and 4.6 percent, giving a forecast range of possible demand between 263,000 and 362,000 tons in the year 2000.

Boron is a ubiquitous element, generally occurring at concentrations of 10 to 30 ppm in soils, approximately 0.1 ppm in surface waters and about 4.5 ppm in seawater. Departures from these norms occur only in the very limited areas where natural boron sources are concentrated, such as in certain arid regions of California. Air monitoring data do not measure boron concentration as there is no evidence that it is present in the atmosphere to a significant degree (2).

According to the U.S. Bureau of Mines (1), neither the production nor use of boron have serious adverse effects on the environment. Boron compounds are mined in a remote desert area where tailings and waste dumps do not encroach on residential, industrial or farm land. Atmospheric pollution is not a problem, although some processing odors and dust are produced, according to the Bureau of Mines. The boron content of streams draining agricultural land where boron compounds have been used to eradicate weeds may become a problem, according to the Bureau of Mines (1).

According to the Office of Toxic Substances of EPA, virtually all boron-containing materials and products requiring boron compounds in their manufacture constitute potential sources of boron contamination. Little is done to check the boron content of surface and river waters in the U.S. Through the use of boron compounds in detergents and other products, the boron concentration in surface waters can be expected to increase. Since sodium perborate is widely used in Europe in place of the American-type of chlorine-based bleach, the trend toward a build-up of boron concentration in surface waters should be more pronounced there. By following the build-up of boron in European waters it may be possible to establish the point where the agricultural ecology of the U.S. will be upset (2). This trend would not be affected by sewage treatment since conventional sewage treatment removes little or no boron.

TOXICITY

Boron compounds, in general, are toxic to humans and animals when ingested. Most boron compounds are moderately toxic by inhalation, while certain compounds, such as the high-energy borane fuels, are highly toxic when inhaled. Dusts of some boron compounds are considered merely nuisance dusts; others, such as dusts of boric acid and boron oxide, are moderately toxic, according to Durocher (3). Borane poisoning in animals has produced listlessness, incoordination, convulsions, coma, and death. Similar symptoms have been observed in humans exposed to the boranes, although none have died. The incidence of acute borane intoxication in humans has been low, and adequate data on exposure, concentrations, and reactions are lacking.

Boron is ingested in the normal human diet, primarily from fruits and vegetables, at the rate of 10 to 20 mg/day and excreted at the same rate with little or no accumulation in the body (2). As regards acute toxicity, the minimum lethal dose of boric acid or borates for man is not established. Intakes of up to four grams per day have been reported in an adult without incident, but single doses of 18 to 20 grams in adults have been claimed to be fatal. Intravenous doses in excess of 20 grams have been used in neutron capture therapy of brain tumors without permanent effect (2).

The most common intoxications from boric acid result from oral intake by infants or from absorption through the skin or bladder after treatment for infections. The symptoms of boric acid poisoning, regardless of the route of administration, include nausea, vomiting, headache, diarrhea, erythema, hypothermy, restlessness, weariness, desquamation, renal injury, and death from circulatory collapse and shock within five days. The clinical picture of boric acid poisoning was indistinguishable from toxic epidermal necrolysis in the newborn infant. An autopsy of an infant fatally poisoned by treatment of oral cavity sores with boric acid solution and treatment of consequent erythema with boric acid powder revealed congestion and edema of the brain, myocardium, lungs and other organs, with fatty infiltration of the liver. Several cases of fatal poisoning due to boric acid irrigation of the bladder after prostatectomy have been reported. The National Clearinghouse for Poison Control Centers reported 166 cases of boric acid poisoning in 1972, 117 for children less than 5 years old (2).

Infants have been accidentally poisoned in hospitals from formulas prepared with boric acid. Since no antidote exists, but equally effective treatments do, many workers feel boric acid should not be used for medical purposes. Some evidence exists, however, that boric acid poisoning by absorption through the skin is a serious possibility only in patients with kidney damage. Experiments conducted with rabbits showed that the biological half life of boric acid was significantly prolonged in rabbits with kidney damage.

Borax is also absorbed through the skin and produces similar symptoms. Boric acid toxicity may be detected by the analysis of borates in the urine by the polyvinyl alcohol-borate-iodine reaction since borates are rapidly excreted by the kidney in largely unchanged form. Hypoacidity in adults and decreased fecal enterokinase activity in children were observed in 288 persons drinking high boron water for prolonged periods (2).

Decaborane in low doses is known to have severe disruptive effects upon behavior. The symptoms include depression, catatonia, muscular fasciculations, and occasional convulsions. Little or no central nervous system damage has been reported as a result of toxicological research on borane fuels (2).

Boron trifluoride is corrosive to the eyes, skin, and mucous membranes, and will cause burns on the skin similar to, but not as penetrating as, hydrogen fluoride. Boron trichloride and boron tribromide are also very corrosive. These effects are apparently due to hydrolysis of the boron trihalides to the halogen acids and not to boron. Organic derivatives of boric acid potentiate the sedative action of hypnotics.

As regards chronic toxicity, workers engaged in the packaging of boron fertilizers complained of poor appetite, nausea, and loss of weight. Aerosols and inadequate shielding in the boron carbide industry may create a risk of pneumoconiosis and pneumosclerosis (2). Concentrations of boron compounds in the fumes and aerosol of boric acid production were two to five times the permissible levels, sulfuric acid fumes were above normal in 64 percent of the samples, and high temperature and humidity prevailed. Medical examinations of 291 workers revealed alterations of upper respiratory tract mucosa, arthralgia and arthropathies. The percentage of gynecological diseases was rather high. Men engaged in boric acid production showed weakened sexual activity, decreased seminal volume, low sperm count and motility, and increased seminal fructose.

Inhalation of the dust from datolite ore [a mixture of $2CaO{\cdot}B_2O_3{\cdot}2SiO_2{\cdot}H_2O$ and $Ca_2FeMn{\cdot}Al_3B(OH)Si_4O_{15}$ containing 9.5 percent boron], ore concentrate [15 to 17 percent boron], or silicon dioxide irritated the lungs and inhibited phagocytic activity in industrial workers (2). The daily ingestion of 25 grams of boric tartrate over a period of 20 years in one case resulted in chronic boric acid poisoning. The symptoms included cachexia, dermatitis, alopecia, hypoplastic anemia and gastric ulcer. All symptoms disappeared when the drug was with-

FIGURE 7: MATERIAL FLOW THROUGH THE ECONOMY SHOWING PRIMARY EMISSION SOURCES—BORON

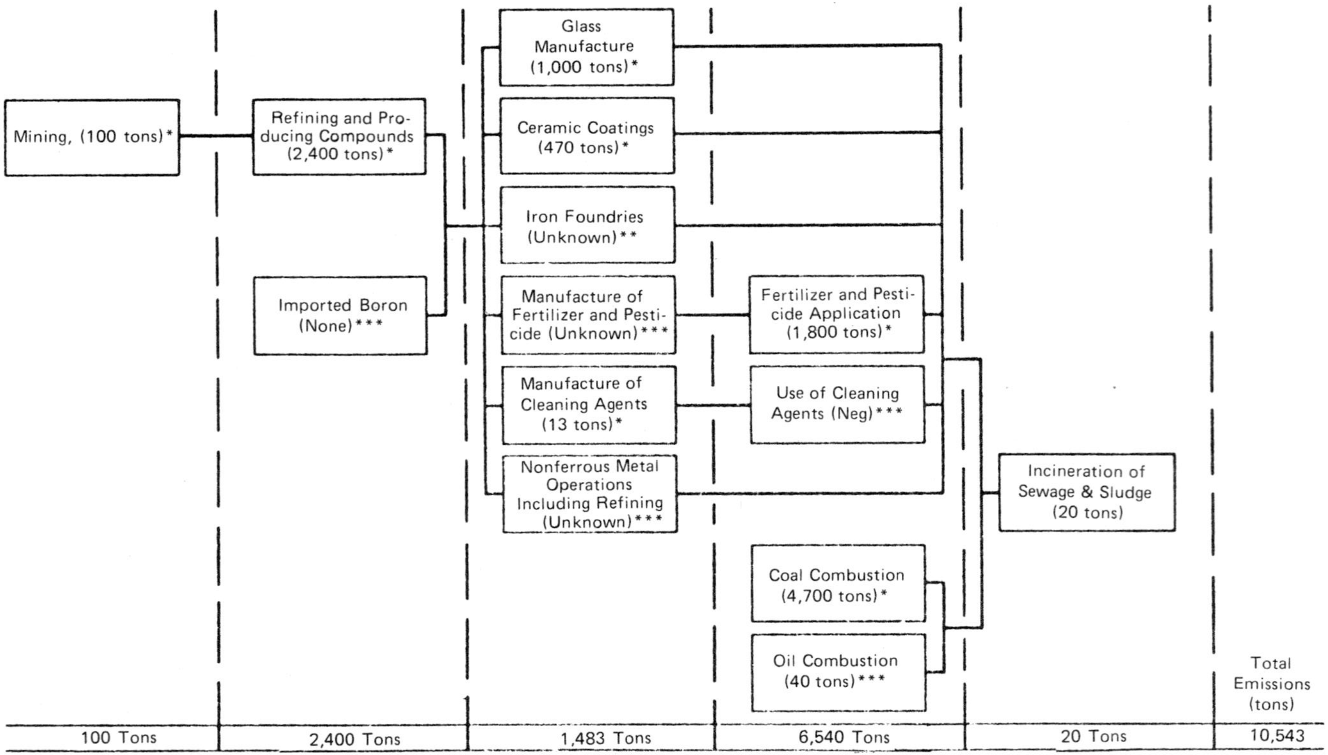

**National Inventory of Sources and Emissions—Boron*, W.E. Davis and Associates, June 1972.
***Preliminary Air Pollution Survey of Boron and its Compounds*, Litton Systems, Inc., October 1969. ***MITRE.

Source: Reference (4)

drawn, but reappeared when therapy was resumed because of epilepsy. No medical evidence of chronic effects of boron trifluoride has been found among workmen exposed to small amounts for periods up to 7 years. However, in persons subjected to boron trifluoride, disturbances of vascular permeability, enhanced tendon reflexes, joint diseases and atrophic changes in the nasal mucosa were observed (2).

EXTENT OF EXPOSURE

The air pollution aspects of boron and its compounds have been reviewed by Durocher (3). The flow of boron in the United States has been traced and charted for the year 1969. The consumption was 85,864 tons, while domestic production totaled 171,361 tons. Imports and exports were 803 and 86,084 tons, respectively. Emissions to the atmosphere during the year were 10,543 tons as shown in Figure 7. Similar data are shown in Table 13.

TABLE 13: BORON EMISSIONS

Source	Amount, tons	Pollutant, %
Borax mining	100	1.05
Glass manufacturing	1,000	10.55
Boron chemicals	2,400	25.32
Ceramic coatings	470	4.96
Soaps and detergent manufacturing	13	0.14
Use of pesticides, herbicides, and fungicides	1,800	18.99
Sewage and sludge incineration	20	0.21
Power plant boilers		
Pulverized coal	2,655	28.01
Stoker coal	304	3.21
Cyclone coal	91	0.96
Industrial boilers		
Pulverized coal	118	1.25
Stoker coal	413	4.36
Cyclone coal	59	0.62
Residential/commercial boilers		
Coal	37	0.39
Total	9,480	100.02

Source: Reference (4)

Nearly 22 percent of the emissions resulted from the processing of boron compounds, more than 34 percent from the manufacture and use of various end products, and about 43 percent from the combustion of coal. Emission estimates for processing and the manufacture of end use products are based on unpublished data obtained from industrial sources.

The emission factors presented in Table 14 are the best currently available. They were determined through a combination of methods which are listed on the following page.

They are, [1] direct observation of emission data and other related plant processing and engineering data; [2] estimates based on information obtained from literature, plant operators, and others knowledgeable in the field; [3] calculations based on experience and personal knowledge of metallurgical processing operations; and [4] specific analytical results where available.

TABLE 14: EMISSION FACTORS

Mining	1 lb/ton boron mined
Processing	28 lb/ton boron processed
End Product Uses of Boron	
Glass manufacture	70 lb/ton boron processed
Ceramic coatings	80 lb/ton boron processed
Other Emission Sources	
Coal	18 lb/1,000 tons of coal burned
Sewage and sludge	55 lb/1,000 tons of sewage and sludge burned

Source: Reference (5)

Details of airborne boron emissions are given below.

Glass – The principal sources of particulate matter in the furnace exhaust gases are raw materials entrained in combustion gases and materials from the melt. Particulates expelled from the furnace are the result of many physical and chemical reactions that occur during melting. One factor which appears significant is the furnace production rate; as the latter is increased the quantity of particulates increases disproportionately.

Ceramic Coatings – The principal emissions of boron from the production and use of ceramic coatings occur during the manufacture of the frit. The smelting operation causes the most significant dust and fume emissions. There are also dust emissions from handling the raw materials as they are received, stored, measured, and mixed prior to charging the smelter. The atmospheric emissions of boron during frit smelting consist primarily of oxides in vapor form and in particulates. The magnitude of emissions varies considerably depending upon the amount of boron in the charge, the combination of ingredients, and the type of smelter.

Agricultural Chemicals – Emissions to the atmosphere occur principally when the materials are applied as sprays or dusts. Too many factors are involved (air movement, drop size, nozzle type, etc.) to give an accurate estimate of emissions.

Soaps and Detergents – Boron compounds in the particulates emitted to the atmosphere are reported to be sodium perborate, sodium tetraborate, and sodium borohydride; the particle size ranges from 1 to 10 microns.

Mining of borax using explosives and its subsequent shipping and handling created a serious dust problem. A three-year, $10 million program to cut down dust emissions at Boron, California was concluded in 1972. Nevertheless, the operation is still not satisfactory and efforts to improve it are continuing. The primary emission from the processing of boron is sodium borate in various degrees of hydration, primarily from 1 to 3 mols of water. Boric oxide, boric acid, and sodium pentaborate are also emitted as these compounds are being produced. Particulate sizes range from less than 1 to 45 microns. Boron emissions vary considerably from plant to plant, ranging from 100 to 2,000 mg/kg of product (2).

Despite its widespread industrial use, information on industrial waste levels of boron and on treatment processes for removal of boron from industrial wastewaters is almost nonexistent in the literature surveyed. Waggott (6) has reported boron levels up to 5.5 mg/l in raw domestic sewage. A strong correlation between anionic detergent and boron concentrations was observed, indicating that the major source of the boron in this wastewater was sodium perborate, used as bleach in household washing powders.

This source, all of which may be assumed to enter the environment shortly after production, amounts to 15 percent of the domestic consumption (which is roughly half of the total production). Thus, based on about 600,000 mt total production in 1972 as B_2O_3, approximately 1.4×10^4 mt of boron may be considered to enter the water per year through the use of boron-containing laundry products (2).

Fluoborate solutions are sometimes used for plating of cadmium, copper, lead, nickel, tin, zinc and a number of alloys (7). They are used only for special purposes such as deposition of lead on battery bolts and terminals, plating of tin-lead alloys on bearings, or in cases where high-speed deposition is required for continuous wire or strip plating. Lowe has given fluoborate plating solution concentrations (as BF_4) of 125,000 to 265,000 mg/l (7). This is equivalent to boron concentrations of 15,800 to 33,500 mg/l. Direct application to the land or crops as fertilizer or herbicides accounts for about 7 percent of domestic usage. This may also be considered as entering the environment immediately upon production. Again, based on 1972 production, approximately 7×10^3 mt of boron appears to enter the environment per year from this source (2). Total boron released to the environment based on 1972 production and use figures may be summarized by:

Source	Amount, metric ton
Laundry products	14,000
Agricultural chemicals, fertilizers	7,000
Mining and processing	3,000
Glass and ceramics	1,500
Combustion of coal	4,000
Miscellaneous	2,500
Total	32,000

Of the estimated total of 32,000 mt/yr, most ends up in the waters because of the relatively high solubility of all the compounds (2).

DETECTION

To detect boron in airborne contaminants collected in air filters a small portion of the filter is rolled up into a cylinder and placed in a hollow graphite electrode.

This portion of the air filter is directly excited by the condensed spark discharge in an oxygen atmosphere. The sample burns, producing the spectra. A photographic recording is used to interpret the spectra and, when improved precision is required, line intensities are measured with a microphotometer. The limits of detection are 0.1 to 1.0 micrograms. The results obtained from the rapid emission spectrographic method are compatible with results obtained with the more conventional technique of ashing the filter, mixing it with a spectrographic buffer, and exciting it in a DC arc (2).

In a colorimetric method of analysis, about 1,000 liters of air passes at a rate of 2 to 3 l/min through four collecting tubes chilled in a methanol-dry ice bath. The collected boron is dissolved in 10 ml of water and added to 1.5 ml of curcumin reagent solution. The solution is evaporated until the odor of phenol disappears. The residue is dissolved in 10 ml ethanol and absorbance is measured at 548 millimicrons (2).

Several devices for detecting pentaborane in air are available, but they are nonspecific (i.e., they detect other compounds with similar chemical properties). A coulometric borane monitor was developed by Braman et al (8), which was based on the oxidation of boranes with electrolytically generated iodine. This instrument, with a sensitivity of approximately ± 0.02 ppm (40 $\mu g/m^3$) pentaborane in air, unfortunately was susceptible to overloading when excessive concentrations of electroactive materials present in the air desensitized it.

A boron hydride monitor that converted hydrides to boric acid by pyrolysis and then determined the presence of boric acid by colorimetry was described by Fristom, Bennet, and Berl (9). However, this monitor was not automated. Kuhns, Forsythe, and Masi (10) have outlined a hand-held device for automatic detection of boron hydrides in air, based on the reduction of boron hydrides with triphenyltetrazolium chloride. These devices provided sensitivity in the 0.1 ppm range, but were not continuous monitors and were subject to interference from all oxidizing or reducing material in air.

Hill and Johnston (11) reported on ultraviolet spectrophotometric detection of decaborane, but this method proved unsuitable for use with other boranes. Braman and Gordon (12) and Braman (13) reported on a direct-reading, automatic, temperature-insensitive instrument capable of detection of pentaborane down to concentrations of 0.05 ppm (100 $\mu g/m^3$).

This device used the flame emission principle of detection; in the presence of boranes, a small hydrogen flame takes on a greenish cast, the intensity of which is proportional to the borane concentration. This device can be operated continuously over an 8-hour period using a self-contained hydrogen supply and storage batteries. Development of an atmospheric monitoring device to detect and record several toxic components, including pentaborane, is reported by the Mine Safety Appliances Company (14). This device also features pyrolysis to form boric oxide from borane, followed by detection of the boron by using n-amylamine reagent. The company claims a detectability of 0.01 ppm (20 $\mu g/m^3$), with rapid response and continuous operation.

In a comprehensive 1964 review of analytical methods for determining the presence of boron, Nemodruk and Karalova (15) listed the following methods. Photometric methods are widely used, and are especially suitable for determining micro amounts, at concentrations of 0.05 to 0.20 μg, these methods produce 90% or better accuracies. Coulometric titration is said to provide determination of amounts as low as 0.2 μg. Spectral analysis is one of the most widely used methods in metallurgy, capable of determining 5 x 10^{-5}% concentrations of boron. Fluorimetric analysis is also used, with sensitivity to 0.1 μg of boron. Radioactive methods are also available, based on the ability of boron's nuclei to absorb thermal neutrons; this method is considered simple, and highly efficient, is capable of providing 100 determinations in 8 hours, and has sensitivity to 0.1 μg.

Monitoring data on the boron content of U.S. waters is limited. For this reason, results of a survey conducted in Europe are of interest. A study was made for the National Swedish Environmental Protection Board, Uppsala, of boron in Swedish and Norwegian fresh waters. The study began in 1970 and took a year to complete. Regional variations due to differences in geology, land use, and population density were found, but there were only small seasonal variations. The enrichment of boron in aquatic plants was very small compared to that of phosphorus and nitrogen. Sewage water showed a lower boron content than that of England. The mean concentration of boron in Swedish and Norwegian fresh waters was low and close to that of other rivers of the world. The median boron content in 355 samples was approximately 12 μg/l (2).

Boron may be monitored in water by the following instrumental techniques (16), according to LBL-1, Vol. 2: direct reading emission spectrometry; potentiometric. Boron may be measured in water by the atomic absorption technique down to a detection limit of 2.6 ppm. An analysis developed by P. Cohen and W.D. Fletcher (17) permits determining the boron concentration in an aqueous solution of boron-containing material. The analyzer has means for adding a polyhydroxyl compound to increase the electrical conductivity of the solution and means for controlling the solution temperature, cell means for continuously measuring electrical conductivity of the solution and for comparing the conductivity of the solution with that of an established standard.

In the past, analyses usually have been based upon manual titrimetric or colorimetric methods. Such methods, however, have not been suitable because of time delays of the order of at least 15 to 20 minutes. Presently available commercial analyzers are generally unsuitable either because of their insensitivity to small changes in boron concentration, or to excessive delays inherent in their mode of operation. Moreover, commercial boron analyzers of the automatic type have been very expensive and require excessive attendance and maintenance. Thus, there has been a need for accurate and fast automatic methods for analyzing the boric acid content of water.

It has been found that a reliable automatic boron analysis may be provided which is simple, relatively low in cost of manufacture, rapid, and sensitive to small changes in boron concentration. The analysis is based upon the increase in solution electrical conductivity which occurs when a polyhydroxyl compound is added to a solution containing boric acid or a borate salt.

Colorimetric and spectrographic methods have been used to analyze boron in fresh water in concentrations of less than 1 ppm. One of the most useful methods is "the identical pH" method with which a sensitivity of ±0.05 ppm boron can be obtained if a 500 ml sample is used. With this method separation or special sample treatment is not required. Detection of the boron by electrometric instead of colorimetric technique is recommended. For use with more than 1 ppm boron, the standard mannitol borate procedure may be used after removing interfering bicarbonate and metal ions (2). The analysis of boron in sea water represents a special case. A high degree of precision and accuracy is required because the ratio of boron to the total salt content is almost constant throughout the world. Only bicarbonate ion removal is required for the mannitol borate procedure. Polyhydroxy organic matter causes low results by complexing the boron (2).

ENVIRONMENTAL STANDARDS

The permissible boron limit in domestic water supplies and in irrigation waters is 1.0 ppm boron. According to the office of Toxic Substances of EPA (2), proposed criteria according to type of crop to be irrigated are as follows:

Crop Species	Critical Boron Concentration, ppm
Sensitive	0.3 to 1.0
Semitolerant	1.0 to 2.0
Tolerant	2.0 to 4.0

Sensitive plants are blackberry, lemon, elm, cherry, peach, persimmon, fig, strawberry, lupine, grape, violet, pansy, kidney bean, cowpea, Jerusalem artichoke, larkspur, and zinnia. The presence of boron in great enough concentration to cause crop damage in some potential irrigation water in parts of Western United States has prompted proposals for criteria for the boron content of irrigation water. No official standard has been set yet.

There is also concern about boron concentration in processes where boron dust may be excessive, such as in the mining and handling of borax ore. In the U.S.S.R. the recommended maximum permissible concentration of calcium borate dust in the air of working locations is 4 to 6 mg/m^3 of air (2). Other standards for boron compounds in air are (2): boron oxide, TLV, 10 mg/m^3 (this value was recommended in view of the very low toxicity to test animals of boron oxide aerosols); boron trifluoride, TLV, 1 ppm by volume, approximately 3 mg/m^3; and boron tribromide, TLV, 1 ppm by volume, approximately 10 mg/m^3.

HANDLING PROCEDURES

There are no particular handling procedures for boron oxide and salts except to avoid their ingestion. As with the other metals, volatile compounds such as hydrides offer much more complex handling problems, however. The only boron compounds covered here are those which are included in "Commodity List of Explosives and Other Dangerous Articles" (2). They are boron trichloride, boron trifluoride, decaborane and pentaborane.

Boron trichloride is classified by the Interstate Commerce Commission as a corrosive liquid, requiring a white label. The maximum quantity allowed in containers by rail express is one liter. International Air Transport Association [IATA] regulations limit its transport to one liter carried in cargo planes only. Boron trifluoride is a nonflammable gas, requiring a green label. It is restricted to 140 kg in one outside container by rail express. IATA regulations permit the transport of up to 140 kg in cargo planes only. Decaborane is classified as a flammable solid requiring a yellow label. Rail transport is limited to 12 kg in one outside container. IATA regulations permit cargo planes only to carry up to 12 kg of decaborane. Pentaborane is a flammable liquid, requiring a red label. It is not accepted for rail express. IATA regulations prohibit its transport in both passenger and cargo planes.

REMOVAL FROM AIR

Normal entrapment and precipitation procedures applicable to most particulate pollutants should be effective in controlling or reducing the amount of boron compound dusts emitted into the atmosphere. No specific methods for abatement of atmospheric borane compounds have been identified, but strict regulatory practices are used to prevent accidental emissions. No information has been found on the economic costs of boron air pollution or on the costs of its abatement. Additional information on atmospheric emissions of boron is available in a report by W.E. Davis and Associates (5).

REMOVAL FROM WATER

Despite its widespread industrial use, information on industrial waste levels of boron, and on treatment processes for removal of boron from industrial wastewaters is almost nonexistent in the literature surveyed (18). Waggott (6) has reported boron levels up to 5.5 mg/l in raw domestic sewage. In the case of irrigation water, a standard of one ppm boron has been established. Normally this limit is attained through dilution with other waters. The manufacturers of the Amberlite IRA-943 resin have made recent improvements and claim that deboration of the average irrigation water can be effected at costs well below three cents per 3.8 cubic meters.

A process has been developed by R.W. Goeldner (19) for the removal of boron from water. It involves evaporation, and recondensing the vapor. The distillation process appears not particularly effective, as it is reported that upon evaporation of a waste containing 21,000 to 22,000 mg/l of boron, the recondensed vapor still contained 50 to 80 mg/l of boron. Upon passing this water through a six foot column containing ceramic Raschig contact rings, the condensed vapor had a boron content of 2 to 3 mg/l. No costs were reported for the process.

SOLID WASTE DISPOSAL

The landfill disposal of boron trifluoride has been discussed by P.W. Powers (20). The recommended procedure involves chemical reaction with water to form boric acid and fluoboric acid. The fluoboric acid is reacted with limestone forming boric acid and calcium fluoride. The boric acid may be discharged into a sanitary sewer system while the calcium fluoride may be recovered or landfilled.

REFERENCES

(1) R.T. MacMillan, in *Mineral Facts and Problems,* Washington, D.C., U.S. Bureau of Mines (1970).

(2) Versar, Inc., *Preliminary Investigation of Effects on the Environment of Boron, Indium, Nickel, Selenium, Tin, Vanadium and Their Compounds, Vol. 1—Boron,* Report EPA-56/2-75-005A, Washington, D.C., Environmental Protection Agency (August 1975).

(3) N.L. Durocher, *Air Pollution Aspects of Boron and Its Compounds,* Report No. PB-188,085, Springfield, Va., Nat. Tech. Information Service (1969).

(4) A.J. Goldberg, *A Survey of Emissions and Controls for Hazardous and Other Pollutants,* Washington, D.C., Environmental Protection Agency (February 1973).

(5) W.E. Davis and Associates, *National Inventory of Sources and Emissions, Sect. II—Boron,* Report PB-210,677, Springfield, Va., Nat. Tech. Information Service (May 1972).

(6) A. Waggott, "An Investigation of the Potential Problem of Increasing Boron Concentrations in Rivers and Water Courses," *Water Research,* 3, 749-765 (1969).

(7) W. Lowe, "The Origin and Characteristics of Toxic Wastes, with Particular Reference to the Metal Industries," *Water Pollution Control,* London, 1970, pp 270-280.

(8) R.S. Braman, D.D. DeFord, T.N. Johnson and L.J. Kuhn, "A Coulometric Borane Monitor," *Anal. Chem.* 32 (September 1960).

(9) G.R. Fristom, L. Bennett and W.G. Berl, "Integrating Monitor for Detecting Low Concentrations of Gaseous Boron Hydrides in Air," *Anal. Chem.* 31 (October 1959).

(10) L.J. Kuhns, R.H. Forsythe and J.F. Masi, "Boron Hydride Monitoring Devices," *Anal. Chem.* 28 (November 1956).

(11) W.H. Hill and M.S. Johnston, *Anal. Chem.* 27, (1955).

(12) R.S. Braman and E.S. Gordon, "The Design of a Portable Instrument for the Detection of Atmospheric Pentaborane and Other Toxic Gases, Institute of Electrical and Electronic Engineers," *Trans. Instrumentation and Measurement,* Vol. IM-14 (1965).

(13) R.S. Braman, *Research and Development of an Automatic Beryllium and Boron Monitor,* Armour Research Foundation of Illinois, Institute of Technology Report ARF3202-3 (1962).

(14) Mine Safety Appliances Company, Final Report, *Development of an Atmospheric Monitoring System,* (1961).

(15) A.A. Nemodruk and Z.K. Karalova, *Analytical Chemistry of Boron,* Academy of Sciences of the U.S.S.R. (Jerusalem: Silvan Press, 1964).

(16) Lawrence Berkeley Laboratory, *Instrumentation for Environmental Monitoring—Water,* Publ. No. LBL-1, Vol. 2, Berkeley, Univ. of Calif. (February 1, 1973).

(17) P. Cohen and W.D. Fletcher; U.S. Patent 3,468,764; September 23, 1969; assigned to Westinghouse Electric Corporation.

(18) J.W. Patterson and R.A. Minear, *Wastewater Treatment Technology,* 2nd Edition, Chicago, Illinois Institute for Environmental Quality (February 1973).

(19) R.W. Goeldner; U.S. Patent 3,480,515; November 25, 1969; assigned to Aqua-Chem Inc.

(20) P.W. Powers, *How to Dispose of Toxic Substances and Industrial Wastes,* Park Ridge, N.J., Noyes Data Corp. (1976).

CADMIUM

Cadmium is found in nature, most generally in low concentrations associated with a similar metal, zinc. It is a soft silver-blue metal, familiar to us as the anticorrosion coating often used on outdoor fittings such as gate latches, roofing nails and screws. It is present as an impurity in the more common galvanized zinc coatings (0.03%) and is also used in many other everyday items such as paint, some pottery pigments and plastics. Automobile tires also contain cadmium which is introduced as an impurity in the zinc used in the production of tires.

Cadmium is a by-product of zinc smelting, and to a smaller extent, cadmium can also be recovered from lead and copper ores. The U.S. production of cadmium in 1972 was 8.1 million pounds while 2.37 million pounds were imported and 12.5 million pounds were consumed.

TOXICITY

Health problems from cadmium were suddenly brought into focus when high cadmium intake was identified as the probable cause of the terrible Itai Itai disease, which has been prevalent in the Minimata area of Northern Japan since World War II, as pointed out by C.R. Parker and D.P. Sandoz (1).

This disease, really poisoning, resulted in gross deformities, pain as bones snapped under weight of the body and in over fifty cases, death. Since that time, investigations have revealed numerous other illnesses which appear to be related to intake of lesser amounts of cadmium. The tragic deaths and deformities experienced since World War II by more than 200 people in Northern Japan have been attributed to cadmium from mining wastes which polluted the water supply, resulting in contamination of rice-growing areas. The damage caused by cadmium in this case was increased by lower than normal amounts of calcium and protein in the diet, which appear to increase the proportion of cadmium retained in the body.

As pointed out by the Office of Toxic Substances of EPA (2), cadmium has unquestioned chronic toxicity leading to serious pathological consequences when ingested in quantities only 3 to 13 times greater than present average intake rates. However, there is a lack of definitive cause-and-effect evidence, or even of a precise analytical projection, between cadmium dissemination and chronic diseases in the United States.

Cadmium and cadmium compounds are toxic substances by all means of administration, producing acute or chronic symptoms varying in intensity from irritations to extensive disturbances resulting in death according to Athanassiadis (3). However, despite increasing use of this metal and increasing attention to its toxic nature, the exact manner in which it affects human or animal organisms is not yet known. Cadmium is toxic to practically all systems and functions of the body, and is absorbed without regard to the levels of cadmium already present, thereby indicating the lack of a natural homostatic mechanism for the control of organic concentrations of cadmium.

Inhalation of cadmium fumes, oxides and salts often produces emphysema, which may be followed by bronchitis. Prolonged exposures to airborne cadmium frequently cause kidney damage resulting in proteinuria. Cadmium also affects the heart and liver. Statistical studies of people living in 28 U.S. cities have shown a positive correlation between heart diseases and the concentration of cadmium in the urban air.

Cadmium may also be a carcinogen. While there is little evidence to support this conclusion from studies of industrial workers, animal experiments have shown cadmium may be carcinogenic. No data were found that indicated deleterious effects produced by airborne cadmium on commercial or domestic animals. However, experiments with laboratory animals have shown that cadmium affects the kidneys, lungs, heart, liver and gastrointestional organs, and the nervous and reproductive system according to Athanassiadis (3).

Many workers regularly exposed to cadmium dust and fumes show symptoms of lung disease similar to emphysema, and as many as 25% in one survey showed symtoms of kidney stones. In cases of inhalation of large quantities of cadmium, severe and irreparable lung damage has resulted. The metal tends to accumulate in several organs of the body, notably the kidneys and the liver.

Excessive concentrations (200 μg/g) cause severe kidney damage, resulting in breakdown of kidney functions and excessive excretion of protein. This can result from an intake of only 2 to 3 times the average intake of a U.S. adult, which is well within the bounds of possibility in areas of contamination or for a heavy smoker.

There is also concern since one statistical survey of cadmium in the air in U.S. cities showed a correlation between cadmium levels and increased arteriosclerotic heart disease, high blood pressure and decreased life expectancy. Much more work is required to firmly establish the connection, but there are now enough indicators to cause concern. Statistical data are reinforced by knowledge that experimental animals have shown marked biochemical changes, including liver damage and decreased life expectancy, while their organs contain lower cadmium concentrations than those normally found in adult humans.

Thus, the health effects of cadmium, both proven and probable, are increased blood pressure, increased incidence of arteriosclerotic disease and reduced life expectancy. As intake increases, kidney damage, protein excretion, anemia and emphysema may result. Chronic high exposures can result in liver and lung damage, degenerative bone disease and death.

Cadmium, usually considered a toxic metal, and under certain circumstances a nephrotoxic one, has been found in the kidneys of all adult human beings from many cities of the world, according to Cothern (4). It is virtually absent in newborns and accumulates with age. There is no known biological function for cadmium in man (5). A marked correlation (r = 0.76) exists between the concentration of cadmium in the air and death rates from hypertension and arteriosclerotic heart diseases (6). Zinc showed a similar correlation with r = 0.56.

Feeding the environmental toxicant cadmium (75 mg/kg) to young Japanese quail for 4 weeks produced growth retardation, severe anemia, low concentrations of iron in the liver and high concentrations of cadmium in the liver. Dietary ascorbic acid supplements (0.5 and 1.0% by weight) almost completely prevented the anemia and improved the growth rate but did not markedly alter concentrations of iron or cadmium in the liver (7). It has been found that 0.1 ppm of cadmium is toxic to some aquatic insects and 0.01 to 10 ppm is toxic to fish (8).

Cadmium in concentrations of 5 ppm were given via their drinking water to mice. This level showed an innate toxicity in terms of survival of male mice (9). Hypertension was induced in rats by adding 5 ppm cadmium to their drinking water. Females were more susceptible at all ages. Cadmium was found in the livers and kidneys of all rats. Along with the hypertension in older rats went salt (sodium chloride) hunger (10), (11). Rats fed 10 ppm of cadmium in water showed in 2 to 4 months chronic arterial hypertension (12). Thus, cadmium is a serious health hazard in water supplies in the ppm range.

Cadmium is extremely toxic to fish. Concentrations between 0.008 and 0.01 are lethal to 50% of a test batch of trout (13). Concentrations of $Cd(OAc)_2$ between 25 and 50 ppm produced marked osteoporosis (14). In concentrations between 0.01 and 10 ppm the fish all overturned in a few days except in the lowest concentrations.

In the upper range hyperplasia and necrosis of the epithelium of secondary gill lamellae were evident. Death probably resulted from interference in respiration. Gill damage was less pronounced in the low concentrations. There were morphological and histological changes in the internal organs especially the spleen, which was nodular in appearance, congested and structurally changed. The liver was also affected. The heart, skeletal muscle and brain had degenerated (15).

The toxicity of cadmium has also been discussed by Luckey et al (16) and at more length by Tucker (17) and by Schroeder (18). It has also been discussed in exhaustive detail by L. Friberg et al (19). Finally the National Environmental Research Center has done a recent summary report on cadmium (20), including its toxicology.

EXTENT OF EXPOSURE

We ingest cadmium from a number of sources, the air, our food, and tobacco.

Studies indicate that the average American adult ingests a total of 50 to 60 micrograms per day (compared with 15 to 20 in Sweden and 80 micrograms in Japan). The typical adult cadmium intake in the U.S. is estimated to be 50 micrograms from food; 1 to 10 micrograms from air, depending on local air content; 1 to 10 micrograms from water; and 2 to 4 micrograms from tobacco for heavy smokers.

Fortunately, most of this is passed through the body with only about 2 micrograms per day retained although this amount may rise to 3 to 4 micrograms per day for a person in a polluted environment or for a heavy smoker. However, even 2 micrograms per day stored in the body daily add up to an appreciable amount over a period of years (1).

Table 15 is a quantitative summary of where cadmium is found, produced, converted, used, and inadvertently altered. Of the cadmium in the commercial flow in the United States, this summary table presents estimates of the amounts dissipated in end products, and of the amounts accessible to the environment via air, water and land discharges.

TABLE 15: CADMIUM EMISSION ESTIMATES*

Source	Airborne Emissions	Waterborne Effluents	Land-Destined Wastes	Total Emissions
Zinc ore mining and beneficiation	0.2	~0	250	
Primary zinc industry	102	10 (1971-72) 2.0 (1977) 1.3 (1983)	~0	
Total: Extraction, refining & production	102	~7 (1974-75) ~2 (1980)	250	359 (1974-75) 354 (1980)
Electroplating shops	~1	10.5 (1972) 4.0 (1977) 0 (1983)	73.5 (1972) 80 (1977) 0 (1983)	
Pigment manufacture	9.5	0.75	16.5	
Stabilizer manufacture	2.7	~0	~0	
Alloy manufacture	2.3	~0	~0	
Battery manufacture	0.7	0.3	8.4 (1973) 11.4 (1977) 9.1 (1983)	
Total: Industrial conversion	15	~8 (1974-75) ~3 (1980)	~102 (1974-75) ~75 (1980)	125 (1974-75) 93 (1980)
Secondary nonferrous metals	2.2	~0	20	
Iron and steel industry	10.5	~0	330	
Galvanized products	~0	~0	40	
Rubber tire wear	5.2	~0	~0	
Incineration	16	~0	70	
Total: Consumption & disposal of Cd-containing products	34	~0	460	494
Phosphate fertilizers	~0	~0	100 (1975) 130 (1980)	
Phosphate detergents	~0	10.2	~0	
Coal combustion	80 (1974) 80 (1980)	~0 ~0	370 (1974) 680 (1980)	

(continued)

TABLE 15: (continued)

Source	Airborne Emissions	Waterborne Effluents	Land-Destined Wastes	Total Emissions
Diesel & fuel oil combustion	50	~0	~0	
Lubricating oils	0.8	~0	~0	
Sewage sludge	20	~0	250	
Total: Inadvertent sources	151	10	720 (1974-75) 1,060 (1980)	881 (1974-75) 1,221 (1980)
Grand Totals	300	25 (1974-75) 15 (1980)	1,500 (1974-75) 1,800 (1980)	1,800 (1974-75) 2,100 (1980)

*Units: metric tons per year as elemental cadmium.

Source: Reference (2)

Of the total cadmium released to the environment through the activities of man, 20% is accountable to the primary nonferrous metals industry (including ore mining and beneficiation); 30% is accountable to the conversion, use, and disposition of cadmium in our economy; while the remaining 50% is inadvertent (derived from fossil fuels, phosphate fertilizers, and sewage sludge) and not at all linked to the production or use of cadmium metal and its derivatives.

Of the total cadmium released to the environment through the activities of man, only 15% is in the form of air pollution from stationary sources, and only 1% is in the form of water pollution from point sources. These quantities appear to be decreasing, according to a comparison of 1968 emission levels with 1974-75 levels. This decrease reflects investments in abatement equipment already made and reflects the concerted efforts made by the government and private sectors to abate the readily-identifiable sources of pollution.

Of the three most important exposure routes for cadmium, exposure via breathing ambient air is the least significant, except in the vicinity of point sources. The major route is via food intake. Smoking may be an additional means of exposure, according to the National Environmental Research Center of EPA (20).

The airborne emissions data in Table 15 are at considerable variance with earlier data published by Goldberg (21) and shown in Table 16 and Figure 8. The discrepancies in some instances are not as bad as they seem. Thus, iron and steel industry emissions while of different magnitudes appear as land destined wastes in Table 15.

As noted by EPA (2), the lower estimates in Table 15 are based on newer data specific to the scrap grades used in steel making; they also reflect effective air pollution abatement practices.

The area of copper, lead and zinc smelting represents an unexplained discrepancy. The more recent data (2) seem to have ignored the categories of copper and lead smelting which made substantial contributions in earlier surveys (21).

The metals industry is the major source of emissions of cadmium into the atmosphere. Cadmium dusts and fumes are produced in the extraction, refining and processing of metallic cadmium.

FIGURE 8: MATERIAL FLOW THROUGH THE ECONOMY SHOWING PRIMARY EMISSION SOURCES

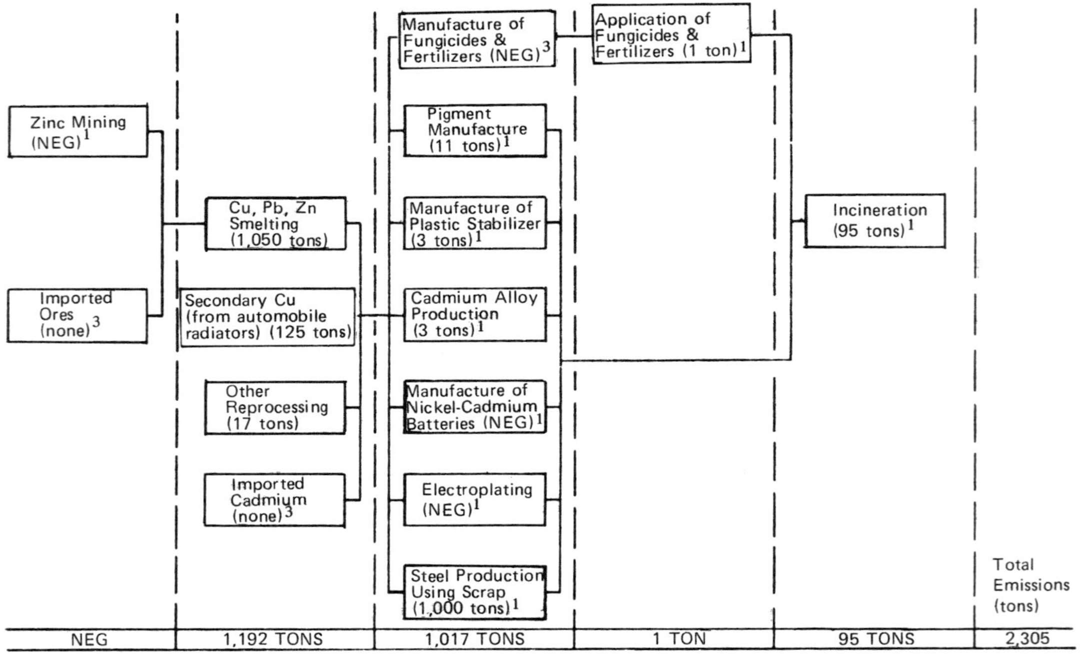

[1] *National Inventory of Sources and Emissions — Cadmium,* W.F. Davis and Associates, February 1970
[2] *Preliminary Air Pollution Survey of Cadmium and its Compounds,* Litton Systems, Inc., October 1969
[3] MITRE

Source: Reference (21)

TABLE 16: CADMIUM EMISSIONS

Source	Amount in Tons	% This Pollutant
Copper mining	Neg	Neg
Zinc mining	<1	0.01
Lead mining	Neg	Neg
Primary copper		
Roasting	229	7.59
Reverberatory furnace	94	3.12
Converters	270	8.95
Material handling	59	1.96
Primary zinc		
Roasting	666	22.07
Sintering	284	9.41
Distillation	90	2.98
Material handling	Neg	Neg
Primary nickel	Neg	Neg
Primary lead		
Sintering	66	2.19
Blast furnace	19	0.63
Reverberatory	3	0.10
Material handling	Neg	Neg
Secondary copper		
Sweating furnace	70	2.32
Blast furnace	55	1.82
Iron & Steel		
Blast furnace	1,000	33.14
Nonferrous alloys		
Furnaces	3	0.10
Material handling	Neg	Neg
Cadmium paint pigments	11	0.36
Cadmium-barium plastic stabilizers	3	0.10
Cadmium-nickel batteries	<1	0.01
Miscellaneous cadmium products	<1	0.02
Use of pesticides, herbicides and fungicides	<1	0.01
Fertilizer application	<1	0.02
Incinerators	95	3.15
Total	3,018	100.06

Source: Reference (21)

Since cadmium is generally produced as a by-product in the refining of other metals, such as zinc, lead, and copper, plants refining these materials are sources of cadmium emissions as well as of the basic metal. Also, because cadmium is present in small quantities in the ores of these metals, cadmium emissions may occur inadvertently in the refining of the basic metal.

Common sources of cadmium air pollution also occur during the use of cadmium. Electroplating, alloying, and use of cadmium in pigments can produce local contaminations of the atmosphere. Also, since cadmium is added to pesticides and fertilizers, the use of these materials can cause local air pollution. In 1964, the average concentration of cadmium in the ambient air was 0.002 $\mu g/m^3$, and the maximum concentration was 0.350 $\mu g/m^3$.

It is difficult to get a realistic idea of natural levels of cadmium in the air, since so much has been added by man. However, it is generally agreed that natural levels are 1 ng/m^3 or less, such as are found in country areas far removed from pollution and in a few clean cities (1). In contrast, yearly averages as high as 65 ng/m^3 have been measured for Chicago while concentrations above 1,000 ng/m^3 have been measured near smelters and other sources of cadmium emissions.

Cadmium taken through the lungs is about 60 times more poisonous than when taken through the digestive tract. Yet a large proportion of the population continues to expose itself to an additional source of airborne cadmium from burning tobacco. It is not widely known that tobacco contains between 1 and 2 micrograms of cadmium per gram, that is, there are 1 to 2 micrograms of cadmium in every cigarette.

Refining cadmium-bearing ores resulted in the release to the atmosphere of more than two million pounds per year in the United States. Cadmium is almost always found with zinc, but in much lower concentrations. We use enormous quantities of zinc, often for corrosion protection as on galvanized water pipes, downpipes and other steel products. During scrap metal recovery processes in the U.S. an estimated 2.3 million pounds of cadmium went up the chimney in 1968.

The total usage of cadmium is around 12 to 13 million pounds per year in the U.S., and 31 million pounds in the world. Apart from the corrosion protection, the metal has a host of uses such as in alloys and nickel-cadmium batteries. In plastics it is a valuable stabilizer, assisting in the manufacturing process and extending the lifetime of the product.

It is also used as a pigment in some paints. During the manufacture of these goods some cadmium escapes to the atmosphere. Many such products end their useful lifetime in incinerators, releasing much of the metal to the air. The total cadmium emitted to the atmosphere was estimated to be 4.6 million pounds in the U.S. in 1968. Most of this eventually settled onto the soil to be taken into plants by roots, then into food products. Some settled in water catchment areas and was swept into our water supplies, and some we breathed directly.

It is known that large quantities of cadmium find their way into our water supplies from electroplating and coating installations and from other industries using cadmium. The exact amounts are not known, but are probably large. We can get an idea of the extent of this problem from the knowledge that silt dredged from the bottom of Lake Erie contained 130 parts per million of cadmium. This is 3 times more than soils immediately adjacent to factories which are known to emit large quantities of cadmium, and is many times larger than the normal soil levels of several parts per million.

The principal sources of aqueous cadmium wastes have been identified as: [1] rinse water and dragout from the electroplating industry; [2] waste sludges from paint manufacture; [3] paint residue left in used paint containers; and [4] wash water from the manufacture of nickel-cadmium batteries (22).

The electroplating industry is the largest user of the cadmium metal and is at the present also the largest source of cadmium wastes. The cadmium wastes are generated primarily as the result of the rinsing operations, and the typical aqueous waste stream contains 100 to 500 ppm cadmium, along with other heavy metals, cyanides, and metal surface cleaning agents. Large volumes of liquid waste streams having a much higher cadmium content, however, have also been reported. An example of such a waste stream contains 1.5% cadmium cyanide, 8.5% sodium cyanide, 3% sodium hydroxide, and traces of other metals. The total amount of cadmium wastes as cadmium from the electroplating industry has been estimated to be 1.44 million pounds per year.

Cadmium wastes are generated in paint manufacture as a result of the kettle washings and equipment cleanup. The two most important cadmium pigments are then cadmium sulfide and the cadmium sulfoselenide. Cadmium in these forms is found in the solvent-based waste paint sludges, which typically contain 27.5% pigments, 25.0% binders, and 47.5% organic solvents. It is estimated that 5,100 pounds of cadmium are lost through 37 million pounds of solvent-based paint sludges per year. An additional estimated 35,000 pounds of cadmium are lost as paint residues left in used paint containers each year.

The major source of cadmium waste in the manufacture of nickel-cadmium batteries (sintered-plate type) is the wash water that is used to remove excess material from the plates. The typical wastewater effluent contains cadmium hydroxide, potassium hydroxide, and potassium nitrate, often in fairly high concentrations due to the cost of the deionized water normally used. Cadmium loss as cadmium from battery manufacture amounts to 3,700 pounds per year.

Potential industrial sources of dissolved cadmium are reported by McKee and Wolf (23) and Santaniello (24) to be: metallurgical alloying, ceramics, electroplating, photography, pigment works, textile printing, chemical industries and lead mine drainage. Of the industries listed above, only the electroplating industry is discussed to any extent in the literature with regard to cadmium-bearing wastes and the treatment thereof. High concentrations of cadmium in lead mine drainage have been reported, however (23). Cadmium wastewater concentrations reported in the literature are summarized in Table 17.

A report on groundwater contamination from plating wastes gives a concentration of 1.2 milligrams per liter cadmium in a groundwater recharge basin receiving the plant wastes. Private wells in the area reported 0.3 to 0.6 mg/l, while one test well contained 3.2 milligrams per liter of cadmium (25).

TABLE 17: CADMIUM CONCENTRATIONS REPORTED FOR INDUSTRIAL WASTEWATERS

Process	Cadmium Concentration, mg/l
Automobile heating control manufacturing	14 - 22*
Plating rinse waters (large installations)	15 av, 50 max
Plating rinse waters	
0.5 gph dragout	48
2.5 gph dragout	240
Plating bath	23,000
Lead mine acid drainage	1,000

*24-hour composite on days when plating baths were dumped. Cadmium levels much reduced due to dilution when solution dragout was the only contributing source.

Source: Reference (26).

DETECTION

Dust and fumes of cadmium compounds may be collected by any method suitable for collection of other dusts and fumes; the impinger, electrostatic precipitator, and filter are commonly used. The National Air Sampling Network uses a high-volume filtration sampler.

Samples are taken from ambient air by filtration through porous media of low, known cadmium content such as glass-fiber filters or membrane filters (20). The filter is held by a flexible gasket in a holder that is a part of a device made to exclude debris, facilitate maintenance, regulate the time of sampling, and permit estimation of the air volume. The sample of particulate matter so collected is prepared for subsequent analysis by combustion (use of a low-temperature asher will eliminate loss of cadmium by volatilization) and dissolution of cadmium in the residue by the use of acids. Wet digestion with acids is also a reliable technique.

Water may be sampled by continuous withdrawal of a sample from a pipe or stream, but the more common method is batch sampling. Clean, acid-washed borosilicate containers should be used, and the samples analyzed as soon as possible. When only one set of samples is to be taken from a stream or lake, it is best to sample at the middle depth. Cadmium can be lost from solution in trace quantities by absorption on container walls, even those made of resistant glass.

The problem of collecting representative samples is complicated by the multitude of source configurations that exist. Cadmium is emitted from both stack and diffuse sources. Because stacks constitute a relatively well-defined source configuration, the possibility of obtaining reliable stack emission data is much better than in the case of diffuse sources. For stack sources, EPA recommends (20) a sampling train consisting of probe, filter holder, impingers, and associated metering system for isokinetic sampling.

The volume flow rate of stack effluents also must be obtained in order to derive pollutant mass emission rates. For diffuse sources, accurate emission data are virtually impossible to obtain by methods currently employed. The usual practice is to do sampling and analysis for ambient concentrations in the vicinity of such sources and to use such data in conjunction with estimated rates of thermal diffusion, turbulent diffusion, deposition, etc. to derive approximate emission rates.

Except in the case of continuous monitors, it is most convenient to consider environmental sampling and analysis as independent of the origin of the sample, with procedural differences consisting of variations in methods of sample preparation.

Methods available for analysis of collected cadmium emission samples include atomic absorption, atomic fluorescence, anodic stripping voltametry, ultraviolet and visible spectroscopy, polarographic procedures, and titrimetric analysis. Multielemental techniques sometimes used are spark-source mass spectrography and optical emission spectrography. X-ray fluorescence appears promising as a rapid, accurate, nondestructive, multielemental technique (20).

In general, the analytical methodology employed is determined by resource considerations and how the data are to be used. Potentially, any method of analysis capable of determining cadmium in the concentrations to be found in environmental samples should be acceptable, but in practice the number of methods in extensive use has been restricted to the classical procedures of colorimetric dithizone (CD), optical emission spectroscopy (OES), neutron activation analysis (NAA), atomic absorption spectroscopy (AAS), and anodic stripping voltametry (ASV).

The dithizone method consists of the formation of a colored dithizone cadmium complex, separation of interferents by closely controlled multiple extractions with various reagents, spectrophotometric determination of the colored complex, and estimation by reference to known responses. The CD method is slow, tedious, and requires excellent technique, but gives reliable results.

The OES procedure uses expensive instrumentation and consists of exciting the cadmium atoms in a sample by electric arc or spark, separation of the emitted light that is characteristic of cadmium by means of a prism or grating, elimination of other lines by means of selective slits, and measuring the light intensity on film or by phototube. The line intensity is compared with the response from standards, and the cadmium content is estimated. The film approach is semi-quantitative, whereas the photometric procedure is more quantitative.

The NAA approach involves irradiation of a sample with neutrons. Cadmium is detected by identification of the radiation emitted by activated cadmium atoms; chemical separation may be necessary to remove interferents. NAA costs, comparatively, are very high, but the sensitivity is good, and the accuracy compares favorably with that of OES.

The AAS procedure consists of exciting vaporized sample atoms in a flame positioned in the path of light from a lamp that has a cathode made of the metal of interest, cadmium in this case. The cadmium atoms in the ground state in the beam will absorb this characteristic radiation. The attenuation of light by the sample is compared with that caused by known cadmium standards, and the cadmium content of the sample is estimated.

Using a deuterium background corrector, atomic absorption spectroscopy meets the requirements for sensitivity, accuracy, precision, ease of handling, speed, and relatively low-cost equipment. AAS is the most frequently used analytical method and is rapidly becoming the accepted method to determine cadmium at trace levels. The sensitivity of AAS can be increased by using dithizone as a chelating agent or by using a heated graphite atomizer.

Interferences in the analysis of cadmium by AAS have been reported to be caused by anions such as $B_2O_4^{=}$, $CO_3^{=}$ and HCO_3^-. Phosphate in concentrations above 0.1 molar (M) could decrease the absorption, and sodium chloride above 0.01 M could increase absorption. These interferences can be removed, if necessary, by extraction of the cadmium into an organic solvent before analysis.

The ASV method consists of measuring the current flow per unit time at varying voltages, which are characteristic of the deposition of metals. Most of the cadmium experience with ASV has been with blood samples for which rapid, cheap, accurate results are claimed.

Atomic fluorescence spectroscopy has high sensitivity, but this method should be held in reserve until it is more widely accepted. Cadmium in air can be estimated continuously by using an excitation source consisting of graphite rods energized by induction with a radio frequency coil (20). At present, the procedure's sensitivity does not permit ambient air analysis, and the equipment is too bulky and complex to be portable. The method in general use consists of examination of a solution obtained by processing a particulate matter sample on a filter of glass fiber or organic membrane. The detection is by OES when a multi-element survey is required. AAS is used when cadmium values alone or data of high accuracy are desired.

In the National Air Surveillance Network, particulate matter samples are sectioned and composites are added in a low-temperature asher using 50 to 100 milliliters (ml) of oxygen per minute at 1 torr[2] with an induction coil energized with 250 watts. At this combustion temperature (about 150°C) cadmium is returned essentially quantitatively, whereas half of the cadmium in samples oxidized in a muffle furnace is lost at between 500° and 550°C.

The residue is extracted using a mixture (4:1 by volume) of redistilled nitric and hydrochloric acid and concentrated. The solution is then freed of silica by centrifugation and brought up to volume with redistilled nitric acid. The solution is analyzed directly by OES, which has a detection limit of 0.3 microgram per milliliter (μg/ml) of sample and a relative standard deviation of 8% at 0.8 μg/ml. For special studies, this solution is diluted 10:1 with water prior to analysis by AAS. The minimum level detectable by AAS is 0.04 μg/l, and the standard relative deviation is 2% at or about 0.5 μg/ml.

Water samples are best analyzed directly by AAS, using the graphite furnace technique. If interferents are expected, the samples may be taken to dryness, and the residue digested with nitric acid. If the cadmium concentration is not detectable, evaporation or extraction with a chelating agent and organic solvent will serve to concentrate the cadmium in the solution analyzed. Friberg et al (19) describe measurement of cadmium in water very well.

ENVIRONMENTAL STANDARDS

As regards air pollution, British emission standards for cadmium (as the element) are 39 milligrams per cubic meter at STP and Australian standards are 23 milligrams per cubic meter at STP. The recommended maximum allowable concentration for cadmium in air has been set at 0.1 milligram per cubic meter of air by the American Standard Association. The American Conference of Governmental Industrial Hygienists has set a standard of 0.05 milligram per cubic meter for cadmium oxide and for cadmium dusts and salts.

The permissible concentration of cadmium in domestic water supplies is 0.01 ppm (27). The permissible cadmium concentration in irrigation water on a continuous use basis is 0.005 ppm and is 0.05 ppm on a short-term use basis in fine-textured soil.

Effluents to storm sewers or streams should contain less than 0.05 ppm of cadmium according to earlier standards. Now, however, new toxic pollutant effluent standards for cadmium are being promulgated by the EPA as outlined in the

Federal Register for December 27, 1973. The new proposed standards say that for streams having a flow of less than 10 cubic feet per second or lakes having an area of less than 500 acres, there shall be no cadmium discharged. In faster flowing streams or larger lakes, the limit is 40 μg/l (about 0.04 ppm) except that the allowable limit is only 10% of this amount if the flow of the stream is less than 10 times the waste stream flow. The proposed allowable discharge into salt water bodies is 320 μg/l (about 0.32 ppm).

HANDLING PROCEDURES

OSHA has not yet issued procedures specific to the handling of cadmium and its compounds. In view of its toxicity, obviously care should be taken to avoid inhalation or ingestion of cadmium compounds, however.

REMOVAL FROM AIR

Air pollution control procedures are employed at some metal refinery plants in order to recover the valuable cadmium that would otherwise escape into the atmosphere. Electrostatic precipitators, baghouses, and cyclones are effectively used for abatement. However, little information has been found on the specific application of these procedures for the purpose of controlling cadmium air pollution, according to Athanassiadis (3).

The procedures for recovering cadmium from exhaust in a copper extraction plant collected significant quantities of valuable cadmium, at the same time reducing local air pollution levels.

Cadmium emitted into the atmosphere will generally be in the form of particulate matter, usually as the oxide, but also as the sulfide or sulfate. Because the boiling point of cadmium is fairly low, 767°C (1410°F), the metal may be vaporized in high-temperature processes and condensed into particles as the process off-gases are cooled. This method of formation would result in very fine particles in the micrometer and submicrometer range.

The exact size distribution of cadmium-containing particles from these various processes has, unfortunately, not been clearly defined. The limited information available indicates that 40% of the particulate mass may be smaller than 2 micrometers in diameter.

In processes where a large fraction of the cadmium-containing particulate is formed by the vaporization/condensation method, it is possible that a significant amount of cadmium may be contained in particulate matter even smaller than 0.1 micrometer.

The shape of the condensed particles is not known, either, although there are indications that the particles might not always be spherical. There is evidence that the percentage concentration of cadmium in fly ash is higher in the finer particles than in the coarser particles.

It is evident, therefore, that the control of atmospheric cadmium emission requires the ability to capture fine particles, as pointed out by the National

Environmental Research Center of EPA (20).

Common sources of cadmium air pollution occur during the use of cadmium. Electroplating, alloying and use of cadmium in pigments can produce local contaminations of the atmosphere. Also, since cadmium is added to pesticides and fertilizers, the use of these materials can cause local air pollution.

Emissions resulting from production and utilization of cadmium-containing products can be effectively controlled by using baghouses, electrostatic precipitators and filters (28) as noted above.

Conventional controls are again utilized in the abatement of cadmium emissions from combustion sources. A study which considered the control of various trace elements by an electrostatic precipitator indicated a removal efficiency range of 97 to 99%.

The dust and fume generated by smelting furnaces can be controlled with similar equipment. Efficiencies of these control systems for normal particulate matter can be more than 99% by mass. However, the collection efficiency for cadmium fumes may be much less than this value due to the high fraction of small particles in the fume. Emissions from machining operations can be controlled by filters, cyclones, and baghouse systems. Similarly, cadmium emissions can be controlled in the manufacture of paint (28).

REMOVAL FROM WATER

Waterborne cadmium emissions are generally in the form of suspended particulates, although cadmium may sometimes be present in a soluble form such as cadmium sulfate. These aqueous emissions may result directly from various processing steps involving cadmium (such as the aqueous beneficiation of zinc ores or the spills, washdowns, and rinsings from cadmium electroplating operations) or they may result from leaching and washing of smelter slag heaps by rainwater (20).

If the cadmium is present in the form of suspended particulates, the aqueous emissions that result directly from process operations involving cadmium may be controlled by employing settling ponds or thickeners. Filtering or centrifuging the aqueous wastes might also be considered.

If the cadmium is present as a soluble compound, it might be removed by precipitation, followed by removal of the resulting solids. Alternatively, the techniques of ion exchange, solvent extraction, or electrolytic deposition might be employed.

Methods of cadmium wastewater treatment currently in use are essentially those employed for most heavy metal wastes encountered in the plating and metal processing industries. These methods include precipitation (generally considered a destructive treatment), and ion exchange. Although not mentioned specifically in the cadmium literature, electrolytic recovery and evaporative recovery processes as used for copper, silver and nickel waste recovery are technically feasible, but only for a concentrated waste or after a concentration step such as ion exchange.

The lack of specific cadmium data in the literature may be related to in-house dilution through central waste collection systems. Cadmium plating rinse streams are likely only a small part of total waste flow in a large plating operation, and resultant dilution can be assumed to reduce cadmium to very low levels.

Cadmium forms an insoluble and highly stable hydroxide at alkaline pH. Jenkins et al (29) report that freshly precipitated cadmium hydroxide leaves approximately 1 mg/l of residual cadmium ion in solution at pH 8, but that this is reduced to 0.1 mg/l at pH 10. Weiner (30) substantiates this behavior by reporting that cadmium is relatively soluble at pH 8 and requires higher pH for effective removal in the absence of iron. However, he indicates that coprecipitation with iron hydroxide at pH 8.5 effects complete removal.

Weiner (30) further reports that in the presence of complexing agents (e.g., cyanide) it is impossible to precipitate out the cadmium ion. In the event that appreciable complexing agents are in the waste solution, pretreatment to destroy the agents would be required.

Cadmium plating wastes normally contain cyanide. Removal of cadmium from spent plating baths or the dilute plating rinse waters is therefore dependent upon prior cyanide removal. Fortunately, cyanide breakdown is relatively rapid and easy, as has been demonstrated for zinc and copper cyanide baths (31) (32). Costs would be expected to be essentially those of cyanide oxidation.

In the absence of cyanide, unit operations involved in cadmium removal are similar to those of the conventional tertiary treatment sequence of coagulation settling, and sand filtration. The latter step has become of increasing importance for complete removal of nonsettling and slow settling metal hydroxide flocs. DiGregorio has provided approximate cost data for this sequence of operations (33).

Hanson and Zabban (34) have reported on lime precipitation of cadmium from a waste stream at the Rochester, Minnesota IBM Machine manufacturing plant. An effluent pH of 9.0 and cadmium concentration of 0.54 mg/l was reported for the lime precipitation process. Although sand filtration was not used, the process incorporated coagulant aid addition (Separan NP-10) at 1 to 2 mg/l. The effluent contained only 2 mg/l of suspended solids, indicating extremely good solids removal by sedimentation.

Linstedt et al (35) have reported extremely high cadmium removal for a pilot plant lime coagulation, settling process. Treating combined sewage effluent from a secondary treatment plant, these workers achieved 94.5% cadmium removal for the trace quantities present (35).

A proprietary hydrogen peroxide oxidation-precipitation system called the Kastone process has been developed, which simultaneously oxidizes cyanides and forms the oxide of cadmium (rather than the hydroxide). Cadmium oxide is claimed to be much easier to remove from solution than cadmium hydroxide (36). This process is reported as suitable for small plating operations, as solids removal is readily accomplished with slight modification to simple filtration apparatus already present in most shops.

Ion exchange can be used as a polishing treatment or recovery process. The concentrated solution obtained upon regeneration of the ion exchange resin is more suitable to economical recovery procedures than the dilute waste streams from which the cadmium was removed. The recovery value of cadmium is estimated as $1.20 to $6.00 per 1,000 gallons, at a solution concentration of 50 to 250 mg/l of cadmium (33).

Mattock (37) states that the extra costs of ion exchange can be offset in one-half to two years when product recovery value is substantial, as is the case with cadmium. Odlan and Hester (38) on the other hand, indicate that ion exchange is unsuitable for recovery of cyanide solutions. No specific applications of ion exchange to cadmium cyanide solutions were found. Linstedt et al (35) have reported 99% reduction of trace quantities of cadmium by a pilot plant cation exchanger, and 99.9% removal by cation plus anion exchange. The waste stream originated from a secondary sewage treatment plant.

Sagakami et al (39) have developed a method of treating an industrial waste liquid containing powder of metals such as cadmium or yttrium and/or their compounds comprising the steps of: [1] mixing the waste liquid with a liquid containing a water-soluble high molecular compound such as polyvinyl alcohol, [2] evaporating water in the mixture by heating with hot blast for concentration, and [3] cooling the dehydrated mixture to form a solid film of the high molecular compound enclosing the waste metals and/or their compounds.

SOLID WASTE DISPOSAL

As described by P.W. Powers (40), cadmium and cadmium compounds are considered as waste stream constituents requiring centralized treatment for the following reasons: [1] the extremely high degree of toxicity of all cadmium compounds; [2] the nondegradable nature of the toxic cadmium component of all cadmium compounds; [3] cadmium compounds are presented in sizable quantities as wastes from electroplating, paint manufacture, battery manufacture, and the metals industry; [4] a significant portion of cadmium wastes is contributed by small plating shops where treatment is either technically or economically infeasible; and [5] the cadmium hydroxide waste sludges resulting from the treatment of cadmium wastes should be disposed of in California Class 1 type landfills only and are not being adequately handled at present.

The only adequate method for the disposal of concentrated cadmium wastes is coagulation with lime, then sedimentation followed by sand filtration. The effluent from this process would probably have to be treated further (for example, adsorption with activated-carbon or ion exchange) to reduce the cadmium concentration to a level in compliance with the U.S. Public Health Service recommendation for public drinking water (0.01 mg/l).

The cadmium hydroxide sludge produced in this process can be dried and placed in an approved chemical landfill area of the California Class 1 type. Cadmium hydroxide is not very soluble (0.00026 g/100 cc), so contamination of water supplies from the landfill operation should not be a problem. If the cadmium hydroxide sludge is relatively pure, it can be dissolved in sulfuric acid and the cadmium metal recovered by zinc dust precipitation.

As pointed out by the Office of Toxic Substances of EPA (2), the great bulk of the total cadmium released to the environment is in the form of land-destined solid wastes, slimes and sludges.

Two categories may be defined for these materials. Less than 10% of this cadmium (the first category) is in a relatively small total volume of wastes, at relatively appreciable concentrations (0.1% or greater).

Many of these wastes are the residuals from air and water pollution control; and there is considerable progress in either disposing of these wastes in an environmentally adequate manner at relatively low cost or in processing these wastes to reclaim the cadmium and other materials.

More than 90% of this land-destined cadmium (the secondary category), however, exits in an enormous total waste volume and at relatively low concentrations (less than 0.1%).

These concentrations are still up to 4 orders of magnitude higher than natural cadmium concentrations in soil; the only apparently feasible posture is the dispersing of this cadmium on our lands in the least-damaging manner.

While the emissions of cadmium to air and water are not expected to increase (improved pollution abatement should at least compensate for industrial growth), the release of cadmium to the land is expected to grow by 20% by 1980.

The primary source of these land-destined wastes, all increasing in volume, are the combustion of coal, the use of phosphate fertilizers, the isolation of sewage sludge, and the production of steel. One method of solid waste disposal is always recovery or recycling. The recovery of cadmium from scrap batteries has been discussed by D.A. Wilson et al (41).

ECONOMIC IMPACT OF CONTROLS

No information has been found on the economic costs of cadmium air pollution or on the costs of its abatement according to Athanassiadis (3). A scarcity of operating data and costs specifically related to cadmium treatment thus exists in the literature.

However, costs associated with the various treatment procedures for other nonferrous heavy metals should apply to cadmium solution of similar general characteristics, due to the similarity of treatment methods. Recovery processes would be expected to be more attractive than with other nonprecious metal systems, due to the high value of cadmium metal.

Recently, a summary of control alterations for cadmium emissions and their economic consequences has been published by the Office of Toxic Substances of EPA (2). Table 18 presents this data, listing the industry segment affected, the control alternative, the feasibility, the effectiveness and the costs. Cost projections have the year for the cost estimation indicated.

TABLE 18: SUMMARY OF CONTROL ALTERNATIVES FOR CADMIUM

No.	Industry Segment Affected	Control Alternative	Feasibility	Effectiveness	Costs
1	Cd electroplating, and Cd producers	Ban use of cadmium for electroplating	Most uses have substitutes; some do not	Would reduce Cd consumption by 2,700 kkg per year now; 3,700 kkg per year in 1985	$49 MM/yr now; $64 MM/yr in 1985
2	Ni-Cd batteries, and Cd producers	Ban use of cadmium for batteries	Many end-items depend on Ni-Cd power sources	Would reduce Cd consumption by 200 kkg/yr now; 1,600 kkg/yr in 1985	$24 MM first year; $10 MM/yr later; $34 MM/yr in 1985
3	Cd pigments	Ban use of cadmium for pigments	No real substitutes for many uses, except white/black pigments	Would reduce Cd consumption by 700 kkg/yr now; 600 kkg/yr in 1985	$8 MM/yr now; $9 MM/yr in 1985
4	Cd heat stabilizers and Cd producers	Ban use of cadmium for heat stabilizers for PVC	Good-adequate substitutes should be available soon	Would reduce Cd consumption by 1,100 kkg per year now; 500 kkg per year in 1985	$9 MM/yr now; $0.4 MM/yr in 1985
5	All Cd-consuming industries and Cd producers	Total ban on use of cadmium	Sum of comments for alternatives 1, 2, 3, 4	Would eliminate Cd consumption of 5,600 kkg per year now; 9,200 kkg per year in 1985	$115 MM first year; $101 MM/yr later; $159 MM/yr in 1985
6	All Cd-consuming industries except batteries, and Cd producers	Ban use of cadmium except for Ni-Cd batteries	Sum of comments for alternatives 1, 3, 4	Could eliminate Cd consumption of up to 4,500 to 4,800 kkg/yr, but lower Cd price would discourage recycling	$63 MM/yr now; $73 MM/yr in 1985
7	Ni-Cd batteries	Encourage recycle of Ni-Cd batteries through labeling and/or deposit system	High incentive for reclamation, some already begun	Should decrease imports by 50% in 1985	None (self-supporting)
8	All Cd-consuming industries and Cd producers	Excise tax and equivalent import duty on cadmium		Ineffective-low demand elasticity for Cd	

(continued)

TABLE 18: (continued)

No.	Industry Segment Affected	Control Alternative	Feasibility	Effectiveness	Costs
9	Cd electroplating and Cd products	Seek voluntary large reduction in Cd electroplating demand by government (including military), automotive, aircraft, shipbuilding, and similar industries	Feasible because these industries are concentrated and structured	These industries account for much of Cd plating demand, reduction in Cd consumption should be much of quantities for Alternative No. 1	Corresponding fraction of costs quoted for Alternative No. 1
10	All Cd-consuming industries and Cd producers	Quota on Cd metal (and flue dusts, etc.) imports, up to a complate ban on Cd imports	Higher Cd price would result; increased revenue to Cd producers would be \$41 MM/yr now; \$51 MM/yr in 1985	Total ban would reduce consumption to domestic production levels-3,400 kkg/yr now; 4,700 kkg/yr in 1985. Reduction would be 2,200 kkg/yr now and 4,500 kkg/yr in 1985	\$33 MM in first year; \$28 MM/yr later; \$60 MM/yr in 1985
11	Zinc ore mining and beneficiating	Regulations to control dusting from dried ore tailings with windbreaks and earth or vegetative covering	High	Reduce air emission of ~0.2 kkg/yr by >50%	Moderate
12	Zinc ore mining and beneficiating	Minimize overflow of slime ponds by requiring sizing for extreme rainfall event	High	Prevent gross discharge. Cd in slimes ~250 kkg/yr	High
13	Zinc ore mining and beneficiating	Imposing severe penalties for dam failure in slime ponds	Good-would promote effective construction, maintenance and inspection	Prevent gross discharge. Cd in slimes ~250 kkg/yr	Moderate
14	Zinc ore mining and beneficiating	Require collection and treatment of drainage from slime pond dams or tailings piles	Good	Excellent-prevents direct discharge at ~15 parts per million Cd	Moderate
15	Primary zinc	Continued stringent air pollution control requirements, plus continued inspection and monitoring. Maintain ⩾95% collection efficiency based upon Cd content of zinc ore concentrates	High. All plants already practice good collection (or are being closed)	Reduce Cd emissions from sintering to ~100 kkg/yr from two plants	Very high for two remaining pyrometallurgical plants, but most of investment already made

(continued)

TABLE 18: (continued)

No.	Industry Segment Affected	Control Alternative	Feasibility	Effectiveness	Costs
16	Primary zinc	Promulgation of effluent guidelines (water pollution control). Treatment and control, and costs, accomplish other objectives in addition to Cd removal		Reduce Cd discharge (in kkg/yr) from 10 (1972) to 2 (1977) to 1.3 (1983)	Capital: \$1.52 MM (1977) + \$1.05 MM (1983). Annual: \$0.46 MM (1977) + \$0.45 MM (1983)
17	Primary zinc	Regulating the land-disposal of Cd-bearing flue dusts, sludges, and other residuals	High. The industry practices widespread reclamation of wastes	Effective safeguard against leaching or migration	Low
18	Cd electroplating	Promulgation of effluent guidelines (water pollution control). Treatment and control and costs, accomplish other objectives in addition to Cd removal	High for 1977; questionable for 1983	Reduce Cd discharge (in kkg/yr) from 10.5 (1972) to 4.0 (1977) to 0 (1983)	Capital: \$37.5 MM (1977) + \$25 M to \$50 MM (1983)
19	Cd electroplating	Pretreatment requirements for wastewater, equivalent to effluent guidelines	Limited by land availability	Would reduce Cd content of sewage sludges	Included in costs for meeting effluent guidelines
20	Cd electroplating	Regulating the land-disposal of Cd-bearing sludges	High. Also high incentive for reclamation	Effective safeguard against leaching or migration of ~80 kkg/yr Cd	\$50,000 per year
21	Cd pigments	Continued stringent air pollution control requirements, plus continued inspection and monitoring. Maintain ≥95% collection efficiency	High. All plants already practice good collection and reclaim dusts	Reduce Cd emissions to an estimated 9.5 kkg/yr	Very high, but investment already made
22	Cd pigments	Promulgation of effluent guidelines (water pollution control)	Most of industry already complies	Reduce Cd discharge to 0.75 kkg/yr	Capital: \$140,000 Annual: \$60,000
23	Cd pigments	Regulating the land-disposal of Cd-bearing sludges	High. Also high incentive for reclamation	Effective safeguard against leaching or migration of ~16.5 kkg per year Cd	\$19,000/yr

(continued)

No.	Industry Segment Affected	Control Alternative	Feasibility	Effectiveness	Costs
24	Ni-Cd batteries	Regulating wastewater discharges	High. Same treatment and control technology as other industries. Most already comply	Reduce Cd discharge to 0.3 kkg/yr	Moderate
25	Ni-Cd batteries	Regulating the land-disposal of Cd-bearing sludges	High. Also high incentive for reclamation	Effective safeguard against leaching or migration of ~10 kkg/yr Cd	\$2,700/yr (1973) to \$4,400/yr (1983)
26	Secondary nonferrous metals	Continued stringent air pollution control requirements, plus continued inspection and monitoring. Maintain ⩾95% collection efficiency	High. Much of industry already complies	Reduce Cd emissions to an estimated 2.2 kkg per year	Very high, but investment already made
27	Secondary nonferrous metals	Regulations to control dusting and wind dispersion of collected flue dusts	High. Use bin with dust control	High	Moderate
28	Secondary nonferrous metals	Regulating the land-disposal of flue dusts	High. Much of industry reclaims dusts	Effective safeguard against leaching or migration of ~20 kkg/yr	Low
29	Iron and steel	Continued stringent air pollution control requirements, plus continued inspection and monitoring. Maintain ⩾95% collection efficiency	High. Much of industry complies	Reduce Cd emissions to an estimated 10.5 kkg per year Cd	Very high, but investment already made
30	Iron and steel	Regulations to control dusting and wind dispersion of collected flue dusts	High for minimills (bins), moderate for primary steel (earth cover)	High for minimills, moderate for primary steel	Moderate
31	Iron and steel	Regulating the land-disposal of flue dusts	Very large quantities involved; incentive for reclamation	Effective safeguard against leaching or migration of ~330 kkg per year Cd	\$140 MM/yr now; \$226 MM/yr in 1985

(continued)

TABLE 18: (continued)

No.	Industry Segment Affected	Control Alternative	Feasibility	Effectiveness	Costs
32	Electric power stations, solid waste incinerators, sewage sludge incinerators	Continued stringent air pollution control requirements, plus continued inspection and monitoring	High for power stations, moderate for incinerators	Reduce Cd emissions to ~120 kkg/yr	Very high
33	Phosphate fertilizers, sewage treatment, zinc ore mining and beneficiating	Regulate the application rate to agricultural land of materials with Cd concentrations greater than 1 ppm. After tillage, the resulting soil should contain ⩽1 ppm Cd, with variations based upon soil pH, soil type, crop type, etc.	Would require educational program, monitoring, labeling. Only moderate success expected because of diversity of users	Would limit Cd in food chain to "normal" range; intended to spread Cd dissipation over wide areas	Moderate
34	Coal gasification and liquefaction	Develop effective Cd removal and recovery technologies as integral part of coal conversion processes; highlight fate of Cd in EIS; develop regulations as technology emerges	High, since most coal conversion R&D is government sponsored	Moderate, because of Cd volatility. Should recover ⩾75% of Cd with control of particulates	Moderate increment to total cost of coal conversion
35	Phosphate fertilizers	Sponsor and encourage research aimed at Cd removal and recovery in the manufacture of phosphoric acid	Possibility using solvent extraction techniques	Would prevent Cd dissipation directly to food chain	Possibly high compared to price of fertilizers
36	Phosphate fertilizers, primary zinc	Banning the use of sulfuric acid with ⩾2 ppm Cd for manufacturing fertilizers	High. Much H_2SO_4 is not smelter-gas acid	Would divert any high-Cd acid to other uses	Low
37	Primary zinc, lead, copper	Regulating the Cd concentration of by-product sulfuric acid from zinc, lead, and copper ores to ⩽2 ppm	High. Two companies have <2 ppm Cd in sulfuric acid	Would recover virtually all Cd, not just divert its use	Moderate
38	Primary zinc	Regulating the Cd concentration of zinc oxide (and other zinc compounds) to ⩽10 ppm	Would require change to French process; involves inequity	Would stop dissipation of ~15 kkg/yr Cd in ZnO	High
39	Primary zinc	Regulating the Cd concentration of all zinc metal to ⩽40 ppm	Involves inequity	Would stop dissipation of ~160 kkg/yr Cd in prime western zinc	$9 MM/yr

Source: Reference (2)

REFERENCES

(1) Parker, C.R. and Sandoz, D.P., *VIA (Varian Instrument Applications)* 7 (2), 18-21, (1973).

(2) Office of Toxic Substances, *Technical and Microeconomic Analysis of Cadmium and Its Compounds,* Report EPA 560/3-75-005, Washington, D.C., Environmental Protection Agency (June 1975).

(3) Athanassiadis, Y.C., *Air Pollution Aspects of Cadmium and Its Compounds,* Report PB 188,086, Springield, Va., Nat. Tech. Info. Service (Sept. 1969).

(4) Cothern, C.R., *Determination of Trace Metal Pollutants in Water Resources and Stream Sediments,* Report PB 213,369, Springfield, Va., Nat. Tech. Info. Service (1972).

(5) Schroeder, H.A. and Balassa, J.J., *J. Chron. Dis.* 14, 236 (1961).

(6) Carroll, R.E., *Journal American Medical Association,* 198, 177 (1966).

(7) Fox, M.R.S. and Fry, B.E., *Science*, 169, 989 (1970).

(8) Warnick, S.L. and Hall, H.L., *J. Water Pollution Control Fed.,* 41, 280 (1969).

(9) Schroeder, H.A., Virton, W.H. and Balassa, J.J., *J. Nutrition* 80, 39 (1963).

(10) Schroeder, H.A., *Amer. J. Physiol.,* 207, 62, (1964).

(11) Schroeder, H.A., Nason, A.P., Prior, R.E., Reed, J.B. and Haessler, W.T., *Amer. J. Physiol.,* 214, 469, (1968).

(12) Schroeder, H.A., Baker, J.T. Hansen, N.M., Jr., and Wise, R.A., *Arch. Envir. Health,* 21, 609, (1970).

(13) Her Majesty's Stationers Office, *Water Pollution Research,* (1967).

(14) Maehara, T., *Nippon Seikeigakagakkai Zasshi,* (1967).

(15) Her Majesty's Stationers Office, *Water Pollution Research,* (1969).

(16) Luckey, T.D., Venugopal, B. and Hutcheson, D., *Heavy Metal Toxicity and Hormonology,* New York, Academic Press (1975).

(17) Tucker, A., *The Toxic Metals,* London, Earth Island Ltd. (1972).

(18) Schroeder, H.A., *The Poisons Around Us,* Bloomington, Indiana University Press (1974).

(19) Friberg, L., Piscator, M. and Nordberg, G., *Cadmium in the Environment: A Toxicological and Epidemiological Appraisal,* Report PB 199,796, Springfield, Va., Nat. Tech. Info. Service (April 1971).

(20) National Environmental Research Center, *Scientific and Technical Assessment Report on Cadmium,* Research Triangle Park, N.C. (July 1975).

(21) Goldberg, A.J., *A Survey of Emissions and Controls for Hazardous and Other Pollutants,* Washington, D.C., Environmental Protection Agency (Feb. 1973).

(22) Ottinger, R.S., Blumenthal, J.L., Dal Porto, D.F., Gruber, G.I., Santy, M.J., and Shih, C.C., *Recommended Methods of Reduction, Neutralization, Recovery or Disposal of Hazardous Waste,* Vol. 1, Report PB-224,580; Springfield, Va., Nat. Tech. Info. Service (Aug. 1973).

(23) McKee, J.E. and Wolf, H.W., *Water Quality Criteria,* 2nd ed., California State Water Quality Control Board, Pub. No. 3-A (1963).

(24) Santaniello, R.M., "Air and Water Pollution Quality Standards, Part 2: Water Quality Criteria and Standards for Industrial Effluents," *Industrial Pollution Control Handbook,* H.F. Lund, ed., New York, McGraw-Hill (1971).

(25) Lieber, M. and Welsch, W.F., "Contamination of Ground Water by Cadmium," *Jour. Am. Water Works Assoc.* 46, 541-547 (1954).

(26) Patterson, J.W. and Minear, R.A., *Wastewater Treatment Technology,* Chicago, Illinois Institute for Environmental Quality (Feb. 1973).

(27) Sittig, M., *Pollution Detection and Monitoring Handbook,* Park Ridge, N.J., Noyes Data Corp. (1974).

(28) McFarland, A.R., ed., *Control Technology for Toxic and Hazardous Air Pollutants,* Chicago, Illinois Institute for Environmental Quality (Dec. 1974).

(29) Jenkins, S.H., Keight, D.G. and Humphreys, R.E., "The Solubility of Some Heavy Metal Hydroxides in Water, Sewage and Sewage Sludge, I. The Solubility of Some Metal Hydroxides," *Int'l. Jour. Air Wat. Pollution,* 8, 53-56 (1964).

(30) Weiner, R.F., "Acute Problems in Effluent Treatment," *Plating,* 54, 1345-1356 (1967).

(31) Eden, G.E., Hampson, B.L. and Wheatland, A.B., "Destruction of Cyanide in Waste Waters by Chlorination," *Jour. Soc. Chem. Ind.,* 69, (8), 244-249, (1950).

(32) Lancy, L.E., "An Economic Study of Metal Finishing Waste Treatment," *Plating,* 54, 157-161 (1967).

(33) DiGregorio, D., *Cost of Wastewater Treatment Processes,* Robert A. Taft Water Research Center Report No. TWRC-6, U.S. Dept. Interior, Washington, D.C. (1968).

(34) Hansen, N.H. and Zabban, W., "Design and Operation Problems of a Continuous Automatic Plating Waste Treatment Plant at Data Processing Division, IBM, Rochester, Minnesota," *Proc. 14th Purdue Industrial Waste Conf.,* pp. 227-249 (1959).

(35) Linstedt, K.D., Houck, C.P. and O'Connor, J.T., "Trace Element Removals in Advanced Wastewater Treatment Processes," *Jour. Wat. Poll. Control Fed.,* 43, 1507-1513 (1971).

(36) "New Process Detoxifies Cyanide Wastes," *Env. Sci. Technol.,* 5, 496-497 (1971).

(37) Mattock, G., "Modern Trends in Effluent Control," *Metal Finishing Jour.,* 14, 168-175 (1968).

(38) Odlan, K. and Hester, T.C., "Profitable Recovery of Plating Wastes by Reconcentration of Reduced Volume Rinses," *Plating,* 43, 1022-1025 (1956).

(39) Sagakami, T., Kakumoto, S., Okawa, K. and Miyashita, K.; U.S. Patent 3,915,691; October 28, 1975; assigned to Matsushita Electric Industrial Co. Ltd.

(40) Powers, P.W., *How to Dispose of Toxic Substances and Industrial Wastes,* Park Ridge, N.J., Noyes Data Corp. (1976).

(41) Wilson, D.A. and Wiegard, B.J., Jr., *Recovery of Cadmium and Nickel from Scrap Batteries,* Report of Investigation No. 7566, Wash. D.C., U.S. Bureau of Mines (1971).

CHROMIUM

Pure chromium is a steel-gray metal. It is prepared from the sesquioxide form by reduction with powdered aluminum and magnesium, by vacuum reduction with carbon or silicon in an electric furnace, or by reduction with dry hydrogen. It is produced from the chloride form by reduction with potassium or sodium, and it can be prepared by electrolysis. The metal has a variety of industrial applications, which include its use in high tensile corrosion resistant alloys and in cutting tools, in plating, in anticorrosion paints, in wood preservatives (chromated zinc chloride), in wools and furs (where it acts as a mordant or fixative for dyeing), and in the photographic industry.

Chromium (Cr, atomic number 24) occurs in nature chiefly as chromite, or chrome iron ore ($FeO{\cdot}Cr_2O_3$), and comprises approximately 0.037% of the earth's crust. Concentrations in soils throughout the world range from traces up to 2.4% chromium, while background atmospheric concentrations of the metal vary from 0.001 to 0.007 $\mu g/m^3$.

TOXICITY

The exposure of industrial workers to airborne chromium compounds and chromic acid mists, particularly the hexavalent chromates, has been observed to produce irritation of the skin and respiratory tract, dermatitis, perforation of the nasal septum, ulcers and cancer of the respiratory tract (1). Chromium metal is thought to be nontoxic. Hexavalent compounds appear to be much more harmful than trivalent compounds, with the toxic effects depending on solubility.

Two effects that appear to be particularly important in relation to air pollution are hypersensitivity to chromium compounds and induction of cancers in the respiratory tract. Exposure of industrial workers in the chromate-producing industry has shown an incidence of deaths from cancer of the respiratory tract which is over 28 times greater than expected. Time-concentration relationships for induction of cancer are not known, according to Sullivan (2).

Hexavalent chromium has been considered a toxic metal for many years. Trivalent chromium is less toxic although it raises the glucose tolerance of rats and rabbits (3). However, it is not clear whether either is an essential element for man. Schroeder et al (3) feels that chromium has not been definitely proven to be a carcinogen. The toxicity of hexavalent chromium to fish is fairly high in that 1.2 to 5.0 ppm is fatal to many species.

Considerable evidence does exist to show that chromium in the atmosphere inhaled by man can be a carcinogen. Hill and Warden (4) suggest that chromyl chloride may play an important role in the formation of lung cancer in chromate workers. Hueper (5) pointed out that the existence of a serious lung cancer hazard for producers of chromates from chromite ore and tar handlers of certain chromium pigments (zinc chromate, barium chromate and lead chromate) is well documented among workers in these plants.

Hueper also reported the development of squamous cell carcinomas of the lung and sarcomas of the lung and of the soft tissues of the thigh of rats which had received intrapleural or intramuscular implants of powdered chromite ore roast suspended in sheep fat. Shelley (6) has pointed out that chromium in welding fumes was the cause of eczematous eruptions of the palms of the workers.

Calcium chromate, sintered calcium chromate and sintered chromium trioxide when introduced in pellet form into the pleural cavity of rats produce cancers, mainly sarcomas, in the majority of animals at the site of implantation within 14 months (5). This effect is related to the solubility of these compounds in the watery medium. It has been reported that as little as 0.7 ppm of chromium in the water can be toxic to some aquatic insects. Thus it can be concluded that chromium may not be an essential element to man, that it may be a carcinogen and that hexavalent chromium is toxic to man at fairly low levels. The toxicity of chromium salts including toxic dosage levels for various chromium salts has been discussed by Luckey et al (7).

EXTENT OF EXPOSURE

Air pollution aspects of chromium and its compounds have been reviewed by Sullivan (2). Sources and estimates of chromium-containing emissions as of 1970 have been presented by GCA Corporation (8) as shown in Table 19.

These 1970 estimates vary significantly from earlier EPA estimates as shown in Table 20 (9). The earlier estimates were much higher in the areas of inadvertent sources (coal and oil combustion) and the GCA Corporation estimates are much higher in the areas of emissions from metallurgical processing and refractory manufacture. Figure 9 shows material flow of chromium through the economy (9).

Chromium occurs in nature as both the trivalent (Cr^{+3}) and the hexavalent (Cr^{+6}) ion. Hexavalent chromium present in industrial wastes is primarily in the form of chromate ($CrO_4^{=}$) and dichromate ($Cr_2O_7^{=}$). Chromium compounds are added to cooling water to inhibit corrosion. They are employed in manufacture of paint pigments, in chrome tanning, aluminum anodizing and other metal cleaning, plating and electroplating operations. In the metal plating industry, automobile parts manufacturers are one of the largest producers of chromium-plated metal parts.

TABLE 19: SOURCES AND ESTIMATES OF CHROMIUM-CONTAINING EMISSIONS IN 1970

Source	Uncontrolled Emission Factor lb/ton	Uncontrolled Emission Factor kg/10^3 kg	Reliability Code	Production Level (tons/yr)	Percent Cr in Emissions	Reliability Code	Emissions of Cr Before Controls (tons/yr)	Estimated Level of Emission Control	Emissions of Cr After Controls (tons/yr)
Mining									
None in U.S.A.							0		0
Refining									
Ferrochromium									
Electric furnace	(200-830) 500[a]	(100-415) 250[a]	C	375,500	22	B	20,600	40%	12,360
Material handling	10	5	B	375,500	65	C	1,220	32%	830
Electrolytic chromium	0.048	0.024	C	9,000	51	A	neg	95%	neg
Refractory									
Noncast	150	75	C	60,300	b		4,500	64%	1,630
Electric cast	225	112	C	6,700	b		754	77%	173
Chemical Processing									
Dichromate	30	15	C	61,000	b		920	90%	92
Other chemicals	-	-	-	-	-	-	-	-	24
Steel and Alloys									
Chromium steels	25	12	C	189,000	b		2,362	78%	520
Cast iron	75	38	C	5,000	b		188	99%	2
Super alloys and alloys	25	12	C	12,000	b		150	78%	33
General steel making	N.A.	N.A.		N.A.	N.A.		N.A.	N.A.	100[c]
Inadvertent Sources									
Coal combustion	N.A.	N.A.		33,800,000[e]	0.026	A	8,700	82%	1,564
Oil combustion	N.A.	N.A.		287,000[d]	0.13	B	370	0	370
Cement production	N.A.	N.A.		934,000[d]	0.03	C	N.A.	N.A.	280
Incineration	N.A.	N.A.		931,000[d]	0.017	C	N.A.	N.A.	158
Asbestos	N.A.	N.A.		6,579	0.15	C	10	99%	0
Total							39,774	54%	18,136

[a]Intermediate value
[b]Emission factor multiplier equal to tons of Cr processed or handled annually.
[c]Needs further investigation.
[d]Emissions, after control.
[e]Emissions, before control.

NOTE: Reliability code is:
A = Excellent
B = Above average
C = Average
D = Below average
E = Poor

Source: Reference (8)

FIGURE 9: MATERIAL FLOW THROUGH THE ECONOMY SHOWING PRIMARY EMISSION SOURCES OF CHROMIUM

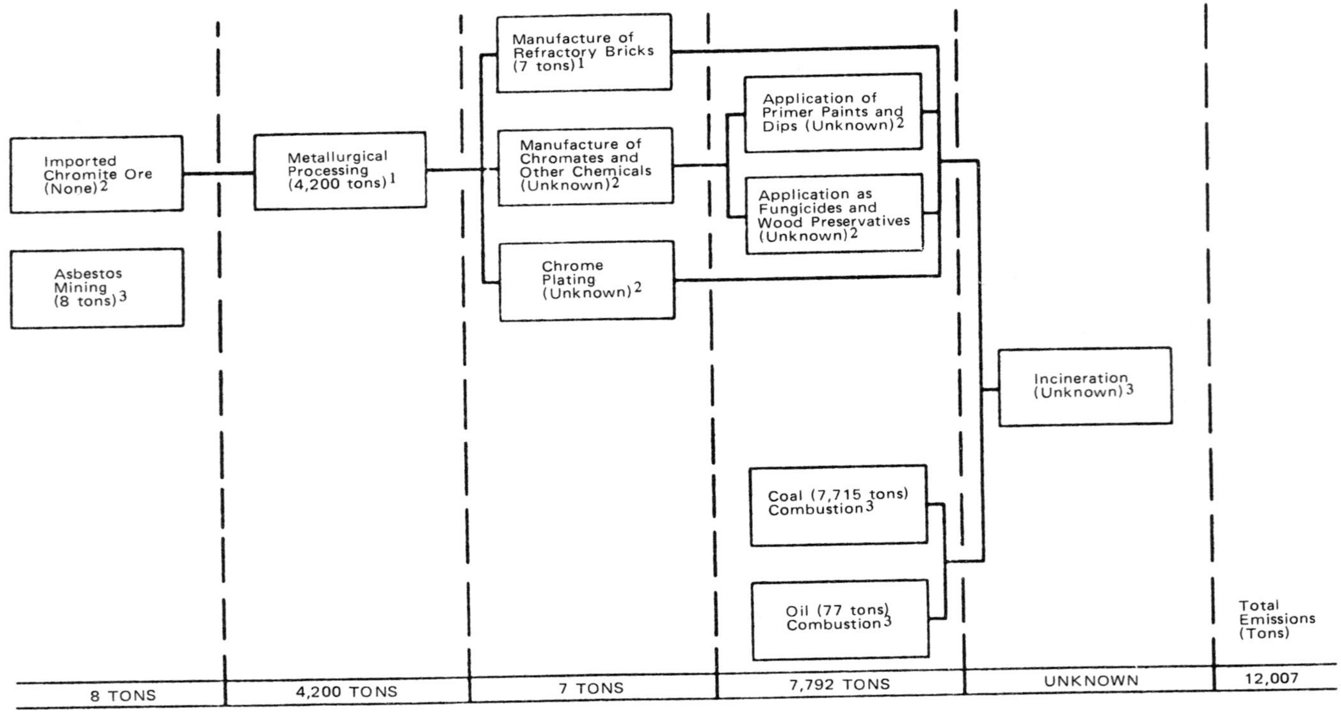

[1] *Control Techniques for Emissions Containing Chromium, Manganese, Nickel, and Vanadium,* Battelle.
[2] *Preliminary Air Pollution Survey of Chromium and Its Compounds,* Litton Systems, Inc., October 1969.
[3] MITRE

Source: Reference (9)

TABLE 20: CHROMIUM EMISSIONS

Source	Amount in Tons	Percent This Pollutant
Asbestos mining	8	0.07
Kraft pulp mill recovery furnace	Neg	Neg
Sulfite pulp mill	Neg	Neg
Primary chromium production	4,200	34.98
Asbestos products	Neg	Neg
Refractory brick production	7	0.06
Installation of asbestos material	Neg	Neg
Spray-on fire proofing	Neg	Neg
Use of insulating cement	Neg	Neg
Power plant boilers		
Pulverized coal	5,571	46.40
Stoker coal	640	5.33
Cyclone coal	192	1.60
All oil	22	0.18
Industrial boilers		
Pulverized coal	247	2.06
Stoker coal	864	7.20
Cyclone coal	123	1.02
All oil	17	0.14
Residential/commercial boilers		
Coal	77	0.64
Oil	38	0.32
Total	12,006	

Source: Reference (9)

Frequently, the major source of waste chromium is the chromic acid used in such metal plating operations. Table 21 is a compilation of typical sources and concentrations of chromate wastewaters.

TABLE 21: HEXAVALENT CHROMIUM WASTEWATER SOURCES AND TYPICAL CONCENTRATIONS

Industrial Source	Chromium (VI) Concentration, mg/l Average	Range
Leather tanning	40	
Wood preserving		0.23 - 1.5
Cooling tower blowdown	31.4	
Cooling tower blowdown		8 - 10.7
Cooling tower blowdown		10 - 60
Bright dip rinse		1 - 6
Bright dip bath		10,000 - 50,000
Bright dip bath		20,000 - 75,000
Bright dip bath		200 - 600
Anodizing bath	173	
Anodizing bath		15,000 - 52,000
Anodizing rinse	49	
Anodizing rinse		30 - 100

(continued)

TABLE 21: (continued)

Industrial Source	Chromium (VI) Concentration, mg/l Average	Range
Plating	1,300	
Plating	600	
Plating		100,000 - 270,000
Plating		60 - 80
Electroplating	140	
Electroplating	41	15 - 70

Source: Reference (10)

Hexavalent chromium treatment frequently involves reduction to the trivalent (Cr^{+3}) form prior to removing the chromium from the industrial waste. Thus trivalent chromium in industrial waste may result from one step of the waste treatment process itself, that of chemical reduction of hexavalent chromium. Industries which employ trivalent chromium directly in manufacturing processes include glass, ceramics, photography and textile dyeing.

In addition to the predominant hexavalent form, some few mg/l of trivalent chromium may be encountered in plating wastes, even prior to chemical reduction of Cr^{+6}. For example, Anderson and Iobst (11) report for one process waste a hexavalent chromium concentration ranging from 0.0 to 18.0 mg/l and a total chromium concentration of 2.9 to 31.8 mg/l. By difference, their data indicate a trivalent chromium concentration in this one waste stream of 2.9 to 13.8 mg/l. These same authors report for an acid bath waste stream a hexavalent chromium content of 122 to 270 mg/l and trivalent levels of 37 to 282 mg/l.

Thus, according to these authors, trivalent chromium represents a major component of this acid waste, even before reduction of hexavalent chromium. After reduction, the trivalent content of this latter waste stream would be approximately 160 to 550 mg/l. Evaluation of similar plating rinse data presented by Germain, et al (12) indicates trivalent chromium levels of 2.5 to 15 mg/l prior to treatment.

The principal sources of chromium wastes in solid, semisolid or concentrated liquid form have been identified as: [1] rinse water and dragout from the metal finishing industry; [2] filter residue from sodium dichromate manufacture; [3] waste sludges from paint manufacture; [4] paint residue left in used paint containers; and [5] surplus chemicals under the custody of Department of Defense. In addition, significant amounts of chromium compounds in highly diluted forms (ppm level) are also lost through the wastewaters from the textile industry, the leather industry and from the blowdown of cooling towers (13).

In the metal finishing industry, hexavalent chromium compounds are formulated for use as cleaning agents, oxidizing agents, surface preparation agents, as well as the chemicals used to electroplate the decorative chrome surface. The wastes are generated from the extensive washing of the metal parts as well as spills, tank leakages, and the periodic draining of the metal treatment tanks. The composition of the wastes varies widely but the chromium levels (present mostly as hexavalent chromium) are normally between a few ppm and a few percent.

Other constituents present in the wastes may include copper, zinc, cadmium, nickel, cyanides, grease, oils, acids, organic additives and cleaning agents. The amount of chromium compounds discharged through the waste streams from the metal finishing industry is estimated to total 62 million lb (calculated as sodium dichromate) per year.

Chromium wastes from sodium dichromate manufacture are generated in the processing of the chromium ores. The waste consists primarily of a gangue residue containing about 4% trivalent chromium as Cr_2O_3 (dry weight basis), and some iron, aluminum and calcium in varied forms depending on the process employed. Approximately 18 million lb of chromium (as Cr_2O_3) are lost as wastes from sodium dichromate manufacture each year.

The major chromium-containing pigments include chrome green, chrome oxide green, chrome yellow, chrome orange, zinc yellow and molybdate orange. With the exception of chrome oxide green (Cr_2O_3), all the other chromium pigments contain chromium in the hexavalent form and are normally only used in solvent-based paints. It is estimated that a total of 140,000 lb of chromium are lost through 37 million lb of waste solvent-based paint sludges each year, and a total of 10,000 lb of chromium lost through 26 million lb of wastewater-based paint sludges each year. An additional estimated 1 million lb of chromium are lost as paint residues left in used paint containers each year.

Chromium compounds under the custody of Department of Defense and awaiting disposal include 34,200 lb of anhydrous sodium chromate, 3,000 lb of potassium dichromate, 2,000 lb of chromium trioxide and smaller quantities of potassium chromate and sodium dichromate dihydrate.

DETECTION

Dusts and fumes of chromium compounds may be collected by any method suitable for collection of other dusts and fumes; the impinger, electrostatic precipitator and filters are commonly used. The National Air Sampling Network uses a high-volume filtration sampler. Chromic acid mists may be collected in an impinger using water or caustic solutions.

Methods of collection of chromic acid [chromium (VI)] mist in air have included absorption in distilled water and alkaline solutions, using impingers or sintered-glass bubblers and filtration with absorbent paper (14).

Of the methods of collection, filtration offers the greatest collection efficiency and ease of collection of breathing zone samples. The AA type of membrane filter has a 0.8 μ pore size and provides a highly retentive matrix for particulates. The use of scrubbing liquids is inconvenient for personal breathing-zone sampling and is thus not recommended.

The concentration of chromic acid mists in air can be estimated by a direct field method described by Ege and Silverman (15)(16)(17). This is a spot-test method using phthalic anhydride and sym-diphenylcarbazide.

The Mine Safety Appliances Co. (18)(19) has a lightweight sampler for estimating quickly the chromic acid mist concentrations in the atmosphere. The operating principle of this instrument is based on the method described above by Ege and Silverman (15)(16)(17).

The use of the colorimetric field analysis technique involving a grab sample and visual analysis must be considered to be only semiquantitative, and useful only for that purpose.

Emission spectroscopy has been used by the National Air Pollution Control Administration for chromium analysis of samples from the National Air Sampling Network (20)(21). The samples are ashed and extracted to eliminate interfering elements. The minimum detectable chromium concentration by emission spectroscopy is 0.0064 $\mu g/m^3$ for urban samples and 0.002 $\mu g/m^3$ for nonurban samples. The different sensitivities result from different extraction procedures for urban samples (22).

Analytical methods have included titration of liberated iodine with standardized sodium thiosulfate solution, colorimetry with hematoxylin of sym-diphenylcarbazide, and field analysis by means of an impregnated filter paper, based on the colorimetric reaction between chromium and sym-diphenylcarbazide and comparison with permanent standards (14).

The iodide-thiosulfate method is subject to interferences from a wide variety of compounds with its nonspecific iodide reaction and the color definition is subject to a slight error. The hematoxylin method is suggested only as a check for very small amounts of chromium and is a visual colorimetric method (14).

West (23) indicates that the ring oven technique for the determination of chromium is sensitive to 0.15 μg and has a normal range of 0.3 to 1.0 μg. The chromium is oxidized to the hexavalent state, and the method, therefore, measures total chromium. Diphenylcarbazide is used as the indicator.

The colorimetric diphenylcarbazide method does not react with trivalent chromium but produces a color with only the hexavalent form (present in chromic acid). However, cyanides, organic matter and other reducing agents, iron, copper and molybdenum at concentrations above 220 ppm and vanadium above 4 ppm, interfere and must be separated or complexed before this method may be expected to provide chromic acid analytical data of an acceptable degree of accuracy and precision (14).

The atomic absorption spectrophotometric method, applied directly, determines the total chromium and cannot make the desired distinction between the hexavalent chromium in chromic acid and the trivalent forms of chromium which may be present in the collected sample. Hence, it is necessary to separate the hexavalent from the trivalent chromium compounds by extracting the chelated complex of hexavalent chromium with ammonium pyrrolidine dithiocarbamate into methyl isobutyl ketone and then applying the atomic absorption spectrophotometric method to the extract for a specific determination of hexavalent chromium. It is not subject to the interferences which affect the diphenylcarbazide method. The atomic absorption method is therefore recommended (14).

Thompson et al (20) have reported that the National Air Pollution Control Administration uses atomic absorption to supplement results obtained by emission spectroscopy. The method has a minimum detectable limit of 0.002 $\mu g/m^3$ based on a 2,000 m^3 air sample.

Hexavalent chromium may be measured in water by a Technicon CSM-6 in a normal range of 0 to 5 ppm and down to a detection limit of 0.05 ppm. Conventional flame atomic absorption may be used to measure chromium down to a detection limit of 7 ppb.

A colorimetric analysis technique for chromates and dichromates has been described by V.M. Marcy and S.L. Halstead (24). According to these investigators, the best known and most widely used indicator for chromate is 1,5-diphenylcarbohydrazide (hereinafter called DPCH), which forms a pink or reddish-violet color in the presence of chromate ion. The intensity of color formed in a test solution treated by a known amount of DPCH is proportional to the concentration of chromate ion in it. No more useful indicator for chromate has been found. A great disadvantage to the use of DPCH as an indicator, however, has been its instability in virtually all media heretofore used as a solvent for it.

In research leading to this process, it was found that the use of glycerin or related polyhydric alcohol as the solvent for DPCH enables the preparation of an indicator solution for chromate having hitherto unrealized stability. An excellent indicator solution can be prepared, for example, by dissolving 2.5 grams of DPCH in 750 ml of glycerin heated to about 110°C. 50 ml of 85% orthophosphoric acid is then added and the mixture diluted to 1 liter with distilled water.

A process developed by C.A. Noll, E.C. Feddern and L.J. Stefanelli (25) permits determining the chromium concentration in an aqueous medium containing either or both trivalent chromium or hexavalent chromium. The method generally entails acidifying the aqueous medium, oxidizing any trivalent chromium present in the aqueous medium to hexavalent chromium, and adding to the medium a composition comprising a water-soluble cobaltous salt and a chelating agent of the acetic acid derivative type. The color intensity of the resulting medium is then measured and compared to the intensity of samples of aqueous medium containing known quantities of chromate in its hexavalent state which have been treated in the same manner. Chromium may be monitored in water by the following instrumental techniques: chemiluminescence, electrochemical, spark source mass spectrometry or x-ray fluorescence (26).

An apparatus described by R.S. Saltzman (27) is one for quantitatively determining the amount of hexavalent chromium (Cr^{+6}) in a process stream by comparing the optical density of the process stream with a reference sample produced by reducing Cr^{+6} to Cr^{+3} by contacting it with a sulfur dioxide reducing agent.

ENVIRONMENTAL STANDARDS

The American Conference of Governmental Industrial Hygienists (28) has set the following threshold limit values for chromium in the workroom environment.

	Mg/m^3
Chromates, certain insoluble forms, (Pb, Zn), and chromate-chromite ore (as Cr)	0.1
Chromic acid and chromates (as CrO_3)	0.1
Chromium, soluble chromic and chromous salts (as Cr)	0.5

The permissible concentration of hexavalent chromium in domestic water supplies is 0.05 ppm. The amount of chromium which can be tolerated in irrigation waters is 5.0 ppm for continuous use and 20.0 ppm for short term use in fine textured soil. The maximum allowable concentration of chromium in an aqueous effluent discharged to a storm sewer or stream should be 0.5 ppm for hexavalent chromium and 1.0 ppm for trivalent chromium.

According to a recommended standard for occupational exposure to chromic acid by NIOSH (14), occupational exposure to chromic acid shall be controlled so that no worker is exposed either to:

1. A concentration of chromic acid greater than 0.05 mg as chromium trioxide per cubic meter of air determined as a time-weighted average exposure for an 8-hour workday, 40-hour work week; or
2. A ceiling concentration in excess of 0.1 mg as chromium trioxide per cubic meter as determined by a sampling time of 15 minutes.

The standard is designed to protect the health and safety of workers for an 8-hour day, 40-hour week over a working lifetime. Compliance with the standard should prevent adverse effects of occupational exposure to chromic acid. Chromic acid is defined to mean chromium trioxide [chromium (VI) oxide, or chromic acid anhydride] and aqueous solutions thereof.

NIOSH has recently issued new criteria for occupational exposure to Cr^{+6} (73). In this document, NIOSH recommends that chromium (VI) in the workplace be considered carcinogenic unless an employer can demonstrate that only the noncarcinogenic chromium (VI) compounds are present: monochromates and bichromates (dichromates) of hydrogen, lithium, sodium, potassium, rubidium, cesium and ammonium or chromium (VI) oxide (chromic acid anhydride).

In those instances where the recommendations for carcinogenic chromium (VI) apply, NIOSH recommends that exposure be limited to 0.001 mg/m^3. If only the noncarcinogenic compounds are present, NIOSH recommends 0.05 mg/m^3 as a 15-minute ceiling and 0.025 mg/m^3 for up to a 10-hour workday, 40-hour work week.

HANDLING PROCEDURES

According to the NIOSH (73), medical surveillance shall be made available as outlined below for all workers with occupational exposure to carcinogenic or noncarcinogenic chromium (VI), including maintenance personnel periodically exposed during routine maintenance or emergency repair operations.

A. Replacement and annual medical examinations shall include:
   1. A comprehensive or interim work history.
   2. A detailed medical history including information on conditions indicating the inadvisability of further exposure to chromium (VI), e.g., potential skin or pulmonary sensitization, a skin or mucous membrane condition that may be exacerbated by chromium (VI), smoking habits, and history of liver or kidney disease.
   3. Examination of the skin for evidence of dermatitis or chrome ulcers, and of the membranes of the upper respiratory tract for irritation, bleeding, ulcerations or perforations.

4. An evaluation of the worker's ability to use negative or positive pressure respirators.
5. Urinalysis.

B. For workers with occupational exposure to carcinogenic chromium (VI), preplacement and annual medical examinations shall include 14" x 17" chest x-rays. Other tests, including sputum cytology and liver function studies, shall be considered by the responsible physician.

C. For workers with occupational exposure to noncarcinogenic chromium (VI) and not to carcinogenic chromium (VI), preplacement medical examinations shall include 14" x 17" chest x-rays. Thereafter, x-ray examinations shall be offered at 5-year intervals and annually after age 40. Other tests, such as liver function studies, may be considered by the responsible physician.

D. Medical examinations shall be made available to all workers with signs or symptoms of skin or upper respiratory tract irritation likely to have been the result of exposure to chromium (VI).

E. If clinical evidence of adverse effects due to chromium (VI) is developed from these medical examinations, the worker shall be kept under a physician's care until the worker has completely recovered or maximal improvement has occurred.

F. Initial annual examinations for presently employed workers shall be offered within 6 months of the promulgation of a standard incorporating these recommendations.

G. The medical representatives of the Secretary of Health, Education and Welfare, of the Secretary of Labor and of the employer shall have access to all medical records. Physicians designated and authorized by any employee or former employee shall have access to that worker's medical records.

H. Medical records shall be maintained for all employees with occupational exposure to carcinogenic or noncarcinogenic chromium (VI) and for maintenance personnel with periodic exposure. Preplacement x-rays and x-rays for the 5 years preceding termination of employment and all medical records with pertinent supporting documents shall be retained at least 30 years after the individual's employment is terminated.

As regards labeling, NIOSH (73) specifies that

A. Except for shipping and storage containers for lithium chromate, lithium bichromate, sodium chromate, sodium bichromate, potassium chromate, potassium bichromate, rubidium chromate, rubidium bichromate, cesium chromate, cesium bichromate, ammonium chromate, ammonium bichromate, and chromium (VI) oxide (chromic acid anhydride), as dry solids or concentrated solutions, all shipping and storage containers for chromium (VI) shall bear the following label in addition to, or in combination with, labels required by other statutes, regulations or ordinances:

CHEMICAL NAME
SYNONYMS
DANGER! EXTREME HEALTH HAZARD
MAY CAUSE IRRITATION, RASH OR EXTERNAL ULCERS
INHALATION MAY CAUSE CANCER

Keep container closed
Avoid contact with skin and eyes
Avoid breathing dust or solution spray
In case of contact, immediately flush eyes with
plenty of water for at least 15 minutes

Call a physician
Flush skin with water
Wash clothing before reuse

B. All shipping and storage containers for lithium chromate, lithium bichromate, sodium chromate, sodium bichromate, potassium chromate, potassium bichromate, rubidium chromate, rubidium bichromate, cesium chromate, cesium bichromate and ammonium chromate, the hydrates of these compounds, high purity aqueous solutions of these compounds, and dry mixtures containing only these materials shall bear the same label except that "Inhalation May Cause Cancer" shall be deleted and "Extreme Health Hazard" shall be replaced by "Moderate Health Hazard."

C. Because of the flammable characteristics of ammonium bichromate (dichromate), shipping and storage containers for dry forms of this compound shall bear the following label in addition to, or in combination with, labels required by other statutes, regulations or ordinances:

AMMONIUM BICHROMATE
DANGER! HIGHLY FLAMMABLE
MODERATE HEALTH HAZARD
MAY CAUSE IRRITATION, RASH OR EXTERNAL ULCERS

Keep away from heat, sparks and open flame
Keep container closed
Avoid contact with skin and eyes
Avoid breathing dust or solution spray
In case of contact, immediately flush eyes with
plenty of water for at least 15 minutes
Call a physician
Flush skin with water
Wash clothing before reuse

D. All storage containers of chromic acid or chromium (VI) oxide (chromic acid anhydride) shall bear the following label in addition to, or in combination with, labels required by other statutes, regulations or ordinances:

CHROMIUM TRIOXIDE
(CHROMIC ACID)
DANGER! POWERFUL OXIDIZER
CONTACT WITH OTHER MATERIAL MAY CAUSE FIRE
MAY CAUSE DELAYED BURNS OR EXTERNAL ULCERS

Keep container closed
Do not get in eyes, on skin, on clothing
Do not breathe dust or mist from solutions
In case of contact, immediately flush skin or eyes
with plenty of water for at least 15 minutes
For eyes, get medical attention immediately
Wash clothing before reuse
Use fresh clothing daily
Take showers after work, using plenty of soap

E. In areas where there is occupational exposure to carcinogenic chromium (VI), the following warning sign shall be posted in readily visible locations, particularly at the entrances to the area.

WARNING
CANCER-SUSPECT AGENT
USED IN THIS AREA
UNAUTHORIZED PERSONS KEEP OUT

The sign shall be printed both in English and in the predominant language of non-English-speaking workers, if any, unless they are otherwise trained and informed of the hazardous areas. All illiterate workers shall receive such training.

F. In areas where airborne chromium (VI) comprises only lithium chromate, lithium bichromate, sodium chromate, sodium bichromate, potassium chromate, potassium bichromate, rubidium chromate, rubidium bichromate, cesium chromate, cesium bichromate, ammonium chromate or chromium (VI) oxide (chromic acid anhydride), their hydrates or mixtures containing only these chromium (VI) materials, the warning sign shall read as follows:

WARNING
CHROMATES, BICHROMATES OR CHROMIC ACID ANHYDRIDE
USED IN THIS AREA
UNAUTHORIZED PERSONS KEEP OUT

The sign shall be posted in readily visible locations, particularly at the entrances to the area. The sign shall be printed both in English and in the predominant language of non-English-speaking workers, if any, unless they are otherwise trained and informed of the hazardous areas. All illiterate workers shall receive such training.

G. In areas where airborne chromium (VI) contains ammonium bichromate, or where ammonium bichromate is stored, manufactured or used, the following shall be added to the warning sign in (E) or (F) above:

FLAMMABLE SUBSTANCE

As regards protective clothing, NIOSH (73) specifies that:

A. Coveralls or other full-body protective clothing shall be worn in areas where there is occupational exposure to chromium (VI). Protective clothing shall be changed at least daily at the end of the shift and more frequently if it should become grossly contaminated.

B. Impervious gloves, aprons and footwear shall be worn at operations where solutions of chromium (VI) may contact the skin. Protective gloves shall be worn at operations where dry compounds of chromium (VI) are handled and may contact the skin.

C. Eye protection shall be provided by the employer and used by the employees where eye contact with chromium (VI) is likely. Selection, use and maintenance of eye protective equipment shall be in accordance with the provisions of the American National Standard Practice for Occupational and Educational Eye and Face Protection, ANSI Z87.1-1968. Unless eye protection is afforded by a respirator hood or facepiece, protective goggles or a face shield shall be worn at operations where there is danger of contact of the eye with dry or wet compounds of chromium (VI) because of spills, splashes or excessive dust or mists in the air.

D. The employer shall ensure that all personal protective devices are inspected regularly and maintained in clean and satisfactory working condition.

E. Work clothing shall not be taken home by employees. The employer shall provide for maintenance and laundering of protective clothing.

F. The employer shall ensure that precautions necessary to protect laundry personnel are taken when soiled protective clothing is laundered.

Engineering controls shall be used wherever feasible to maintain airborne carcinogenic and noncarcinogenic chromium (VI) concentrations below those recommended by NIOSH (73) above. Compliance with the permissible exposure limits by the use of respirators is only allowed when airborne chromium (VI) concentrations are in excess of the workplace environmental limit because required engineering controls are being installed or tested, when nonroutine maintenance or repair is being accomplished, or during emergencies. When a respirator is thus permitted, it shall be selected and used in accordance with the following requirements:

A. For the purpose of determining the type of respirator to be used, the employer shall measure the airborne concentration of chromium (VI) in the workplace initially and thereafter whenever process, worksite, climate or control changes occur which are likely to increase the airborne concentration of chromium (VI); this requirement does not apply when carcinogenic chromium (VI) is present.

B. The employer shall ensure that no worker is overexposed to chromium (VI) because of improper respirator selection, fit, use or maintenance.

C. A respiratory protection program meeting the requirements of 29 CFR 1910.134 and 30 CFR 11 which incorporates the American National Standard Practices for Respiratory Protection Z88.2-1969 shall be established and enforced by the employer.

D. The employer shall provide respirators in accordance with Table 22 when appropriate, and shall ensure that the employee uses the respirator provided.

E. Respirators described in Table 22 shall be those approved under the provisions of 29 CFR 1910.134 and 30 CFR 11.

F. The employer shall ensure that respirators are adequately cleaned, and that employees are instructed on the use of respirators assigned to them and on how to test for leakage.

G. Respirators specified for use in higher concentrations of airborne chromium (VI) may be used in workplaces with lower concentrations of airborne chromium (VI).

H. Where an emergency may develop which could result in employee injury from chromium (VI), the employer shall provide an escape device as listed in Table 22 where appropriate.

TABLE 22: RESPIRATOR SELECTION GUIDE FOR PROTECTION AGAINST CHROMIUM (VI)

Carcinogenic Chromium (VI)

Self-contained breathing apparatus with positive pressure in full facepiece

Combination supplied air respirator, pressure-demand type, with auxiliary self-contained air supply

Noncarcinogenic Chromium (VI)

Multiples of TWA Limit	Respirator Type
Less than or equal to 10X	Half-mask respirator with replaceable high efficiency filter(s), or Type C supplied-air respirator, demand type (negative pressure), with half-mask facepiece

(continued)

TABLE 22: (continued)

Multiples of TWA Limit	Respirator Type
Less than or equal to 100X	Full facepiece respirator with replaceable high efficiency filter(s), or Type C supplied-air respirator, demand type (negative pressure), with full facepiece, or Self-contained breathing apparatus in demand mode (negative pressure), with full facepiece
Less than or equal to 200X	Powered air-purifying (positive pressure) respirator with high efficiency filter(s)
Greater than 200X	Self-contained breathing apparatus with positive pressure in full facepiece, or Combination supplied-air respirator, pressure-demand type, with auxiliary self-contained air supply
Emergency (no concentration limit)	Self-contained breathing apparatus with positive pressure in full facepiece, or Combination supplied-air respirator, pressure-demand type, with auxiliary self-contained air supply
Evacuation or Escape (no concentration limit)	Self-contained breathing apparatus in demand or pressure-demand mode (negative or positive pressure), or Gas mask, Type N, with high efficiency filter, and mouthpiece respirator with high efficiency filter(s)

Note: A high efficiency filter is defined as a filter having an efficiency of at least 99.97% against 0.3 μm DOP (Dioctyl Phthalate).

Source: Reference (73)

NIOSH (73) specifies that at the beginning of employment or assignment for work in a chromium (VI) area, employees with occupational exposure to chromium (VI) shall be informed of the hazards, relevant signs and symptoms of overexposure, appropriate emergency procedures, and proper conditions and precautions for the safe use of chromium (VI).

Instruction shall include all information applicable to the specific chromium (VI) product or material to which there is exposure. This information shall be posted in the work area and kept on file, readily accessible to the worker at all places of employment where chromium (VI) is involved in unit processes and operations.

A continuing educational program shall be instituted to ensure that all workers have current knowledge of job hazards, proper maintenance procedures, and cleanup methods, and that they know how to use respiratory protective equipment and protective clothing correctly.

Information shall be recorded on U.S. Department of Labor Form OSHA-20 "Material Safety Data Sheet" or a similar form approved by the Occupational Safety and Health Administration, U.S. Department of Labor.

As regards work practices, NIOSH (73) specifies the following:

A. Control of airborne contamination.
   Emission of airborne particulates (dust, mist, spray, etc.) of chromium (VI)

shall be controlled at the sources of dispersion by means of effective and properly maintained methods such as fully enclosed operations and local exhaust ventilation. Other methods may be used if they are shown to effectively control airborne concentrations of chromium (VI) within the limits of the recommended standard.

B. Control of contact with skin and eyes.

1. Employees working in areas of possible contact of skin or eyes with chromium (VI), dry or wet, shall wear full-body protective clothing, including neck and head coverings, and gloves, in accord with Section 4(a)(73).
2. Clean protective clothing shall be put on before each work shift.
3. If, during the shift, the clothing becomes wetted with a solution, slurry, or paste of a chromium (VI) material, or grossly contaminated with a dry form of such material, it shall be removed promptly and placed in a special container for garments for decontamination or disposal. The employee shall wash the contaminated skin area thoroughly with soap and a copious amount of water. A complete shower is preferred after anything but limited, minor contact. Then, clean protective clothing shall be put on before resuming work. When working directly with chromium (VI) oxide, with unsealed containers of chromium (VI) oxide, or with chromium (VI) oxide in other than fully enclosed operations, protective devices and clothing shall be removed and the arms, hands and face thoroughly washed after working with chromium (VI) oxide, and at 30-minute intervals when working with chromium (VI) oxide for extended periods of time.
4. Minor areas of skin (principally the hands) contaminated by contact with chromium (VI) shall be washed immediately and thoroughly with an abundance of water. Water shall be easily accessible in the work areas from low-pressure, free-running hose lines or showers.
5. If chromium (VI) comes into contact with the eyes, the eyes should be flushed with a large volume of low-pressure flowing water for at least 15 minutes. Medical attention shall be obtained without delay but not at the expense of thoroughly flushing the eyes.

C. Procedures for emergencies, including firefighting, shall be established to meet foreseeable events. Necessary emergency equipment, including appropriate respiratory protective devices, shall be kept in readily accessible locations. Only self-contained breathing apparatus with positive pressure in the face-piece shall be used in firefighting. Appropriate respirators shall be available for use during evacuation.

D. Special supervision and care shall be exercised to ensure that the exposures of repair and maintenance personnel to chromium (VI) shall be within the limits prescribed by this standard.

E. Prompt cleaning of spills of chromium (VI).

1. No dry sweeping shall be performed. Wet methods or dry vacuuming shall be used as appropriate.
2. Wet spills and flushing of wet or dry spills shall be channeled for appropriate treatment or collection for disposal. They shall not be channeled directly into the sanitary sewer system.

F. General requirements.

1. Good practices of housekeeping shall be observed to prevent or minimize contamination of areas and equipment and to prevent build-up of such contamination.
2. Good personal hygiene practices shall be encouraged.
3. Equipment shall be kept in good repair and free of leaks.
4. Containers of dry chromium (VI) shall be kept covered insofar as is practical.

As regards sanitation facilities, NIOSH (73) specifies the following:

A. Washing facilities.
Emergency showers and eye-flushing fountains with adequate pressure of cool water shall be provided and be quickly accessible in areas where there is potential of skin or eye contact with chromium (VI). This equipment shall be frequently inspected and maintained in good working condition. Showers and washbasins shall be provided in the employees' locker areas. Employees exposed to chromium (VI) shall wash before eating or smoking during the work shift.

B. Food facilities.
Food storage, preparation, and eating shall be prohibited in areas where chromium (VI) is handled, processed or stored. Eating facilities provided for employees shall be located in nonexposure areas. Washing facilities should be accessible nearby.

C. Employees shall not smoke in areas where chromium (VI) is handled, processed or stored.

D. Clothing and locker room facilities.
Locker room facilities shall be provided in a nonexposure area for employees required to change clothing before and after work. The facilities shall provide for the separate storage of street clothing and clean work clothing from soiled work clothing. Showers and wash basins should be located in the locker area to encourage good personal hygiene. Covered containers should be provided for work clothing discarded at the end of the shift or after a contamination incident. The clothing will be held in these containers until removed for decontamination or disposal.

As regards monitoring and housekeeping requirements NIOSH (73) specifies that: workers are not considered to have occupational exposure to chromium (VI) if, on the basis of a professional industrial hygiene survey, [a] the airborne concentration of carcinogenic chromium (VI) is sufficiently low that a sampling volume greater than 1.0 m^3 is necessary in order to collect 0.5 μg of carcinogenic chromium (VI) and [b] the airborne concentration of noncarcinogenic chromium (VI) is not greater than half the recommended limit of 25 μg Cr(VI)/m^3. All samples of airborne chromium (VI) shall be analyzed by the chemical analytical method specified (73); if samples can be demonstrated to contain only noncarcinogenic chromium (VI), other methods of chemical analysis equivalent to the method specified (73) may be used. Records of these surveys, including the basis for concluding that there is no occupational exposure to chromium (VI) shall be maintained until a new survey is conducted.

In workplaces where chromium (VI) is handled or processed, surveys shall be repeated annually and when any process change indicates a need for reevaluation. Requirements set forth below apply to areas in which there is occupational exposure to chromium (VI).

Employers shall maintain records of workplace environmental exposures to chromium (VI) based upon the following sampling, analytical and recording schedules:

A. In all monitoring, samples representative of the exposure in the breathing zone of employees shall be collected by personal samplers.

B. An adequate number of samples shall be taken in order to permit construction of TWA exposures for every operation or process. Except as otherwise determined by a professional industrial hygienist, the minimum number of representative TWA determinations for an operation or process shall be based on the number of workers exposed as provided below:

Number of Employees Exposed	Minimum Number of Employees Whose Individual Exposures Shall Be Determined
1 - 20	50% of the total number of exposed employees
21 - 100	10 plus 25% of the excess over 20 exposed employees
Over 100	30 plus 5% of the excess over 100 exposed employees

C. The first determination of the workers' exposures to airborne chromium (VI) shall be completed within 6 months after the promulgation of a standard incorporating these recommendations.

D. A reevaluation of the exposures of workers to airborne chromium (VI) shall be made within 30 days after installation of a new process or process changes.

E. Samples of airborne chromium (VI) shall be collected and analyzed at least every 2 months for those work areas with occupational exposure to carcinogenic chromium (VI) and at least every 3 months if the airborne chromium (VI) is noncarcinogenic.

F. A reevaluation of the workers' exposures to airborne chromium (VI) shall be repeated at 1-week intervals when the airborne concentration has been found to exceed the recommended workplace environmental limit. In such cases, suitable controls shall be instituted and monitoring shall continue at 1-week intervals until 3 consecutive surveys indicate the adequacy of controls.

G. Records of all sampling and analysis of airborne chromium (VI) and of medical examinations shall be maintained for at least 30 years after the individual's employment is terminated. Records shall indicate the details of [1] type of personal protective devices, if any, in use at the time of sampling, and [2] methods of sampling and analysis used. Each employee shall be able to obtain information on his own exposure. In the event that the employer who has or has had employees with occupational exposure to carcinogenic chromium (VI) ceases business without a successor, he shall forward their records by registered mail to the Director, National Institute for Occupational Safety and Health.

H. A regulated area shall be established and maintained where:
   1. Carcinogenic chromium (VI) is manufactured, reacted, repackaged, stored, handled or used; and
   2. Airborne concentrations of carcinogenic chromium (VI) are in excess of the permissible exposure limit in Section 1.

I. Access to the regulated areas designated by Section 8h shall be limited to authorized persons. A daily roster shall be made of authorized persons who enter; these rosters shall be maintained for 30 years.

REMOVAL FROM AIR

Chromium compounds polluting the air from metallurgical uses are assumed to be in either the trivalent or zero-valent state. Effluent particulate matter from an oxygen lanced open hearth furnace contains from 0.11 to 0.16% chromium, while a typical dust analysis of an electric arc furnace reveals that chromium, as Cr_2O_3, accounts for about 1.1% of the dust. Particulate chromium effluxing from processes associated with the manufacture of iron and steel are in the submicron size range. Nearly 90% of all chromium fumes are less than 1 μm (29).

Chromium emissions from coal-fired power plants have been found to range from 220 to 2,200 $\mu g/m^3$ depending upon the type of boiler firing. Fly ash collection reduces this range to 18 to 500 $\mu g/m^3$.

Chromium particles of a mean size of 0.12 μm have been reported to efflux from glass furnaces of several types. Chromate-producing plants have been pointed out as possible sources of chromium pollution. Other possible sources include paint manufacturing, the production of fungicides, cement production and Cr-containing corrosion inhibitors whose use finds application in road salt. It has been established that the metal emanates in the solid phase from these processes (29).

Particulate chromium emissions can be minimized through use of standard dust collecting apparatus. Chromium particles lie in the size range from 0.01 μm to about 100 μm, with the majority typically in the submicron range.

Collection efficiencies of fabric filters can be expected to be greater than 90% for particulate chromium. In ferroalloy manufacturing baghouse collection efficiencies have been as great as 99% for chromium pollutants. A six-section bag filter using Terylene fabric has been used successfully on an electric arc steel furnace to limit particulate emissions to below 0.05 grains per scf. A heat exchanger was used to cool effluent gas from 1500° to 100°C. Fiber glass bag collectors preceded by a U-tube cooler have successfully removed over 95% chromium-containing particulate material from the waste gas of a silicon metal furnace (29).

Electrostatic precipitation is widely used in the iron and steel industry. Particulate chromium collection efficiencies have reportedly been as high as 98%. In the open hearth system similar efficiencies have been experienced. Studies on trace element emissions from coal-fired power plants have shown uncontrolled chromium emissions to range from 0.0007 to 0.0087 grains per acf, varying with the chromium content of the coal burned. Application of a single stage electrostatic precipitator reduced chromium effluent by 95 to 99%. Particulate trace element emissions from municipal incinerators have been reduced similarly through electrostatic precipitation, although total chromium emitted from combustion of refuse is considered to be negligible.

Scrubbers have been employed for reduction of trace pollutants in plating operations, in certain chromium industries, and in several steelmaking processes. A high energy venturi-type scrubber has recorded a particle removal efficiency of 93% on one 20 Mw FeCrSi furnace and 98% on a similar ferroalloy unit. Collection efficiencies in the removal of metallic fumes from electric arc steelmaking furnaces have been found to be greater than 90%.

Single stage spray scrubbers have been successfully employed in removing chromic acid aerosols from waste gases emanating from certain plating processes. Recovered acid is either recycled or rinsed, where in the latter case the acid rinse water is subsequently freed of chromates either by direct precipitation or by chemical reduction of chromates to a trivalent state and then chemical removal. Packed towers also have been applied to reduce trace element emissions from plating processes, but performance data are not given (29).

Chromic acid mists emanate from certain plating processes and can be dealt with using wet scrubbers. Alternative methods include direct mist separation. On such a control system, the chromic acid mist is led into a casing with two lines of drop separator profiles made of polyvinyl chloride. The liquid particles are removed at these profiles and flow downward. Removal efficiencies are greater than 99%.

Some chrome plating facilities have installed moisture-extractor vanes in the hood-duct system to eliminate acid mists. Most exhaust systems employ slot hoods to capture the mists discharged from the plating solutions. Other commercially available mist inhibitors use a surfactant in the plating solution which reduces surface tension and suppresses the mist. Mist eliminators used to remove mercury mist from waste gas chlor-alkali manufacture can be adopted to ameliorate the chromic acid problem. One available system employs a special fiber filter which is intermittently flushed with water. Acidic wastewater is then freed of chromium in a manner similar to the cleansing of wastewater from a wet scrubbing device (29).

REMOVAL FROM WATER

The various techniques which are applicable to the removal of chromium from water are:

[1] Reduction of hexavalent chromium and subsequent precipitation of the trivalent chromium ion. This is the most common method.
[2] Ion exchange to treat chromate and chromic acid wastes.
[3] Evaporative recovery.
[4] Chemical precipitation to give insoluble chromates.
[5] Solvent extraction.
[6] Reverse osmosis.
[7] Biological purification.

Some examples of each type of treatment are cited in the following paragraphs.

Reduction

The standard reduction treatment technique is to lower the waste stream pH to 3.0 or below with sulfuric acid and convert the hexavalent chromium to trivalent chromium with a chemical reducing agent such as sulfur dioxide, sodium bisulfite or ferrous sulfate. The trivalent chromium is then removed, usually by precipitation with lime (30)(31). The reduction of hexavalent to trivalent chromium is not 100% effective, and the amount of residual nonreduced hexavalent chromium depends upon the allowed time of reaction, pH of the reaction mixture and concentration and type of reducing agent employed. Treatment of chromic acid waste does not normally require pH adjustment with sulfuric acid, as the pH of the chromic acid waste itself is sufficiently low for the reduction reaction to proceed.

The treatment of a metal electroplating waste containing 140 mg/l of hexavalent chromium has been reported (32). Reduction was carried out at pH 2.5 to 2.8, employing sodium bisulfite as the reducing agent. The process was reported to reduce the hexavalent chromium to concentrations of 0.7 to 1.0 mg/l. The Electric Autolite Company (Sharonville, Ohio) also attempted to use sodium bisulfite to reduce the waste from chrome plating automobile bumpers (33). They abandoned the use of sodium bisulfite as a reductant because of odors and corrosion hazards associated with its use, and now employ sulfur dioxide in its place.

Sulfur dioxide appears to be the most popular reducing agent used in treatment of chromium wastes. Hulse et al (34) have reported its use at a Boeing plant to treat chromium in metal finishing wastes. The waste was adjusted to pH 2.5 with sulfuric acid before reduction. Schink (35) also reports the use of sulfur dioxide to treat chromic acid plating bath and chromic acid etch bath wastes.

At an average waste flow of 40 gpm, and for a 20 to 30 minute treatment period, Schink reported that residual hexavalent chromium was less than 1.0 mg/l. Hexavalent chromium concentration of the raw waste was not reported.

Sulfur dioxide has also been used as the reductant at an IBM plant (36), to treat a waste estimated to contain 1,300 mg/l of hexavalent chromium. The chromate wastes were treated with sulfur dioxide at pH 2. Treatment detention time was approximately 90 minutes. Sulfuric acid was added to the waste, to maintain the proper pH. Chemical analyses by the state Department of Health have shown zero hexavalent chromium in the treated effluent.

An unusual source of sulfur dioxide is employed by the Fisher Body Plant of Elyria, Ohio, to treat the hexavalent chromium in its plating waste (37). Sulfur dioxide is washed out of the power house stack gas, and used as the chemical reducing agent for converting hexavalent chromium to the trivalent form. The process has been in operation since 1959. On occasions when additional reduction is necessary, sodium bisulfite is employed as a supplement. No values for raw or treated waste chromium concentrations are reported, although it is claimed that the process is highly effective. Lacy and Cywin (38) report that the use of sulfur dioxide for reduction imparts an oxygen demand to the waste effluent, unless the effluent is oxygenated by passing air through it.

A process developed by R.P. Selm et al (39) involves the reduction and precipitation of chromates in aqueous industrial wastes by the steps of dissolving flue gas containing carbon dioxide into the wastes to carbonate same, passing the carbonated wastes rapidly through a bed of ferrous chips to activate the bed, recycling the wastes that pass through the ferrous chip bed at least once through the steps of dissolving flue gas containing carbon dioxide therein and again passing the wastes through the activated ferrous chip bed to maintain a rapid flow therethrough for reduction of the chromates in the flow of wastes, and withdrawing a portion of the wastes after passage thereof through the ferrous chip bed for disposal as nontoxic waste.

A process developed by L.E. Lancy (40) provides a method of quickly and effectively conditioning spent cooling water having a toxic hexavalent chromium solution content to make the water innocuous. It comprises providing a mass of hard metal sulfide granules, moving the spent cooling water through the mass in contact with surface portions of the granules and surface-reacting the granules with the hexavalent chromium solution content to fully convert it into trivalent chromium, while inhibiting the forming of a soluble sulfide in the water from the sulfide granules, and while maintaining surfaces of the metal sulfide granules reactive to the hexavalent chromium solution content of the water, and removing the water in an innocuous condition from the mass and discharging it.

A process developed by A. Gloster and H.G. Bocckino (41) involves combining the streams from a cooling tower and an acid plant scrubbing tower to reduce the pollution problems of these industrial wastewaters. The cooling tower stream is characterized by the presence of chromium products in which the chromium is in the hexavalent state, such as chromic acid or a chromate salt which are common corrosion inhibitors in cooling waters. The acid plant scrubbing tower is characterized by an acidic component such as sulfur dioxide dissolved in water. By combining these two streams the hexavalent chromium under acidic conditions is converted to trivalent chromium. The combined streams are then neutralized to precipitate chromium hydroxide and other compounds which are undesirable in the effluent. After settling out the chromium hydroxide and other compounds the waste stream may be discharged without presenting a pollution problem.

A process developed by T.D. Henley and R.F. Reeves (42) is one in which dissolved hexavalent chromium and dissolved iron cyanide $[Fe(CN)_6]$ complex are removed from a liquid waste effluent by adding thereto a reducing agent having an anion capable of reducing the hexavalent chromium to trivalent chromium and a cation which forms a solid or precipitate with the complex iron cyanide. The preferred reducing agent is zinc hydrosulfite (ZnS_2O_4).

A process developed by L.B. Fournier and R.A. Meyer (43) is one in which the hexavalent chromium present in aqueous solutions is chemically reduced to trivalent chromium with hydrogen peroxide. This process is improved if the pH of the aqueous solution is maintained below 1.5 and the hydrogen peroxide is gradually added to the solution to keep the chromate in excess as long as possible.

A process developed by R. Stewart (44) is one in which SO_2 is injected into an essentially unimpeded stream of an acidified aqueous industrial effluent flowing in a simple conduit which is a mere passageway having no baffles or other obstructive or detaining means therein, the stream containing soluble chromium (VI) and having a turbulence characterized by a Reynolds Number of at least about 5,000. The resulting chromium (III) is then precipitated as $Cr(OH)_3$, which is allowed to settle and is subsequently separated from the essentially chromium-free supernatant.

Ion Exchange

Ross, in his text on industrial waste treatment, claims that ion exchange processes can be economically utilized for chromium recovery and elimination of waste (45). Cation exchange can be applied to remove trivalent chromium, as described in a separate report, and anion exchange employed to remove hexavalent chromium in the chromate form normally present in industrial wastes.

When the anion exchange resin is exhausted, it is regenerated (usually with sodium hydroxide), and sodium chromate is eluted from the ion exchange resin. The eluted sodium chromate can be passed through a cation exchange resin to recover purified chromic acid. If chromic acid is not recovered, the concentrated waste yielded by the resin regeneration process must be disposed of. This may be accomplished by reduction to trivalent chromium, followed by lime precipitation.

Waste treatment by ion exchange produces a reusable water, which may provide an economic edge when water costs are high.

Yuronis (46) has reported that ion exchange recovery is economically and technically feasible for wastes containing chromate ion concentrations up to 200 mg/l. This study indicated that chromate wastes of 100 to 500 mg/l were suitable for reduction and precipitation. Wastes with chromate concentration exceeding 500 mg/l were suitable for evaporative recovery.

A process developed by A.J. Saraceno et al (47) is an improved anion exchange process for the removal and recovery of trace quantities of chromates from water. The process utilizes a bed of basic anion exchange resin and the combination of upflow exhaustion of the chromates from the water and downflow regeneration of the resin. Efficient recovery of chromates is accomplished while virtually eliminating chromate leakage during the initial portion of the exhaustion operation. Regeneration can be accomplished without the use of strong, expensive chemicals.

A process developed by L. Sloan et al (48) permits recovery of chromate ions from the blowdown of a cooling system involving cooling towers. In the case of air conditioning or other cooling systems, water is used to absorb heat and is then passed to cooling towers in which cooling is effected by evaporation of a portion of the water. To minimize corrosion chromate ions are added to the recirculating water usually in the form of sodium chromate or chromic acid.

Because of the fact that evaporation takes place during operation, the salts present in the water which is used become concentrated, and to limit the concentration it is necessary to blowdown the system either continuously or intermittently, bleeding off portions of the water which are then replaced by makeup. The blowdown has been wasted to the sewer, but this has two major objections: first, the expensive chromate ions are thus lost, and second, the chromate ions represent a pollution problem and are in some cases required to be removed but in such fashion that recovery is impractical from the standpoint of cost.

Removal by ion exchange has been used, but has been carried out in expensive fashion. A typical blowdown may include twenty to several hundred parts per million of chromate ion, CrO_4. The chromate ions may be accompanied by, typically, 900 ppm of sulfate ions, 300 ppm of chloride ions and 15 to 20 ppm of phosphate ions.

It involves a two-zone anion exchange process, the first of which contains a high concentration of the absorbed ionic material, and a second of which contains a lower concentration of the same absorbed ionic material.

Relatively little information is available in the technical literature on ion exchange treatment of trivalent chromium. This is in part because most industrial waste chromium exists in the hexavalent form, and if recovery is preferred direct ion exchange of the hexavalent chromium, chromate, or dichromate ion is employed. Ion exchange is dependent upon the electron charge on an ion.

Trivalent chromium has a plus 3 charge, and hexavalent chromium in its usual form of chromate or dichromate has a minus 2 charge. Thus different ion exchange resins must be employed to recover each form of chromium. An anion type of resin is required to capture negatively charged forms such as the hexavalent chromium chromate and dichromate ions. A cation type of resin is employed for positively charged (e.g., Cr^{+3}) forms. Ross (45) has pointed out two reasons to recover trivalent chromium by ion exchange. The first is the passage of concentrated chromic acid baths through a cation resin to remove metallic contaminants such as iron, aluminum and trivalent chromium from the chromic acid. This process would purify the chromic acid solution, which would then be reused.

The second reason discussed by Ross is for the complete purification of a plating rinse water, in which this rather dilute waste would be passed through both an anionic and a cationic resin to remove chromium forms of both positive and negative charge. The treated rinse water could then be recycled. No capital cost figures are available for ion exchange treatment of trivalent chromium.

A process developed by J.F. Zievers et al (49) for treating mixed rinse waters used to rinse workpieces removed from chromium and other metal treating baths provides for reducing hexavalent chromium ions carried by the workpieces prior to rinsing and mixing the rinse waters from a number of different rinses prior to passage thereof through cation and anion exchangers. The method and system also provide for the immediate removal of cyanide from the anion regenerant while it has a high pH value and then combining the cation and anion regenerants prior to neutralization.

A process developed by A. Yagishita (50) for reclaiming plating wastes containing chromic acid and the like is one in which the rinse liquid, from one of a plurality of wash tubs into which plated articles are dipped successively to rinse off the plating solution, is circulated through an ion exchange column, where it passes through a resin layer that removes chromic acid and other impurities before the liquid is returned to the wash tub. Periodically, or when the resin becomes saturated with chromic acid, it is regenerated by pumping a solution of caustic soda through the exchange column, thus producing a sodium chromate solution to which barium hydroxide is added to produce barium chromate, which is filtered out of the caustic soda solution before the latter is again used for regeneration purposes.

A process developed by R.B. Smith (51) is one in which chromium values are recovered from wash water following a chemical treatment step in electrotinplating lines in a process by treating in a cation exchange zone, treating in an anion exchange zone, regenerating the anion exchange zone, treating effluent from the regeneration in a cation exchange zone and passing effluent from the cation exchange zone to a storage zone, passing effluent from the regeneration to the storage zone without intermediate cation exchange treatment.

Evaporative Recovery

The evaporative recovery process consists of evaporating metal plating rinse water to drive off the water as vapor and thus concentrate the chromic acid for recovery and reuse. Evaporative recovery can be used on almost all process rinse water systems, with the exception of those which deteriorate with use. Culotta and Swanton (52) reported in a case study of plating waste recovery that rinse water containing only a few milligrams per liter of chromic acid could be concentrated to above 900 mg/l.

In a process developed by A. Yagishita (53) for reclaiming plating wastes containing chromic acid and the like, a plurality of wash tubs are provided into which plated articles are dipped successively to rinse off the plating solution. The tub water from the tub, into which the articles are first dipped, is sucked into a tower heated by steam to concentrate the tub water to plating strength; and this water is returned to the plating tank.

Water is siphoned from the other tubs successively back to the first tub. The vapor in the tower is used in a condenser to heat water flowing to the tub most remote from the first tub, to replenish the system. The tower has an inner container; and the tower proper and the space between this container and the tower proper is held under vacuum.

In another process developed by A. Yagishita (54) for reclaiming plating wastes containing chromic acid and the like, the rinse liquid from the first of a plurality of wash tubs into which plated articles are dipped successively to rinse off the plating solution, is sucked into a tower where it is heated by steam partially to vaporize and concentrate it to plating strength. The vapor generated in the tower is passed through a check valve to an ejector, where it is introduced transversely into a jet of cooling water to be condensed and conveyed with the stream of cooling water to a reservoir. The cooling-water jet develops a vacuum in the tower; and a trap is interposed between the tower and ejector to trap any cooling water which a faulty check valve might otherwise allow to back up into the tower.

Chemical Precipitation

A process developed by R. Richards (55) involves chemically removing chromium ions from industrial waste solutions by direct precipitation with barium carbonate in such solutions acidified with nitric or hydrochloric acid or their salts.

A process developed by G.J. Nieuwenhuis (56) is one in which hexavalent chromium contained in aqueous metal treating wastes is substantially completely removed from the aqueous solution by metering the acid wastes into rinse water, metering stoichiometric amounts of a lead compound such as lead nitrate into the rinse water containing the hexavalent chromium to form an insoluble lead chromate precipitate and an aqueous precipitate residue which is subsequently treated with a caustic cleaning solution to neutralize the acids and precipitate the remaining metals and metal compounds in the aqueous treating solution. The residual aqueous solution is suitable for reuse.

Trivalent chromium may be removed by precipitation as the hydroxide upon addition of lime or caustic soda, or it may be concentrated by ion exchange and recovered. The high cost of chromium as an industrial material dictates that ion exchange and recovery be seriously considered as a waste treatment process. However, the literature indicates that at this time precipitation of (trivalent) chromium hydroxide, and disposal of the resulting sludge, remains the most common practice.

Trivalent chromium can be removed as insoluble chromic hydroxide, $Cr(OH)_3$, by precipitation with caustic soda or lime (31). The chromic hydroxide sludge formed by the process frequently is disposed by landfill. The precipitation process is most effective at pH 8.5 to 9.5 due to the low solubility of chromic hydroxide in that range.

Because a typical metal bearing waste will include other metal ions, as well as chromium to be removed by precipitation, and because the optimum pH of precipitation varies for different metal hydroxides, an average waste stream pH of about 8 seems to achieve the best overall results on a waste containing several metal ions (57). Thus much of chromium treatment reported in the literature occurred at pH 7 to 8. Schink (35) has reported chromium precipitation at an even higher pH of 9, in an electroplating waste. No values of raw or effluent trivalent chromium concentrations were given. An effluent containing less than 1.0 mg/l hexavalent chromium and 2.0 mg/l total heavy metals was achieved via the reduction-neutralization-precipitation treatment sequence (35).

Soda ash has been used to precipitate trivalent chromium (31)(33). However, the most commonly employed chemical for precipitation appears to be lime, $Ca(OH)_2$. Avrutskii (32) has reported reducing the chromium content of an electroplating waste from 140 to 1.0 mg/l. This was achieved by reduction, neutralization and precipitation with lime at pH 7 to 8.

In a detailed discussion of waste treatment at an IBM plant, Hansen and Zabben (36) report that hexavalent chromium (at 1,300 mg/l) in one waste stream was converted to trivalent chromium and precipitated with lime to achieve an effluent containing zero hexavalent and 0.06 mg/l total chromium. As total chromium must equal hexavalent plus trivalent forms, the effluent thus contained 0.06 mg/l trivalent chromium.

Anderson et al (11) report the use of reduction and lime precipitation to treat chromium bearing waste from a General Electric appliance plant. Sedimentation of the precipitate was improved by addition of an anionic polyelectrolyte. Plant effluent was reported to contain no hexavalent chromium and an average of 0.75 mg/l total chromium.

A process developed by S.K. Lee (58) is one in which contaminating anions such as chromates are removed from aqueous media by reacting them indirectly with lead from an anode to produce an insoluble lead compound which is subsequently separated from the aqueous medium. The lead is obtained from an anode containing lead or an insoluble lead compound which reacts with an activating ion, such as acetate or chloride, to produce soluble lead ions, which subsequently react with the contaminant anion, e.g., chromate from a purge stream of cooling tower water or from rinses of chrome-plated articles.

A process developed by G.J. Nieuwenhuis (59) is one in which chromium anions in the parts per million range are removed by contacting the water-laden chromium solution with a water-insoluble lead compound, such as lead oxide, lead carbonate or lead hydroxide, adsorbed in the pores of an essentially inert, porous, particulate matrix such as vermiculite. The lead compound reacts with the chromium in the water to form insoluble lead chromate. The effluent may be passed through an ion exchange column containing an acidic cation exchanger in the hydrogen form for removing any lead ion solubilized during passage through the bed of adsorbed lead compound.

A process developed by J.M. Lalancette and B. Coupal (60) is one in which chromium-polluted solutions are treated by a process which avoids the necessity of reducing chromium (VI) to chromium (III). The process involves forming a chromium sulfide complex by adding a water-soluble ferric salt to the chromium polluted solution and insolubilizing the complex formed by treatment with a compound which when dissolved in water forms sulfide ions and separating the insoluble precipitate by filtration through a bed of peat moss. The peat moss containing the insoluble chromium iron sulfide complex is oxidized under heat to recover the chromium as chromium oxide. If other metals such as cadmium, copper, lead, mercury, iron and zinc are present in the polluted solution they will also be recovered in the form of their oxides except the mercury which will be recovered as such by pyrolysis of the flue gases.

A process developed by W. Izdebski (61) involves treating plating effluents, including at least chromium plating effluent and, in plants having it, zinc plating effluent as well as effluents from plating copper, cadmium or nickel, removing from these effluents the heavy metal ions. Hexavalent chromium ions are removed directly, without first reducing them to trivalent forms, by adding barium acetate. When other plating effluents are treated, the formed barium compounds are used as a coagulant for their hydroxides, which may be formed by mixing any of these other effluents with the formed barium compound and adjusting the pH to the value enabling precipitation of the hydroxides. Sulfuric acid may be used for this pH adjustment, thereby also precipitating out any remaining barium ions in solution.

Solvent Extraction

A process developed by R.R. Dougherty (62) is one in which spent or used aqueous chromic acid etchants are treated to separate chromic acid from by-products of the etching process which are dissolved or suspended in the spent etchants, thereby permitting the by-products of the etching process to be concentrated or isolated for waste disposal and/or recovery of the metal values contained therein.

In a typical process, chromic acid and water are separated from the by-products by acetone extraction. The extract is then stripped of the acetone to produce an aqueous solution of chromic acid which can be reused in the etching process. The raffinate is rich in the by-products of the etching process.

A process developed by N. Texier et al (63) involves treating residual solutions containing hexavalent chromium compounds in order to render them nonpolluting and at the same time providing sodium chromate by the extraction of hexavalent chromium by means of an insoluble organic solvent (such as an alkyl phosphate) and reextraction.

The aqueous solution used for the reextraction of hexavalent chromium from the organic solvent is a sodium chromate solution. A concentrated sodium bichromate solution is obtained. Application is to the treatment of residual waters from processes for the extraction of chromium from ores thereof, or from electroplating installations.

Reverse Osmosis

A process developed by A. Geinopolos et al (64) is one in which process water containing hexavalent chromium is pretreated to bring it to a state of dichromate dominance and the water is then subjected to reverse osmosis using a loose membrane. The membrane allows passage of most of the dichromate ions with the product water while concurrently effecting a significant reduction in the hardness and dissolved solids levels. The softened product water containing dichromate ions is recyclable with the recirculated process water and the wasted fraction contains substantially lower amounts of potentially harmful chromium.

Biological Purification

A process developed by V.I. Romanenko et al (65) involves biological purification of industrial effluents from chromates and bichromates, wherein industrial effluents are mixed with domestic sewage and then the chromates and bichromates contained therein are biochemically reduced under anaerobic conditions.

SOLID WASTE DISPOSAL

Soda ash or other alkaline chemicals are added to trivalent chrome or other heavy metal solutions to adjust the pH to about 9.5, thereby precipitating the insoluble heavy metal hydroxide gels. Aluminum sulfate (alum) or other suitable flocculating agents are also often used to aid in precipitation, settling of the gels and clarification of the effluent. Slaked lime must also be added to precipitate insoluble calcium fluoride, a reaction which requires at least 24 hours. The effluent is neutralized before sewering and the resultant sludge is landfilled (66).

The procedure can be carried out in either batch or continuous processes. Complete equipment systems are available for performing the precipitation on an automated or semiautomated basis. The biggest drawback of this procedure is the removal and disposal of the highly hydrated metal hydroxide sludges. It is not unusual for these sludges to contain upwards of 75 to 80% water by volume.

Precipitation and settling is normally a slow procedure and with a high volume of effluent it is normally necessary to have settling ponds or lagoons in which to allow the slow coagulation process to function. The clear effluent can be removed and the residual sludge dried. No feasible economic process has yet been developed for the reclamation of valuable metals or other credits from this type of sludge since it often contains other organic and inorganic materials which hinder effective purification and recovery.

Three ultimate disposal options for these sludges are: [1] landfill, [2] incineration, and [3] ocean disposal. Landfill is believed to be, by far, the most prevalent final disposal procedure but the problem of potential contamination of surrounding land or water tables due to the leaching of poisonous materials from the sludge must be considered. It is therefore recommended that California Class 1 type landfill be utilized for these wastes. Incineration reduces the sludge to an ash residue which can be more easily handled for ultimate disposal but at the same time could also lead to the formation of poisonous soluble metal oxides. Incineration of course requires additional equipment and a combustible fuel source as well as equipment for air pollution control.

Ocean disposal is currently being practiced to a considerable degree by municipal waste treatment plants located near the coast where the primary and secondary sludges can be either pumped or barged out to submarine disposal areas. This disposal approach has been the subject of much recent debate and it is likely that it will not long remain one of the significant sludge disposal options.

Chromite-contaminated land-destined hazardous wastes are generated by the manufacture of these products in three locations (66). One plant disposes of the chrome ore gangue off-site to a contractor for land and road construction purposes. One uses on-site land dumping and another uses approved private landfill.

The plants and plant complexes manufacturing chrome pigments, iron blues and cadmium colors generate land-disposed hazardous wastes containing lead, chromium, zinc, cyanide and cadmium hazardous constituents (66). Of the eight plants land-disposing such wastes, one disposed on-site by ponding and one by landfill after treatment, which consists principally of neutralization and precipitation. Six other plants use off-site disposal, one to municipal landfill, four to secured private landfill, and one to a land dump. Two remaining chrome pigments plants do not land dispose, one claiming to recover all wastes and the other discharging.

The chromates and dichromates are considered as waste stream constituents for centralized treatment for the following reasons: [1] the high degree of toxicity of the compounds; [2] hexavalent chromium oxides cannot normally be reduced to the nontoxic elemental chromium form; [3] the chromate and dichromate wastes appear in sizable quantities; [4] a substantial portion of the chromate and dichromate wastes are contributed by small plating shops that cannot afford waste treatment equipment; and [5] the disposal of waste chromium hydroxide sludges can be adequately handled at a centralized facility and chromium and heavy metal values present can be more readily recovered if such a need arises.

For the disposal of concentrated hexavalent chromate wastes at national disposal sites, it is recommended that a reductive-precipitative facility be constructed. This scheme is by far the most applicable approach for treating chromate concentrates. The subprocess of electro-reduction might have merit to replace the chemical reduction step of the operation and save on necessary reducing chemicals. The technique is efficient and adequate for large-scale removal of chromates. The effluent can have 1 to 2 ppm levels of Cr^{+3} or lower.

The permanent containerized storage or proper landfilling of chromium waste sludges or their incinerated residues are recommended for ultimate disposition. The landfill sites used must be located over nonwater-bearing sediments or have only unusable groundwater underlying them and must also be completely protected from flooding, surface runoff or drainage, so that all waste materials and internal drainage are restricted to the site. The economics of transportation to the landfill or storage site must be considered in the decision to transport the hydrated sludge or to incinerate it prior to shipment. Such studies will be necessary in site selection and process recommendation for sites in various parts of the nation. The very same facility is usable for many of the toxic metal compounds which would also be likely to need disposal from time to time including cadmium, mercury, lead, copper and arsenic.

A process developed by S. Adachi (67) is one in which water-soluble chromates contained in water-leached residues from roasted products obtained by the alkali oxidizing roasting of chrome ore can be insolubilized and fixed in the residue by mixing lumpy and/or powdery water-leached residue with organic materials capable of being converted into active carbon, roasting the mixture in a gaseous atmosphere of low oxygen concentration at a gas temperature of 400° to 1000°C and a material temperature of 400° to 700°C, and thereafter quenching the roasted product. The organic materials include pulp waste liquor, sawdusts, chaffs, pulp sludge and lignite.

A process developed by C.O. Weiss (68) permits recovering the chromium content from waste sludge produced from the separation of the sludge as precipitated solids from the waste effluent streams of metal finishing processes, the chromium content of the sludge being in the form of barium chromate.

The above is achieved by forming an aqueous slurry of the sludge, adding sulfuric acid thereto, agitating the resulting admixture for a period of time at ambient temperature, separating the insolubles therefrom, adding to the filtrate from the separation an amount of a carbonate selected from the group consisting of Ca^{+2}, Sr^{+2}, Ba^{+2} and Pb^{+2} to precipitate out a substantial portion of the sulfate ion from the reacting environment, adding to the filtrate of the second separation step strontium carbonate to precipitate out the remaining sulfate ion, and separating the insolubles from the reacting environment to obtain a solution containing all the CrO_3 content from the initial extraction.

A process developed by S. Yamada (74) is a process for incineration of a sludge containing chromium substance comprising a sequential combination of a step of incinerating the sludge containing chromium substance into ashes, a step of reducing hexavalent chromium existing within the product of the first step into trivalent chromium by means of reducing gas, and a step of cooling the product of the second step before discharging the same to the outside.

ECONOMIC IMPACT OF CONTROLS

On-site experience with wet scrubbers for the removal of chromic acid mists yields an average control system cost of about $0.50 per cfm of gas treated, although this varies widely with design and construction materials (69). Similar estimates place the cost of packed towers for acid mist removal from plating operations at $1.00 per cfm of waste gas handled (69). As noted earlier, when wet type control systems are employed, wastewater must be cleansed of chromium, typically in its hexavalent form, prior to either recycling or permanent disposal. Table 23 gives the general chemicals cost for removal of hexavalent chromium from wastewater by several processes.

TABLE 23: TREATMENT OF HEXAVALENT CHROMIUM

	Amounts per Pound of Chromium as Cr					
Method	$FeSO_4 \cdot 7H_2O$ $24.00/Ton	H_2SO_4 $32.30/Ton	$Ca(OH)_2$ $19.25/Ton	$Na_2S_2O_5$ $132/Ton	SO_2 $0.05/lb	Total Cost
Ferrous	16.03 lb	6.01 lb	10.52 lb	-	-	-
sulfate	$0.19	$0.09	$0.09	-	-	$0.37
Sodium	-	1.52 lb	2.64 lb	2.88 lb	-	-
metabisulfate	-	$0.02	$0.02	$0.19	-	$0.23
SO_2	-	-	2.64 lb	-	1.85 lb	-
	-	-	$0.02	-	$0.10	$0.12

Source: Reference (20)

Yuronis (46) has provided cost data on the treatment of chromium wastes by reduction. For complete treatment, including pH adjustment, reduction and neutralization, and removal of the trivalent chromium, Yuronis indicates that treatment operating costs range from $55 to $100 per day, dependent on the cost of municipal water to an industry. The basis for his costs is a waste flow of 100 gal/min at 120 mg/l of chromium, and a 16-hour day. Based upon his values, and the cost of water in Chicago (25¢/1,000 gal, 1968), daily operating costs would be about $75 for the waste flow and concentration given. This cost is less than $1.00/1,000 gal of waste treated.

Zievers et al (57) have also published treatment costs for chromium wastes, which include capital and operating costs for the reduction of hexavalent chromium and subsequent pH neutralization to an acceptable level.

Rothstein (71) reported successful ion exchange treatment of a metal finishing waste from Fairchild Camera and Instrument Corp. (Syosset, L.I.) to meet the chromate effluent standard of 0.05 mg/l. This chromate level is equivalent to a hexavalent chromium concentration of 0.023 mg/l. The chromate and chromic acid wastes treated at this plant by ion exchange were segregated from other metal finishing wastes.

Recovery of chromic acid was practiced, and resulted in a five-year savings of $8,342 above the cost which would have been required for chromium reduction and precipitation. Rothstein reports that standard reduction and precipitation of this waste would cost $1.25/lb of chromate removed. This is equivalent to a treatment cost of $1.00/1,000 gallons at a hexavalent chromium concentration of approximately 50 mg/l.

Lacy et al (38) report that the cost of ion exchange treatment depends upon the industrial source of the waste. Waste treatment operating costs are given by these authors as 16 to 24¢/1,000 gallons of waste stream treated. Ross (45) has reported ion exchange costs for a waste stream of 73 gal/min with a chromic acid concentration of approximately 5,000 mg/l (equivalent to 2,300 mg/l of hexavalent chromium).

Operating costs for the waste treatment process totaled $72.24 per day, with a credit of $57.30 per day resulting from chromic acid recovery and reuse of purified water. Net operating costs were, therefore, $14.94 per day. By contrast, reduction and precipitation would have cost $33.19 per day for that waste.

Besselievre (31) has reported ion exchange treatment costs of $17 per day for a chromium waste, versus an estimated reduction and precipitation cost of $63 per day. Keating and Duff in 1955 reported a net profit of $10 per day on an ion exchange chromium recovery system, as well as a recycle of 86,500 gallons of treated wastewater to the plant (72).

Yuronis has reported ion exchange costs of 15¢/lb of chromate removed (46). He estimated capital costs for a plant treating 100 gal/min and a chromium content of 50 to 100 mg/l as approximately $40,000. Daily operating costs would be $65,000, or 68¢ per 1,000 gallons.

Yuronis (46) reported evaporative recovery operating costs of $2.50 to $10.00 per 1,000 gallons. Capital equipment costs for a continuous flow evaporative recovery system were estimated at $60,000.

The treatment costs associated with trivalent chromium removal by precipitation are due to the lime precipitation process and the expense of sludge disposal. In the Midwest, lime costs 2.25¢/lb in 1-ton lots (1968 price). This compares to a cost of 1.75¢/lb on both the east and west coasts (57). Yuronis (46) has given a somewhat higher lime cost of 3.6¢/lb in his discussion of chromate treatment. He estimates a lime cost of approximately $2.50/1,000 gal for chromium precipitation from a waste containing 2,000 mg/l of Cr^{+3}.

The cost for treatment of a stronger or weaker waste would vary directly. For example, a waste containing 1,000 mg/l Cr^{+3} would have a lime cost of $1.25 per 1,000 gallons. Cost of complete treatment from the hexavalent chromium form would include reduction cost, precipitation cost and sludge handling and disposal costs. Reduction costs are approximately $1.00/1,000 gal.

Anderson et al (11) have given details of the costs associated with treatment of a metal bearing waste which contains large quantities of chromium and lesser concentrations of nickel, copper, iron and aluminum. The treatment plant had an initial installation cost of $1,090/1,000 gal capacity per day. In 1967, the plant treated 205,400 gallons of waste per day, at a total cost of approximately 80¢/1,000 gallons.

Costs include chemicals, manpower, utilities, depreciation and taxes. Cost of chromium treatment represent about 20% of total treatment costs. This results in an estimated chromium treatment installation cost of $218/1,000 gallons treated per day and 16¢/1,000 gallons for operating costs. These costs include sludge disposal.

Zievers et al (57) have also discussed sludge disposal and associated costs. Major costs associated with sludge disposal result from vacuum filtration to concentrate the sludge solids. These authors point out that there is a growing emphasis on filtration of metal hydroxide sludges, due to more stringent enforcement of effluent suspended solids standards.

Lacy et al (38) have reported operating costs for ion exchange treatment of 16 to 24¢ per 1,000 gallons. Ross (45) has reported daily operating costs for ion exchange treatment of trivalent chromium as approximately $28/day for a waste flow of 73 gal/min. Assuming a 16-hour day, operating costs are estimated at 40¢/1,000 gallons. If the assumption is erroneous and the process day consists of only 8 hours, operating costs double to 80¢ per 1,000 gallons.

REFERENCES

(1) Booz-Allen Applied Research, Inc., *A Study of Hazardous Waste Materials, Hazardous Effects and Disposal Methods,* Report PB 221 466, Springfield, Va., Nat. Tech. Information Service (July 1973).

(2) Sullivan, R.J., *Air Pollution Aspects of Chromium and Its Compounds,* Report PB 188 075, Springfield, Va., Nat. Tech. Information Service (Sept. 1969).

(3) Schroeder, H.A., Balassa, J.J. and Tipton, I.H., *J. Chron. Dis.* 15, 941 (1961).

(4) Hill, W.H. and Warden, F.X., *Am. Ind. Hyg. Assoc. J.* 23, 186 (1962).

(5) Hueper, W.C. et al, *Am. Ind. Hyg. Assoc. J.* 21, 530 (1960).

(6) Shelley, W.B., *J. Am. Med. Assoc.* 189, 170 (1964).

(7) Luckey, T.D., Venugopal, B. and Hutcheson, D., *Heavy Metal Toxicity and Hormonology,* New York, Academic Press (1975).

(8) GCA Corp., *National Emissions Inventory of Sources and Emissions of Chromium,* Report PB 230 034, Springfield, Va., Nat. Tech. Information Service (May 1973).

(9) Goldberg, A.J., *A Survey of Emissions and Controls for Hazardous and Other Pollutants,* Wash., D.C., Environmental Protection Agency (Feb. 1973).

(10) Patterson, J.W. and Minear, R.A., *Wastewater Treatment Technology,* 2nd Ed., Chicago, Illinois Institute for Environmental Quality (Feb. 1973).

(11) Anderson, J.S. and Iobst, E.H., Jr., "Case History of Wastewater Treatment in a General Electric Appliance Plant," *Jour. Wat. Poll. Control Fed.,* 10, pp. 1786-1795 (1968).

(12) Germain, J.E., Vath, C.A. and Griffin, C.F., "Solving Complex Waste Disposal Problems in the Metal Finishing Industry," presented at Georgia Wat. Poll. Control Assoc. Meeting, September 1968.

(13) Ottinger, R.S., Blumenthal, J.L., Dal Porto, D.F., Gruber, G.I., Santy, M.J. and Shih, C.C., *Recommended Methods of Reduction, Neutralization, Recovery or Disposal of Hazardous Waste,* Report PB 224 580, Springfield, Va., Nat. Tech. Information Service (Aug. 1973).

(14) Nat. Inst. for Occupational Safety and Health, *Criteria for a Recommended Standard... Occupational Exposure to Chromic Acid,* Wash., D.C. (1973).

(15) Ege, J.F., Jr. and L. Silverman, "Stable Colorimetric Reagent for Chromium," *Anal. Chem.* 19:693 (1947).

(16) Silverman, L., and J.F. Ege, Jr., "A Rapid Method for Determination of Chromic Acid Mist in Air," *J. Ind. Hyg. Toxicol.* 29:136 (1947).

(17) Silverman, L. and J.F. Ege, Jr., "Chromium Compounds in Gaseous Atmospheres," U.S. Patent 2,483,108 (1949); *CA–Bull. Cancer Progr.* 44:490 (1950).

(18) "MSA Chromic Acid Mist Detector," *Mine Safety Appliances Co. Bull.* 0811-8 (1959).

(19) "MSA Samplair," in *Catalog of Industrial Safety Equipment,* Mine Safety Appliances Co. Cat. 7-B, Sect. 3, p. 38, Pittsburgh, Pa. (1957).

(20) Thompson, R.J., Morgan, G.B. and Purdue, L.J., "Analyses of Selected Elements in Atmospheric Particulate Matter by Atomic Absorption," Preprint. Presented at the Instrument Society of America Symposium, New Orleans, La. (May 5-7, 1969).

(21) "Air Pollution Measurements of the National Air Sampling Network—Analyses of Suspended Particulates," 1957-1961, *U.S. Public Health Serv. Publ. 978* (1962).

(22) *Air Quality Data from the National Air Sampling Networks and Contributing State and Local Networks,* 1966 ed., U.S. Dept. of Health, Education and Welfare, National Air Pollution Control Administration Publication No. APTD 68-9, U.S. Govt. Printing Office, Washington, D.C. (1968).

(23) West, P.W., "Chemical Analysis of Inorganic Pollutants," Chapter 19 in *Air Pollution,* Vol. II, 2nd ed., A.C. Stern, Ed. (New York: Academic Press, 1968);

(24) Marcy, V.M. and Halstead, S.L.; U.S. Patent 3,216,950; November 9, 1965; assigned to Calgon Corp.

(25) Noll, C.A., Feddern, E.C. and Stefanelli, L.J.; U.S. Patent 3,656,908; April 18, 1972; assigned to Betz Laboratories, Inc.

(26) Lawrence Berkeley Laboratory, *Instrumentation for Environmental Monitoring—Water,* Publ. No. LBL-1, Vol. 2, Berkeley, Univ. of Calif. (Feb. 1973).

(27) Saltzman, R.S.; U.S. Patent 3,941,563; March 2, 1976; assigned to E.I. du Pont de Nemours and Co., Inc.

(28) American Conference of Governmental Industrial Hygienists, *Threshold Limit Values for Chemical Substances in Workroom Air,* Cincinnati, Ohio (1974).

(29) McFarland, A.R., ed., *Control Technology for Toxic and Hazardous Air Pollutants,* Chicago, Illinois Institute for Environmental Quality (Dec. 1974).

(30) MacDougall, H., *Waste Disposal at a Steel Plant: Treatment of Sheet and Tin Mill Wastes,* ASCE Proc. Separate No. 493 (1954).

(31) Besselievre, E.B., *The Treatment of Industrial Wastes,* McGraw-Hill Book Co., New York (1969).

(32) Avrutskii, P.I., "Control of Chromium (VI) Concentration in Waste Waters," *Chem. Abst.* 70, 206-207 (1969).

(33) Avrutskii, P.I., "Liquid SO_2 in New Reduction Role," *Indust. Eng.*, 53, 29A-30A (1961).
(34) Hulse, B.T., Selim, R.P. and Summers, G.E., "Control of Metal Finishing Wastes Using ORP," *Jour. Wat. Poll. Control Fed.*, 32, 975-981 (1960).
(35) Schink, C.A., "Plating Wastes: A Simplified Approach to Treatment," *Plating*, 55, 1302-1305 (1968).
(36) Hansen, N.H. and Zabben, W., "Design and Operation Problems of a Continuous Automatic Plating Waste Treatment Plant at the Data Processing Division, IBM, Rochester, Minnesota," *Proc. 14th Purdue Industrial Waste Conf.*, pp. 227-249 (1959).
(37) Fisco, R., "Plating and Industrial Waste Treatment at the Expanded Plant of Fisher Body, Elyria, Ohio," *Proc. 25th Purdue Industrial Waste Conf.* (1970).
(38) Lacy, W.J. and Cywin, A., "The Federal Water Pollution Control Administration's Research and Development Program: Industrial Pollution Control," *Plating*, 55, 1299-1301 (1968).
(39) Selm, R.P. and Hulse, B.T.; U.S. Patent 3,027, 321; March 27, 1962; assigned to Wilson & Co.
(40) Lancy, L.E.; U.S. Patent 3,294,680; December 27, 1966; assigned to Lancy Laboratories, Inc.
(41) Gloster, A. and Bocckino, H.G.; U.S. Patent 3,810,542; May 14, 1974; assigned to Texas Gulf Inc.
(42) Henley, T.D. and Reeves, R.F.; U.S. Patent 3,819,051; June 25, 1974; assigned to Amchem Products, Inc.
(43) Fournier, L.B. and Meyer, R.A.; U.S. Patent 3,896,209; July 22, 1975; assigned to E.I. du Pont de Nemours and Co.
(44) Stewart, R.; U.S. Patent 3,901,805; August 26, 1975; assigned to Dow Badische Co.
(45) Ross, R.D., (ed.), *Industrial Waste Disposal*, New York, Reinhold Book Corp. (1968).
(46) Yuronis, D., "Metal Finishing Waste Treatment: Comparative Economics," *Plating*, 55, 1071-1074 (1968).
(47) Saraceno, A.J., Walters, R.H., Jones, D.B. and Wiehle, W.E.; U.S. Patent 3,664,950; May 23, 1972; assigned to U.S. Atomic Energy Commission.
(48) Sloan, L., Nitti, N.J. and Pratt, J.B.; U.S. Patent 3,306,859; February 28, 1967; assigned to Crane Co.
(49) Zievers, J.F. and Novotny, C.J.; U.S. Patent 3,681,210; August 1, 1972; assigned to Industrial Filter & Pump Manufacturing Co.
(50) Yagishita, A.; U.S. Patent 3,761,381; September 25, 1973.
(51) Smith, R.B.; U.S. Patent 3,847,757; November 12, 1974; assigned to National Steel Corp.
(52) Culotta, J.M. and Swanton, W.F., "Case Histories of Plating Waste Recovery Systems," presented at 56th Annual Conf. Amer. Electroplaters Soc., Detroit (1969).
(53) Yagishita, A.; U.S. Patent 3,542,651; November 24, 1970.
(54) Yagishita, A.; U.S. Patent 3,616,437; October 26, 1971.
(55) Richards, R.; U.S. Patent 3,371,034; February 27, 1968.
(56) Nieuwenhuis, G.J.; U.S. Patent 3,493,328; February 3, 1970.
(57) Zievers, J.F., Crain, R.W. and Barclay, F.G., "Waste Treatment in Metal Finishing: U.S. and European Practices," *Plating*, 55, 1171-1179 (1968).
(58) Lee, S.K.; U.S. Patent 3,766,037; October 16, 1973; assigned to Andco, Inc.
(59) Nieuwenhuis, G.J.; U.S. Patent 3,791,520; February 12, 1974.
(60) Lalancette, J.M. and Coupal, B.; U.S. Patent 3,835,042; assigned to Universite de Sherbrooke.
(61) Izdebski, W.; U.S. Patent 3,869,386; March 4, 1975; assigned to Schlage Lock Co.
(62) Dougherty, R.R.; U.S. Patent 3,531,262; September 29, 1970; assigned to Control Data Corp.
(63) Texier, N., Cuer, J.P. and Gabriel, M.; U.S. Patent 3,856,917; December 24, 1974; assigned to S.A. Products Chemique Ugine Kuhlmann.
(64) Geinopolos, A., Gupta, M.K. and Katz, W.J.; U.S. Patent 3,625,885; December 7, 1971; assigned to Rex Chainbelt, Inc.
(65) Romanenko, V.I., Kuznetsov, S.I. and Korenkov, V.N.; U.S. Patent 3,941,691; March 2, 1976.

(66) Powers, P.W., *How to Dispose of Toxic Substances and Industrial Wastes,* Park Ridge, N.J., Noyes Data Corp. (1976).

(67) Adachi, S.; U.S. Patent 3,803,032; April 9, 1974; assigned to Nippon Denko Co. Ltd.

(68) Weiss, C.O.; U.S. Patent 3,552,917; January 5, 1971; assigned to M & T Chemicals, Inc.

(69) Paulus, H. and Platt, C., "Air Pollution Controls in the Small Electroplating Shop," *Plating,* 13-14, (Jan. 1972).

(70) Subcommittee on Environmental Pollution Control, American Electroplaters' Society, "A Report on the Control of Chromium in Plating Shop Effluents," *Plating,* 890 (Sept. 1970).

(71) Rothstein, S., "Five Years of Ion Exchange," *Plating,* 45, 835-841 (1958).

(72) Keating, R.J. and Duff, J.H., "Plating Waste Solutions: Recovery or Disposal," *Proc. 10th Purdue Industrial Waste Conf.,* (1955).

(73) National Institute for Occupational Safety and Health, *Criteria for a Recommended Standard. . . Occupational Exposure to Chromium (VI),* Wash., D.C. (1975).

(74) Yamada, S.; U.S. Patent 3,968,756; July 13, 1976; assigned to Chugairo K.K. Ltd.

COPPER

Copper (Cu, atomic number 29) is a reddish-brown ductile malleable metal whose electrical and thermal conductivities are higher than those of any known metal except silver. The metal occurs in quantities averaging 55 ppm in the earth's crust. Copper is found free as native copper (cuprite, 88.8% Cu), in the oxide ores azurite [$2CuCO_3 \cdot Cu(OH)_2$], chryrocolla ($CuSiO_3 \cdot 2H_2O$) and malachite [$CuCO_3 \cdot Cu(OH)_2$], and in the sulfide ores bornite (Cu_5FeS_4), chalcocite (Cu_2S) and chalcopyrite ($CuFeS_2$). The latter is economically the most important copper ore. The metal is widely distributed in the natural environment. Soils contain up to 20 ppm copper, and background ambient air concentrations range from 0.01 to 0.25 $\mu g/m^3$.

Copper is extracted from its ores by crushing, roasting and smelting to produce a matter in which nearly ⅓ of the copper is contained as sulfide. From the molten matte sulfur is normally removed by treatment with air in a converter; iron impurities are removed as an oxide slag by the addition of a siliceous flux, the by-product being brittle blister copper. Further refining takes place in a reverberatory furnace or by an electrolytic process.

Its unique physical properties has led to wide industrial use of copper. The metal is found in numerous alloys which have a variety of general engineering applications. It is used in the electrical industry in fine resistance wires, commutator bars and high conductivity tubes. Copper is used in the construction industry in pipes and roof sheeting. It is used in bronze paints, in insecticides, and has many other applications.

The domestic copper industry faces localized land-use conflicts because a larger percentage of copper will continue to be produced by surface mining. In addition, large quantities of solid waste material are generated in mining, concentrating, and smelting copper ore. Careful planning by copper producers has largely solved objections to surface disturbance and waste storage problems.

Discharge of obnoxious gases and fumes from smelters is a problem that will require increasing attention. Employment of high stacks to disperse the pollutants will undoubtedly need to be supplemented by adoption of practices to remove the objectionable substances. An alternative solution is the more revolutionary change to a hydrometallurgical extractive process replacing the conventional pyrometallurgical method.

TOXICITY

The toxicity of copper salts has been reviewed by T.D. Luckey et al (1) and some of the data from that source are quoted in the following table.

TABLE 24: TOXICITY OF COPPER SALTS

Compound	Animal	Route	Toxic Dose	Dosage Expressed as mg of: Compound/kg	Metal/kg
Copper carbonate	Rat	Oral	LD_{50}	159	81.7
Copper chloride	Rat	Oral	LD_{50}	140	66.1
	Guinea pig	SC	LD_{50}	100	47.2
Copper nitrate	Rat	Oral	LD_{50}	940	318
Copper sulfate	Mouse	IV	LD	50	12.7
	Rat	Oral	LD_{50}	300	76.3
	Guinea pig	IV	LD	2	0.51
	Rabbit	IV	LD	4.5	1.15

Source: Reference (1)

Forage with too much copper can cause fatal hemolytic anemia and hepatic necrosis in cattle and sheep. Copper toxicity is a fundamental cause of Wilson's disease (2). It has been recognized that this disorder is inherited as an autosomal recessive trait and is the only example of significant copper toxicity in man. This has made it clear that man has had to develop a mechanism, transmitted from one generation to the next which prevents copper poisoning in all the rest of us.

The geographic location has influence on the biological priorities related to copper intake. In Florida cattle with 60 ppm Cu on dry weight basis were often showing severe diarrhea, achromotrichia, bone change, and anemia and severe reproduction disturbances. Yet in the Netherlands and in Oregon values of 25 ppm Cu are found in the liver with the animals apparently in normal health (3).

Although not much information appears to be available concerning the effect of copper in water supplies, it is known that 0.01 to 1.7 ppm Cu is the toxic range for fish and as little as 0.027 ppm Cu can be toxic to aquatic insects (4).

Copper in ionic form is very toxic to the photosynthesis of the green alga, *Chlorella pyrenoidosa* and the diatom, *Nitzschiz palea*, in concentrations of copper normally found in natural waters, indicating that copper is not ordinarily present in ionic form but is complexed to organic matter such as polypeptides (5).

Copper concentrations of 0.04 ppm were acutely toxic to chinook salmon fry and concentrations of 0.02 ppm increased mortality and inhibited growth. The maximum acceptable concentration for the flathead minnow is between 0.03 and 0.08 ppm (6). The survival of gammarus, an invertebrate, was markedly decreased between 12.9 and 6.2 ppb. Newly hatched amphipods grew to the adult stage only in concentrations less than or equal to 4.6 ppb (7).

Concentrations between 0.56 and 3.2 ppm resulted in the fatty metamorphosis of the liver, necrosis in the kidney and gross changes in gill architecture in the winter flounder. A concentration of 0.18 ppm produced an extracted appearance in the gill lamellae. The epithelial layer appeared vacuolated, the basilamellar region was reduced in thickness and the lamellar mucus cells were few, while chloride cells appeared in their stead. Under an electron microscope, vacuolation in epithelial vesicles, myelin like figures, various membrane bound vesicles and apical homogeneous layers of reduced thickness were evident as were increased amounts of particulate matter adhering to the external surface of the epithelial cells (8).

EXTENT OF EXPOSURE

Primary sources of copper in industrial waste streams are metal process pickling baths and plating baths. Brass and copper metal working requires periodic oxide removal by immersing the metal in strong acid baths. Solution adhering to the metal surface, referred to as "dragout" is rinsed from the metal and contaminates the waste rinse water. Similarly, metal parts undergoing copper or brass plating drag out some of the concentrated plating solution. Jewelry manufacturers employ copper plating either directly or as a base metal for silver and other precious metal surfaces. Copper is also employed in the alkaline Bemberg rayon process, as cupro-ammonium salts.

Copper bearing acid mine drainage also contributes significant quantities of dissolved copper to waste streams. Table 25 summarizes values reported for copper in various industrial process wastes.

TABLE 25: CONCENTRATIONS OF COPPER IN PROCESS WASTEWATERS

Process	Copper Concentration, mg/l
Plating wash	20-120*
Plating wash	0-7.9
Brass dip	2-6
Brass mill rinse	4.4-8.5
Copper mill rinse	19-74
Metal processing	204-370
Brass mill wash	
Tube mill	74
Rod and wire mill	888
Rolling mill	34
Brass mill bichromate pickle	
Tube mill	13.1
Rod and wire mill	27.4
Rolling mill	12.2
Copper rinse	13-74
Brass mill rinse	4.5
Appliance manufacturing	
Spent acids	0.6-11.0

(continued)

TABLE 25: (continued)

Process	Copper Concentration, mg/l
Alkaline wastes	0-1.0
Typical large plater	
Rinse waters	up to 100 (20 av)
Four plating operations	6.4-88
Automobile heater production	24-33 (28 av)
Silver plating	
Silver bearing	3-900 (12 av)
Acid wastes	30-590 (135 av)
Alkaline wastes	3.2-19 (6.1 av)
Brass industry	
Pickling bath wastes	4.0-23
Bright dip wastes	7.0-44
Business Machine Corp.	
Plating wastes	2.8-7.8 (4.5 av)
Pickling wastes	0.4-2.2 (1.0 av)
Copper plating rinse water	5.2-41
Copper tube mill waste	70 (av)
Copper wire mill waste	800 (av)

*High values from spray rinse

Source: Reference (9)

Soluble copper wastes are of particular concern because of their high degree of toxicity to aquatic organisms (10). The principal sources of soluble copper wastes have been identified as: [1] rinse water and dragout from the electroplating industry; [2] etching solution wastes from printed circuit production; [3] spent copper catalysts; [4] copper pickling liquor; and [5] wastewater from the textile industry.

Copper wastes from the electroplating industry come from several sources, but the most important of these is the rinse water. A typical waste stream has soluble copper concentrations in the range of 50 to 10,000 ppm, alkaline concentrations from 1 to 20% as sodium hydroxide, along with some cyanides, chromium, nickel, and lead. It is estimated that there are 2.1 million lb of copper lost in the waste streams from the electroplating industry each year.

Copper wastes from printed circuit production are generated mainly in the etching process where the copper not comprising the circuit is chemically removed from the circuit board. The waste is generally characterized by a high concentration (1,000 to 50,000 ppm) of soluble copper. The other constituents present in the waste depend primarily on the type of etchant used. Ferric chloride is the major etchant material, but cuprous chloride, ammoniacal chlorite, ammonium persulfate, and chrome/sulfuric acid solutions are also used. It is estimated that 460,000 lb of copper are put through the waste streams from printed circuit production each year.

The chemical industry generates 600,000 lb of copper wastes (as copper) in the form of sludges each year by discarding spent copper catalysts. The composition of the sludge depends on the type of copper catalyst used and the particular segment of the chemical industry utilizing it. A typical waste sludge from aniline production contains 27% cuprous-cupric chloride tar in 65% mixed organics. The latter is mostly diphenyl ether.

Copper wastes are generated from the acid pickling of copper largely as a result of dragout of liquor from the pickling vats and the dumping of the vats when the acid strength is depleted. The typical waste stream from the pickling area contains 0.5 to 1% soluble copper, largely as copper sulfate when sulfuric acid is used. The total amount of copper waste generated from the pickling of copper is estimated to be 8.4 million pounds (as copper) annually.

Copper wastes from the textile industry are generated mainly in the dyeing and finishing of cotton cloth. Wastes from these operations are extremely variable in contaminating matter, although the copper is usually found in the form of copper sulfate, copper ammonium fluoride, and copper ammonium carbonate. It is estimated that 6 million pounds of copper are lost through the waste streams from the textile industry annually (10).

A report has been prepared by W.E. Davis and Assoc. (11) regarding the nature, magnitude and extent of airborne emissions of copper in the United States for the year 1969. Background information concerning the basic characteristics of the copper industry has been assembled and included. Brief process descriptions are given; they are limited to the areas that are closely related to existing or potential atmospheric losses of the pollutant.

The copper emissions and emission factors are based on data obtained from production and reprocessing companies. Additional information was acquired during field trips to inspect the air pollution control equipment and observe processing operations.

Emissions to the atmosphere during the year were some 13,680 tons. About 64% of the emissions resulted from the metallurgical processing of primary copper, and about 20% from the production of iron and steel. The combustion of coal was the only other significant emission source.

The emission factors presented in Table 26 are the best currently available. They were determined through a combination of methods consisting of

[1] direct observation of emission data and other related plant processing and engineering data;
[2] estimates based on information obtained from literature, plant operators, and others knowledgeable in the field;
[3] calculations based on experience and personal knowledge of metallurgical processing operations; and
[4] specific analytical results where available.

The emission factors shown in Table 26 for mining and milling, metallurgical processing, secondary production, and metal fabrication are based on the averages reported by industry.

The degree of accuracy is judged to be plus or minus 30%.

TABLE 26: EMISSION FACTORS

Mining and milling	200 lb/1,000 tons of copper mined
Metallurgical processing	10 lb/ton of copper produced
Secondary production	300 lb/1,000 tons of copper produced
Metal fabrication	1 lb/1,000 tons of copper fabricated
Other emission sources	
Coal	4 lb/1,000 tons of coal burned
Oil	160 lb/million bbl of oil burned
Blast furnaces	22 lb/1,000 tons of pig iron produced
Open-hearth furnaces	51 lb/1,000 tons of steel produced
Basic oxygen furnaces	2 lb/1,000 tons of steel produced
Electric furnaces	7 lb/1,000 tons of steel produced
Foundries	5 lb/1,000 tons of gray iron produced

Source: Reference (11)

DETECTION

Copper may be monitored in air by the following instrumental techniques (12): emission spectroscopy and atomic absorption. Copper may be monitored in water by the following instrumental techniques (13): atomic fluorescence, electrochemical, spark source mass spectrometry and x-ray fluorescence. Copper may be measured using the Technicon CSM-6 in the 0 to 10 ppm range down to a detection limit of 0.1 ppm. Copper may be measured down to a detection limit in the 1 to 5 ppb range by a variety of techniques. Those techniques include: flame atomic absorption, flameless atomic absorption, neutron activation and polarography.

The determination of copper by a spectrofluorometric method has also been reported by K.H. Mancy (14). The use of tetrachlorotetraiodofluorescein-o-phenanthroline as an indicator gave a sensitivity of 0.001 μg/ml with only cyanide ion interfering.

A continuous copper analyzer has been developed by J.J. Morrow and L.L. Dailey (15). It is an analyzer for directly and continuously measuring the concentration of soluble copper in an aqueous solution by adding potassium thiocyanate thereto in a dilute solution to form cupric thiocyanate. The cupric thiocyanate passes through an electrochemical cell where it is quantitatively reduced to cuprous thiocyanate. The resultant electric current is directly proportional to copper concentration and is measured to provide an indication thereof.

ENVIRONMENTAL STANDARDS

The permissible concentration of copper in domestic water supplies is 1.0 ppm; it is desirable, however, that it be virtually absent. The allowable copper concentration in irrigation water is 0.2 ppm on a continuous use basis and is 5.0 ppm on a short term use basis in fine textured soil. The maximum allowable concentration of copper to be discharged to a storm sewer or stream is less than 0.5 ppm.

The threshold limit value set by the American Conference of Governmental Industrial Hygienists for copper dusts and rusts was 1.0 mg/m^3 which is being revised downward to 0.2 mg/m^3 (16).

HANDLING PROCEDURES

There are no procedures for "handling copper." Indeed copper is handled in multimillion ton quantities in a wide variety of industries. All that can be said here is that like other metals, copper intakes by humans should be watched, measured and minimized.

REMOVAL FROM AIR

Recent surveys have revealed a marked contrast between urban and nonurban atmospheric concentrations of copper (17). Quantities found in the air over heavily industrialized areas have been found to range from about 0.03 to 4.00 μg/m^3, in sharp contrast to the nonurban concentrations noted earlier (18). Industrial process sources of atmospheric copper have been identified as fossil fuel combustion, incineration of copper-bearing wastes, manufacture of brass and steel, metal fabrication and mining and milling processes (11).

Problems associated with the control of copper emissions from industrial process sources are similar to those encountered in efforts made to reduce zinc emissions. Copper effluxes largely in oxide form, as particulate CuO. Particle sizes range from about 0.01 μm to greater than 100 μm. The art of controlling copper emissions makes use of electrostatic precipitators, fabric (baghouse) filters, and the scrubbers. In addition, some special processes have been developed for controlling copper emissions, which will be discussed below.

Electrostatic precipitators have found use in abating particulate emissions from all of the copper source processes enumerated above. Collection efficiencies of this apparatus used on a coal-burning power plant have been observed to range from about 91 to greater than 99% for copper. Similar particle size collection efficiencies hence have been experienced in applications to municipal incinerators (20). Precipitator efficiency tests on oxygen steelmaking facilities have shown collection efficiencies of greater than 99% (21). Efficiencies ranging from 94 to 99% have been found to be typical for electrostatic precipitators applied to minimize emissions from various types of steelmaking furnaces (20).

Theoretically, particle collection efficiencies for the baghouse system of fabric filtration are very high, ranging from 95 to nearly 100% for particles having a size distribution comparable to that of CuO. In practice, such efficiencies have been encountered, particularly in certain aspects of steelmaking, brass production and other areas of metallurgy (22). In nearly all cases of its application, the fabric filtration system is preceeded by effluent gas cooling apparatus and/or waste gas precleaning devices.

Typically, these may include heat exchangers, cyclonic separators, water-cooled ducts and spray chambers (23). Application of the baghouse system to power plants has been hindered, not by design efficiency difficulties, but by the absence of convincing evidence that the system can achieve long-term availability and reliability required for power plants without incurring excessive maintenance requirements (24).

For the removal of particles in the micrometer range, the most efficient scrubbers are those of the venturi type, which capture 96 (low energy venturi) to 98 (high energy venturi) percent of all particles in this size range (25). However, for capturing particles in the submicron size range, which hosts a large proportion of metal oxide fumes, scrubbers may not be as efficient as electrostatic precipitators. Efficiency considerations, along with the disadvantages of high power consumption, mechanical wear, and required further separation of collected metallic fumes from the recirculating water have generally discouraged scrubber usage, with the major exception of incinerator applications.

REMOVAL FROM WATER

As with most heavy metal wastes, treatment processes employed for reduction of soluble copper in wastewaters may be of a destructive nature involving precipitation and disposal of resulting solids, or recovery processes which include ion exchange, evaporation and electrolysis. The value of copper metal frequently makes recovery processes more attractive for it than for iron or zinc.

The importance of good internal housekeeping and process modification in reducing overall treatment requirements has been summarized by Domey and Stiefel (26) and Foulk (27). Modification of rinsing techniques for metal parts emerging from pickling baths and plating baths can greatly reduce rinse water volumes, and require treatment plant capacity. Such modifications also increase the concentration of metal in the rinse water and make recovery processes more feasible technically.

As pointed out by E.A. Tomic (28) as little as 2 to 4 micrograms of copper per liter of aqueous solution in streams flowing through aluminum processing equipment causes extensive corrosion damage. Thus, a process has been proposed for the removal of copper from nonaqueous and aqueous solutions employing a polymeric agent especially selective for copper.

The process employs a chelating polymer having a repeating thiosemicarbazide unit of the structural formula:

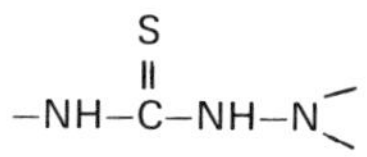

The standard destructive treatment for copper and most other heavy metals is precipitation of the relatively insoluble metal hydroxide at alkaline pH. As most heavy metal wastes are initially acidic, lime is the desired base for pH adjustment, due to its low cost and reaction to form the metal hydroxide.

Copper hydroxide solubility is minimal around pH 9.0 to 10.3 (29). Jenkins, et al (30) have reported a maximum solubility of 0.01 mg/l at pH 10. This value would correspond to the theoretical minimum effluent level which could be achieved by precipitation. Theoretical levels are seldom attained, due to poor settling of colloidal precipitates, slow reaction rates, pH fluctuations, and the influence of other ions in solution. Levels of copper treatment and associated costs reported in the literature are discussed below.

Parsons and Rudolfs (31) reported that copper in a plating bath waste was reduced from 33 mg/l to less than 1 mg/l by precipitation with lime. Nyquist and Carroll (32) report treatment of a waste containing from 204 to 385 mg/l copper, reduced to an average effluent copper concentration of 0.5 mg/l. Treatment also included zinc reduction from initial levels of 55 to 120 mg/l to less than 1 mg/l.

Stone (33) reports that copper precipitation at pH 8.5 to 9.0 provided an effluent concentration of copper ranging from 0.2 to 2.5 mg/l prior to installation of rapid sand filters. After filter installation, more consistent results were obtained, although minimum values were not reduced. The post-filter concentrations of copper ranged from 0.2 to 0.5 mg/l. Costs for construction of the entire waste treatment plant, which treated between 1.5 and 2.0 MGD, was nearly $800,000.

Hanson and Zabban (34) report an effluent concentration of 2.2 mg/l copper after lime treatment and coagulant addition at the Rochester, Minnesota IBM plant. Nemerow (35), in a lab study of copper cyanide waste from a silver plating operation, was able to reduce 14 to 18 mg/l of copper to 0.05 to 0.5 mg/l at pH 8.0. The process involved cyanide destruction with sodium hypochlorite, pH adjustment to pH 6.5 with ferric chloride, further addition of ferric chloride to 200 mg/l for coagulation, and addition of lime to pH 8.0 for metal precipitation.

In-plant operation of this process, also employing sand filtration, produced an effluent containing 0.16 to 0.3 mg/l copper. Untreated waste concentration of copper was, on the average, 30 mg/l. Watson (36) reports effluent copper concentrations of 0 to 1.2 mg/l for metal rinse waters, with the average copper level being 0.15 to 0.19 mg/l. No initial concentrations were given, but the author stated that copper levels were high enough to warrant treatment.

Tallmadge (37) reports that the Scoville Brass Mill of New Milford, Conn., disposes of its rinse waters and waste pickle liquors by lime precipitation. Rinse waters containing 10 to 20 mg/l copper are treated on a continuous basis, while the more concentrated waste pickle liquors are pumped into holding tanks and bled into the waste stream slowly, thus allowing continuous treatment of the high concentrate solutions. Lime is used to neutralize the acid and precipitate copper. The resultant effluent containing 1 to 2 mg/l of residual copper is diluted four-fold with other process waters to yield a final concentration of 0.25 to 0.50 mg/l of copper in the discharge effluent.

A process developed by C.H. Roy (38) is one in which the copper content of aqueous effluent containing the rinsings from the etching of copper containing substrates is reduced to less than about 5 ppm, by the steps of: (a) adjusting the pH of the effluent to no more than about 4; (b) providing in the effluent, while vigorously agitating, hydrogen peroxide and a water-soluble compound such as calcium chloride; and (c) adding to the reaction mixture as required an alkaline material such as lime or a water-soluble salt of carbonic acid, in an amount effective to adjust the pH to at least about 8.

The standard domestic sewage tertiary treatment sequence of chemical coagulation, sedimentation and rapid sand filtration is equivalent to metal precipitation operations, except for neutralization of acids to near neutral pH. Treatment costs should, therefore, be similar for the treatment units.

A second category involves recovery processes. Process modification for waste stream segregation and flow reduction is usually an integral part of recovery processes. In fact, direct recovery can result from such modifications, as reported by Barnes (39). Use of countercurrent rinse copper plating baths allowed direct return of the rinse flow into the plating baths as evaporative makeup water. Rinse water flow was reduced from 480 gal/hr to 16 gal/hr for the copper plating lines. The final rinse tank concentration of copper was reduced to 0.9 mg/l from a previous value of 40 mg/l.

Domey and Stiefel (26) report that copper plating bath rinse water volume could be reduced from 4,100 gal/day to 600 gal/day by rinsing modifications.

Use of evaporative recovery has been practiced for over twenty years (40). In the event that water evaporation from the plating bath is not sufficient to allow direct return of the dilute rinse water, preliminary evaporation of rinse water may be necessary in order to increase its copper concentration.

Bothan and Bryson (41) reported copper removal by ion exchange from 1.02 mg/l down to less than 0.03 mg/l at flow rates of approximately 0.80 gal/ft^2/min. McGarvey (58) reports that "complete" removal of copper was accomplished from alkaline rayon process wastewaters prior to World War II by ion exchange recovery. Ion exchange is capable of achieving very high levels of copper removal, particularly from dilute wastes.

Direct electrolytic recovery of copper metal is possible from relatively concentrated waste solutions. This process generally requires a preconcentration step such as ion exchange or evaporation. Acid leach solution from copper ore is electrolyzed directly by the Inspiration Consolidated Copper Company of Arizona to recover copper. The resultant regenerated sulfuric acid is also recycled (42).

For copper cyanide baths, copper recovery by electrolysis can be accomplished in conjunction with cyanide destruction (43)(44)(45). The process is not feasible for dilute solutions but is appropriate for treatment of spent plating bath solutions or ion exchange concentrates.

A process developed by J.D. Smith (46) utilizes electrodialysis to concentrate the metal to a sufficiently high concentration that convenient recovery of the metal, e.g., nickel or copper, can be completed by handling the metal as an ion in a liquid phase and without the need to plate the metal out on an electrode or precipitate it in any solid form.

A process developed by E.H. Newton et al (47) involves extracting copper as metal from a used etchant solution containing complexed cupric ions as an oxidant while simultaneously regenerating the etchant for further use. The etchant treated may have a pH from below 0 to 13, but preferably has a pH of at least 3 and more preferably has a pH between about 4 and 10. The process comprises electrowinning a portion of the copper from solution under conditions effective for electrowinning but not etching. These conditions include a substantial freedom from oxygen in the vicinity of the cathode, high current density, a substantial freedom from solution agitation during the winning operation and preferably for good efficiency, low solution temperature at the interface of the solution and the cathode, typically below 140°F. The process is economical because in the preferred embodiment, a portion only of the copper in solution is removed, the remaining copper being left in solution and available as a source of cupric ion for reuse of the etchant. The process is an important contribution to pollution abatement efforts as it eliminates the need for dumping copper and other wastes resulting from an etching operation.

A process developed by G.D. Parikh et al (48) involves electrolytically recovering metal from an electrolyte and/or an etchant on specially constructed and configured cathodes. The metal deposits advantageously may be subsequently removed from the cathodes with no mechanical or chemical cathodic stripping operation being required.

In a more specific application, the process relates to a continuous cupric chloride-hydrochloric acid etching system for use in recovering etched copper from articles, such as printed circuit boards, wherein the spent etchant is continuously, or intermittently, electrolytically regenerated and etched copper simultaneously recovered as loose, spongy, fine grain deposits on elongated cathodes preferably having arcuate or curvilinear plating surface profiles. The recovered copper deposits advantageously may be readily removed from the cathodes by passing them through a water spray.

A process developed by W.N. Schjerven, Jr. (49) relates to a method of recovering solid particulate copper from a liquid, and more particularly to a method of recovering such metallic solids from a neutralized metal processing solution, to make the solution suitable for reuse or for discharge into a sewer.

In the manufacture of rolled copper rod which is to be drawn into electrical conductor wire, it is standard practice to dip the copper rod in a pickling solution, which includes sulfuric acid, certain other process enhancing agents and water, to remove copper oxides which have formed on the surface of the rod during its passage through a rolling mill. The rod subsequently is dipped in a series of water rinse tanks to remove acid solution remaining on the rod as a result of the pickling process, and finally in a neutralizing soap dip. Eventually, the water in the rinse tanks becomes contaminated with solid particulate material from the surface of the copper rod, which must be removed from the water before it can be reused in manufacturing operations or discharged into a sewer.

In this regard, these copper solids usually are in the form of copper oxide particles (cupric and cuprous), pure copper fines and small amounts of copper hydroxide, with traces of copper sulfate in solution.

One well-known procedure for removing the copper solids from the rinse water involves introducing the contaminated rinse water into a settling tank in which the copper solids settle to the bottom of the tank and in which, after a preselected period of time, the rinse water is withdrawn from the tank. In a specific known system of this type, the water, instead of being withdrawn from the tank, is permitted to seep out of the tank through pervious vertical side walls and through a filter bed.

In other known processes the effluent from a settling tank is passed through a relatively thick filter bed of a suitable filter material, such as dolomite. The filter material may be pretreated so as to react chemically with any suspended or colloidally dissolved copper salts in the water, so as to remove these contaminants from the water. When the filter bed becomes clogged, it is backwashed to remove the copper contaminants therefrom, and the backwash water and the copper contaminants are returned to the settling tank for resettling.

One disadvantage of these prior known systems is that the recovery process generally is relatively time consuming. Further, settling tank systems require a large amount of space and settling tank systems in which the water is withdrawn after a predetermined settling time, without further treatment, are not particularly suited for the recovery of small solid particles and fines, because of the tendency for these materials to remain in suspension.

In addition, however, systems in which the water subsequently passes through a filter bed also are disadvantageous in that usually a relatively large amount of filtering material is required, there is a tendency for the filter bed to become clogged rapidly when there are a large number of fines and colloidal suspensions in the water, and it is necessary to take the filter bed out of operation periodically when backwashing is required.

Filtration processes and equipment for removing the copper solids from the rinse water also are known in which a filter aid, in the form of a 2 to 3 inch layer of a filtering material, such as diatomaceous earth, is deposited on a unitary supporting filter media, such as an automatically indexable filter paper. The rinse water, after strip acidity has been neutralized, then is passed through the diatomaceous earth and the filter paper, which cooperate to filter out the copper solids.

This filtering process is assisted by providing a vacuum beneath the filter paper, and when the diatomaceous earth and the filter paper become clogged with solid particulate material such that the vacuum increases to a preselected value, a new portion of the paper is automatically indexed into filtering position and a filter aid bed of the diatomaceous earth is formed on the new portion of the paper. The clogged diatomaceous earth and filter paper subsequently may be collected in the same container for disposal, or the diatomaceous earth may be removed from the filter paper, in which case they are collected in separate containers for disposal.

This last mentioned system also is disadvantageous, however, for several reasons. For example, while the copper solids have been removed from the neutralized rinse water, whereby it can be reused in manufacturing operations or discharged into a sewer, the clogged diatomaceous earth and the paper filter media still must be disposed of in a suitable manner, such as by hauling them to a land fill area. Another disadvantage of the system is that there is no practical procedure from an economic standpoint for recovering the copper solids from the diatomaceous earth and reusing the solids in manufacturing, and they must be discarded with the diatomaceous earth as a sludge or "soft scrap."

In the process by Schjerven (49) a filter aid in the form of a relatively thin layer of solid particulate material of the same type as the solid particulate material to be recovered from the liquid, is formed on a unitary filter medium. The liquid is then passed through this filter aid and the filter media, which cooperate to remove solid particulate material from the liquid. Subsequently, the filtered solid particulate material and the solid particulate material forming the filter aid are removed from the filter medium.

More specifically, solid particulate material of the type to be removed from the liquid initially is mixed with a volume of the same type of liquid to form a relatively thick slurry. The slurry then is passed through the filter medium and the material which is filtered out is leveled to a uniform depth on the order of $\frac{1}{16}$ to $\frac{1}{8}$ of an inch to form a filter aid on the filter medium. After the initial operating cycle has been completed, a portion of the solid particulate material which has been removed from the filter medium may be recycled through the process and used to make the slurry for forming the filter aid.

Further, after removal of the solid particulate material from the filter medium, the medium may be incinerated to recover any particulate material embedded therein or adhering thereto.

A process developed by E.B. Saubestre (50) involves the removal of substantially all copper from a substantially cyanide-free alkaline waste solution containing ionic copper. A reducing agent, for example, formaldehyde, for the ionic copper is added to the solution if not already present therein, and usually a complexing agent for the ionic copper. The process provides for the recovery of marketable zero valent metallic copper.

The alkaline waste solution is contacted with a material or metal catalytic to the reduction of the ionic copper by the reducing agent to zero valent metallic copper, and the catalytic contacting continued until the desired amount of copper is precipitated as zero valent metallic copper. The metallic copper precipitate is then separated, for instance, by filtration, from the solution. Exemplary of the catalytic metals for use in catalyzing the reduction of the ionic copper to zero valent metallic copper are palladium, platinum, gold and silver.

A process proposed by Lancy (51) is nonelectrolytic and simply involves adding an excess of caustic soda to the toxic solution and heating the solution, to convert the ammonium and copper content into ammonia gas and copper oxide, and to drive off the ammonia gas and precipitate out or remove the copper oxide from the solution to form a copper-free and ammonia-free supernatant liquid.

A process developed by C.A. Sumner (52) involves the reaction of copper in solution with hydrogen sulfide to form solid metal sulfide crystals. A portion of the metal sulfide is recycled and contacted with the hydrogen sulfide which is adsorbed onto the surface of the solid metal sulfide. When the soluble metal contacts the adsorbed hydrogen sulfide, larger metal sulfide crystals are formed that can readily be recovered by thickening and filtration procedures.

J. Hulsebos (53) developed a process and apparatus for recovering copper values from acidic solutions containing dissolved copper comprising reacting in an intake line and feed pump powdered iron in the size range of up to -40 mesh with the solution to displace the copper as metallic copper. The copper is separated from the solution by settling the metallic copper in the bottom of a polishing tower. The solution is clarified by passing it upwardly through a bed of scrap iron and then centrifuging to remove the particles.

In summary, copper removal by destructive treatment is currently accomplished by precipitation with lime. Effluent concentrations below 0.5 mg/l are achieved on a consistent basis only with proper pH control and either proper clarifier design accompanied by good sludge settling characteristics or final effluent treatment with sand filtration. Final copper concentrations below 0.1 mg/l are reported but do not seem readily attainable on a consistent basis with current precipitation methods. Since copper hydroxide solubility increases at higher pH after passing through a minimum at approximately pH 9 to 10, variations in the waste stream may be a causitive factor in limiting the removal of copper.

Process modification and waste stream separation provide a means of enhanced treatment of copper bearing wastes. When such modification is introduced, recovery processes become more attractive. Recovery of one type or another is practiced in many industrial operations and appears economically attractive due to the value of recovered copper.

SOLID WASTE DISPOSAL

Three ultimate disposal options for copper hydroxide and sulfide sludges, according to P.W. Powers (54) are: [1] incineration, [2] landfill, and [3] ocean disposal.

Incineration

Incineration reduces the sludge to an ash residue which can be more easily handled for ultimate disposal but at the same time could also lead to the formation of poisonous, soluble metal oxides. Incineration also requires additional equipment and a combustible fuel souce as well as equipment for air pollution control.

Landfill

Landfill is believed to be, by far, the most prevalent final disposal procedure but the problem of potential contamination of surrounding land or water tables, due to the leaching of poisonous materials from the sludge, must be considered.

Ocean Disposal

Ocean disposal is being practiced to a considerable degree by municipal waste treatment plants located near the coast where the primary and secondary sludges can be either pumped or barged out to submarine disposal areas. This disposal approach has been the subject of much recent debate and it is likely that it will not long remain one of the significant sludge disposal options.

The recovery of copper from solid wastes, as one disposal option, has been reviewed at some length by Sittig (55).

ECONOMIC IMPACT OF CONTROLS

Recovery of copper is more attractive than some other metals due to its higher value. A 1969 study (56) indicated that the recovery value for copper solutions ranging from 100 to 500 mg/l is \$0.60 to \$3.00/1,000 gal. In brass rinse waters containing 40 to 250 mg/l of copper, the recovery values are given as \$0.20 to \$1.20/1,000 gal. Unless a recovery process is a closed loop operation, however, the wastewater may still require final treatment for residual metal removal.

Cost of evaporative recovery versus destructive treatment can be compared based upon a study by Culotta and Swanton (57). They report the cost of copper cyanide plating solution to be \$0.60/gal. Their cost analysis equations allow calculation of approximate costs for destructive treatment and for evaporative recovery. An economic analysis for zinc cyanide waste demonstrated that at a plating solution dragout rate of greater than 4 gal/hr, recovery had the economic edge over destruction. Copper plating solution has a higher cost than zinc (\$0.60 versus \$0.45/gal), and therefore more favorable recovery economics.

The initial costs of an ion exchange unit depend upon the required resin volume. For dilute systems, flow rate will be the limiting factor, while the resin exchange capacity will dictate the volume for more concentrated systems. Costs of various units and types of resins are given by McGarvey et al (58) as a function of required capacity. These costs are for complete units (valves, resin, etc.).

The economics of electrolytic copper recovery from plating bath solutions compares favorably with other copper cyanide treatment processes. Operating data are presented by Easton (44), and an economic analysis of the process (1967 prices) is given. It has been claimed however that electrolytic recovery of copper from copper etching solutions (including pH neutralization) is too expensive.

REFERENCES

(1) T.D. Luckey, B. Venugopal and D. Hutcheson, *Heavy Metal Toxicity and Hormonology,* New York, Academic Press (1975).

(2) Z.H. Scheinberg, *Proc. Third Annual Conf. on Trace Substances in Envir. Health,* p. 79, Columbia, Mo. (1969).

(3) G.K. Davis, *Proc. Third Annual Conf. on Trace Substances in Envir. Health,* p. 135, Columbia, Mo. (1969).

(4) S.L. Warnick and H.L. Bell, *J. Water Pollution Control Fed.* 41, 280 (1969).

(5) E.S. Nielsen, *Marine Biology* 6, 2 (1970).

(6) R. Mount and C. Stephan, *J. Fishery Reserve Board of Canada* 26 (1969).

(7) J. Arthur and E. Leonard, *J. Fishery Reserve Board of Canada* 27 (1970).

(8) J.J. Baker, *J. Fishery Reserve Board of Canada* 26 (1969).

(9) J.W. Patterson and R.A. Minear *Wastewater Treatment Technology,* Report PB 216,162, Springfield, Va,, Nat. Tech. Info. Service (Feb 1973).

(10) R.S. Ottinger, J.L. Blumenthal, D.F. Dal Porto, G.I. Gruber, M.J. Santy and C.C. Shih, *Recommended Methods of Reduction, Neutralization, Recovery or Disposal of Hazardous Waste,* Vol. I, Report PB-224,580, Springfield, Va., Nat. Tech. Info. Service (August 1973).

(11) W.E. Davis and Assoc., Report PB 210,678, Springfield, Va., Nat. Tech. Info. Service (May 1972).

(12) Lawrence Berkeley Laboratory, *Instrumentation for Environmental Monitoring - Air,* Publ, LBL-1, Vol. 1, Berkeley, Univ. of Calif. (Feb. 1973).

(13) Lawrence Berkeley Laboratory, *Instrumentation for Environmental Monitoring - Water,* Publ. LBL-1, Vol. 2, Berkeley, Univ. of Calif. (Feb 1973).

(14) K.H. Mancy, *Instrumental Analysis for Water Pollution Control,* Ann Arbor, Mich., Ann Arbor Science Publishers (1971).

(15) J.J. Morrow and L.L. Dailey; U.S. Patent 3,857,762; December 31, 1974; assigned to Fischer & Porter Co.

(16) American Conference of Governmental Industrial Hygienists, *Threshold Limit Values for Chemical Substances in Workroom Air,* Cincinnati, Ohio (1974).

(17) A.R. McFarland, Ed., *Control Technology for Toxic and Hazardous Air Pollutants,* Chicago, Illinois Institute for Environmental Quality (Dec. 1974).

(18) P. Harrison, K. Rahn, R. Dorns, J. Robbins, J. Winchester, S. Brar, and D. Nelson, "Areawide Trace Metal Concentrations Measured by Multielement Neutron Activation Analysis." *Journal of the Air Pollution Control Association,* 21, 563-568 (September, 1970).

(19) S. Cuffe, R. Gerstle, A. Orning and C. Schwartz, "Air Pollutant Emissions from Coal-Fired Power Plants: Report No. 1," *Journal of the Air Pollution Control Association,* 14, 355-359, (September, 1964).

(20) S. Oglesby and G. Nichols, *A Manual of Electrostatic Precipitator Technology Vol. II,* Southern Research Institute, Birmingham, Alabama (1970).

(21) J. Smith, "Air Pollution Control in Oxygen Steelmaking," *Jour. of Metals,* p. 634 (Sept. 1961).

(22) L. Hardison and H. Herington, *Study of Technical and Cost Information in the Lime & Secondary Non-Ferrous Metallurgical Industries,* Rye, N.Y., Industrial Gas Cleaning Institute (1970).

(23) J. Fife and R. Boyer, "What Price Incineration Air Pollution Control," *Proceedings of the 1966 National Incineration Conference,* New York (1966).

(24) H. Falkenberry, A. Slack and F. Gartrell, "Control of Fossil Fuel Power Plant Stack Gas Effluents," *Combustion,* p. 10 (October 1972).

(25) W. Ellison, *Control of Air and Water Pollution from Municipal Incinerators with the Unit-Approach Venturi Scrubbers,* New York, Chemical Construction Co. (1970).

(26) W.R. Domey and R.C. Stiefel, "Wastewater Reduction in Metal Finishing Operations," presented at 43rd Annual Conf. Wat. Poll. Control Fed., Boston, Mass (Oct. 5, 1970).

(27) D.G. Foulk, "Metal Finishing Products," in *Chemical Technology,* Vol. 2, *Industrial Waste Water Control,* C. Fred Gurnham, ed., New York, Academic Press (1965).

(28) E.A. Tomic; U.S. Patent 3,196,107; July 20, 1965; assigned to E.I. du Pont de Nemours and Co.

(29) W. Stumm and J.J. Morgan, *Aquatic Chemistry,* New York, Wiley-Interscience (1970).

(30) S.H. Jenkins, D.G. Keight and R.E. Humphreys, "The Solubility of Heavy Metal Hydroxides in Water, Sewage and Sewage Sludge. I. Solubility of Some Metal Hydroxides," *Internat. Jour. Air Wat. Poll.* 8, 53-56 (1964).

(31) W.A. Parsons and W. Rudolfs, "Lime Treatment of Copper Pyrophosphate Plating Wastes," *Sew. Ind. Wastes Eng.* 22, 313-315 (1951).

(32) O.W. Nyquist and H.R. Carroll, "Design and Treatment of Metal Processing Wastewaters," *Sew. Ind. Wastes,* 31, 941-948 (1959).

(33) E.H.F. Stone, "Treatment of Non-ferrous Metal Process Waste of Kynoch Works, Birmingham, England," *Proc. 25th Purdue Industrial Waste Conf.,* pp 848-865 (1967).

(34) N.H. Hansen and W. Zabban, "Design and Operation Problems of a Continuous Automatic Plating Waste Treatment Plant at Data Processing Division, IBM, Rochester, Minnesota," *Proc. 14th Purdue Industrial Waste Conf.,* pp 227-249 (1959).

(35) N.L. Nemerow, *Theories and Practices of Industrial Waste Treatment,* Reading, Mass., Addison Wesley (1963).

(36) K.S. Watson, "Treatment of Complex Metal-Finishing Wastes," *Sew. Ind. Wastes* 26, 182-194 (1954).

(37) J.A. Tallmadge, "Nonferrous Metals," in *Chemical Technology,* Vol 2, *Industrial Waste Water Control,* C. Fred Gurnham, ed., New York, Academic Press (1965).

(38) C.H. Roy; U.S. Patent 3,816,306; June 11, 1974.

(39) G.E. Barnes, "Disposal and Recovery of Electroplating Wastes," *Jour. Wat. Poll. Control Fed.*, 40 1459-1470 (1968).

(40) J.W. Culotta and W.F. Swanton, "Case Histories of Plating Waste Recovery Systems," *Plating,* 57, 251-255 (1970).

(41) G.H. Botham and W.R. Bryson, "Note on the Corrosion of Aluminum Dairy Equipment by Water," *J. Dairy Res.*, 20, 154-155 (1953).
(42) R.W. Simpson and K. Thompson, "Chlorine Treatment of Cyanide Wastes," *Sew. Ind. Wastes Eng.*, 21, 302-304 (1950).
(43) H.L. Pinkerton, "Waste Disposal" in *Electroplating Engineering Handbook,* 2nd ed., A. Kenneth Graham, ed-in-chief, New York, Reinhold Pub. Co. (1962).
(44) J.K. Easton, "Electrolytic Decomposition of Concentrated Cyanide Plating Wastes," *Jour. Wat. Poll. Control Fed.*, 39, 1621-1625 (1967).
(45) American Electroplating Society, "A Report on the Control of Cyanides in Plating Shop Effluents," *Plating,* 56, 1107-1112 (1969).
(46) J.D. Smith; U.S. Patent 3,766,049; October 16, 1973; assigned to Process Research, Inc.
(47) E.H. Newton, J.M. Ketteringham, J.L. Sienczyk and C.M. Isaacson; U.S. Patent 3,783,113; Jan. 1, 1974; assigned to The Shipley Co., Inc.
(48) G.D. Parikh and W.C. Willard; U.S. Patent 3,784,455; Jan. 8, 1974; assigned to Western Electric Co., Inc.
(49) W.N. Schjerven, Jr.; U.S. Patent 3,696,928; October 10, 1972; assigned to Western Electric Co.
(50) E.B. Saubestre; U.S. Patent 3,666,447; May 30, 1972; assigned to Enthone, Inc.
(51) L.E. Lancy; U.S. Patent 3,218,254; November 16, 1965; assigned to Lancy Laboratories, Inc.
(52) C.A. Sumner; U.S. Patent 3,743,705; July 3, 1973; assigned to Stauffer Chemical Co.
(53) J. Hulsebos; U.S. Patent 3,900,314; August 19, 1975; assigned to Cities Service Oil Co.
(54) P.W. Powers, *How to Dispose of Toxic Substances and Industrial Wastes,* Park Ridge, N.J., Noyes Data Corp. (1976).
(55) M. Sittig, *Resource Recovery and Recycling Handbook of Industrial Wastes,* Park Ridge, N.J., Noyes Data Corp. (1975).
(56) Resource Eng. Assoc., *State of the Art Review on Product Recovery,* Report PB 192,634, Springfield, Va., National Technical Information Service (Nov. 1969).
(57) J.W. Culotta and W.F. Swanton, "The Role of Evaporation in the Economics of Waste Treatment for Plating Operations," *Plating* 55, 957-61 (1968).
(58) F.X. McGarvey, R.E. Tenhoor and R.P. Nevers, "Brass and Copper Industry: Cation Exchangers for Metals Concentration from Pickle Rinse Waters," *Ind. Eng. Chem.*, 44, 534-541 (1952).

INDIUM

Indium was first used commercially over 30 years ago as a dental alloy (1). The only available breakdown on the uses of indium is on an international, rather than domestic basis (2). (See Figure 10.) The solder and alloy industries offer the best opportunity for growth for the indium market. This is because indium increases hardness and fatigue resistance and improves corrosion resistance, producing generally high quality solder and braze-bonded connections (1).

The use of indium in semiconductor devices is closely tied to silicon and germanium device production where indium is used to form the alloy junction. The use of germanium devices is declining. They are being replaced by silicon devices. The use of indium-containing solders in silicon device fabrication, therefore, is on the increase.

Substitutes are available for most uses of indium. Continuing research in the transistor field could lead to replacement of the germanium-indium type of transistor. Gallium can be used instead of indium in dental alloys and glass joint seals; however, gallium is more expensive than indium. Since the cost of materials is a relatively small part of the end product cost in most of the uses, adequacy of supply and desirable properties (or fewer undesirable properties) are more likely to determine which element is used.

The U.S. Bureau of Mines prepared (3) a forecast of the requirements for indium by the year 2000. 1968 was used as the base year and the annual growth rate projected for total population was 1.6% for 1968 to 2000. Projecting from an estimated consumption of 18,000 grams of indium in 1968, the demand in the year 2000 will be in the range of 21,000 to 37,000 grams, or a median of 29,000. This estimate was postulated on normal technological development, growth in existing uses, and no new breakthroughs.

A new use developed by Western Electric and now being implemented by the Bell System is indium-plated components (1). Splicing connectors which employ indium-plated contacts are used with both aluminum and copper cables. One type connector is for splicing, bridging and butt splicing.

FIGURE 10: INTERNATIONAL USE OF INDIUM

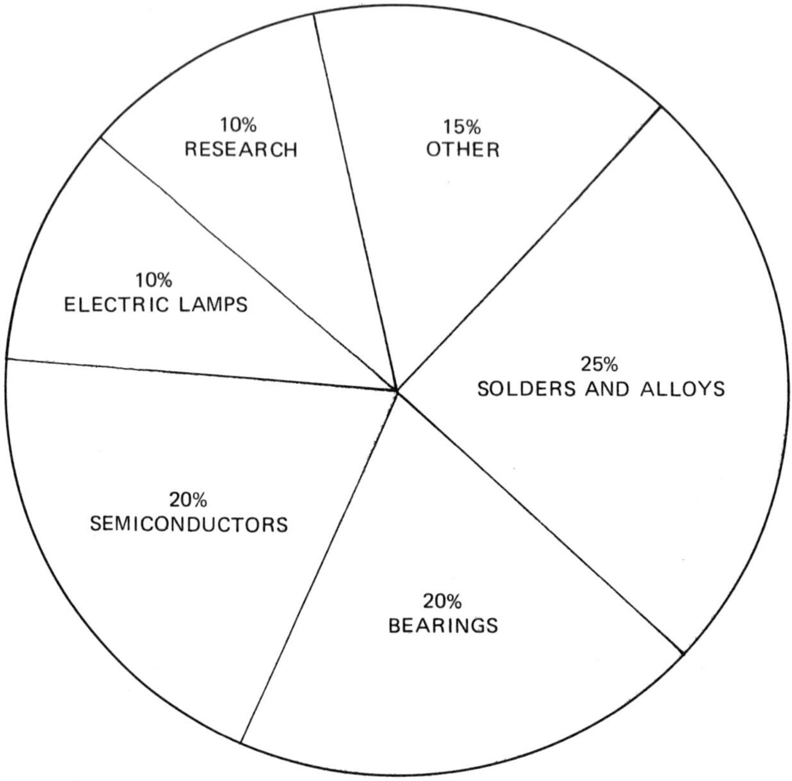

Source: Reference (2)

Special tools have been designed for installing the connectors. Coating of the contacts with indium may present a chance for indium to spill over into the environment. The Bell System will use about a billion of these units. In terms of the quantity of indium involved, this could amount to as much as 3% of the total consumed in the United States. A supply problem is not anticipated. One company supplies indium sulfate plating solution to a number of contractors. They provide the indium-plated components in a sealed system.

TOXICITY

Indium and most of its compounds are considered to be locally a moderately toxic irritant. Indium ingested or inhaled into the system is considered severely toxic. Injections of indium under the skin in quantities from 7 to 30 milligrams per kilogram of body weight and venous injections of indium are toxic.

Ingestion of large quantities of compounds is harmful, and injections of compounds have caused intoxication and degenerative changes in the liver, heart, and kidneys as noted by the U.S. Bureau of Mines (3). Acid foodstuffs, containing indium from indium-coated containers showed some absorption of the element by the body.

As regards occupational exposures, it has been reported that individuals exposed to indium compounds during the production of indium complained of tooth decay, pains in joints and bones, nervous and gastrointestinal disorders, heart pains, and general debility (4).

In other exposures, out of 770 patients injected with indium-113M radiopharmaceuticals for liver scanning, three developed anaphylactoid reactions. These three patients had cancer and an altered general condition. Vascular shock set in immediately or within 20 minutes after intravenous injection and lasted from 10 minutes to an hour.

In a parallel study of coagulation in patients receiving labelled or unlabelled indium compounds, changes in intravascular coagulation were observed. During the first two hours after injection, hypercoagulation with increased platelet adhesiveness occurred. After twenty-four hours, and five days later, lengthened coagulation time was observed with decreases in platelet adhesiveness and fibrinogen concentration (5). The toxic effects of indium can be reduced by the use of carrier-free radioactive isotopes.

The toxicity of indium and some of its compounds has been reviewed by Luckey et al (6).

EXTENT OF EXPOSURE

Fairly routine procedures are used to recover indium as a by-product of primary zinc production. Operational data are not available; however, indium contamination of the environment should not pose a problem. This is because special procedures are required to take care of wastes (lead for instance) and contaminants (such as sulfur dioxide) present or developed during the production of primary zinc.

The environmental hazard from indium appears to result from the use and disposal of radioactive isotopes and not chemical toxicity (1). In a study in Cape Town, South Africa, for example, indium-113 was still detectable as a bolus with an average radioactivity on the order of 10^{-4} microcuries per milliliter even after travelling a considerable distance after being dumped into the city sewage system by a hospital.

The use of carrier-free radioactive isotopes of indium in medical applications essentially eliminates the hazard from chemical toxicity. The use of indium as a lubricant and the use of indium antimonide for semiconductor purposes appear to pose no direct environmental threat.

Indium is recovered commercially as a by-product in processing zinc-bearing ores. Over and above that obtained by design in the procedures for recovering zinc (as well as other nonferrous materials), there is some concentration of indium in the residuals. In this manner indium may get into the environment from flue and Cottrell precipitator dusts, residues from zinc retorts, residues from cadmium recovery, lead blast furnace slags and crude zinc and lead. The amounts of indium involved in these inadvertent sources should not cause problems unless trace amounts of indium should prove harmful. Furthermore, should the supply of indium become tight, some of these could become valuable for recovery of indium (3).

DETECTION

Detection may be carried out as follows (7):

[1] Physical techniques — Indium may be detected by the characteristic blue-violet color it imparts to a flame at 4511.31 and 4101.76 A. As little as 0.01 to 0.02 ppm indium can be detected by this technique. Indium may also be detected spectroscopically by identification of the indium lines at the following wavelengths given in decreasing order of intensity: 4511.31, 4101.76, 3256.09, and 3039.36 A.

[2] Chemical techniques — There are no specific chemical methods available for the detection of indium. Of the methods used, the Alizarin and Quinalizarin Method is recommended as suitable and sensitive. In the method, indium forms a dark red lake with alizarin, and a violet lake with quinalizarin in ammoniacal solution. The effect of aluminum can be masked by fluoride, and that of zinc, manganese, ferrous iron, cobalt, and nickel by cyanide. The limit of detection is 0.01 to 1 microgram of indium, depending on the conditions used.

Quantitative analysis may be carried out by one of a number of classical analytical methods available for the determination of indium, involving gravimetric or titrimetric procedures (7). In the gravimetric methods indium is precipitated as the hydroxide, then ignited to the oxide In_2O_3. Prior to the precipitation of indium, indium must be separated from interfering elements. Two schemes are presented which, although tedious, are used to remove interfering elements.

Other gravimetric methods obtain the indium as the sulfide or hydroxyquinolinate instead of as the oxide. Nongravimetric methods listed are electrodeposition and titrimetric methods. Unfortunately the quantitative methods of analysis are not characterized as to their sensitivity.

ENVIRONMENTAL STANDARDS

The American Conference of Governmental Hygienists has recommended a Threshold Limiting Value of 0.1 mg/m^3 based almost entirely on experiments with animals and the severity of injury from indium salts. No recommendations are available for indium content of effluents, drinking water, etc.

HANDLING PROCEDURES

Indium in the form of dust may pose a moderate fire hazard in the presence of heat or flame (8). Heated indium burns with a nonluminous blue-red flame to form indium sesquioxide. Approved respiratory equipment should be worn when burning indium. Carbon dioxide, dry chemicals, or foam can be used to extinguish an indium or an indium sesquioxide fire (9).

Indium metal requires careful packing in bars, stick, foil, or shot for shipment to prevent pieces from sticking and to avoid cutting or denting of the shapes. Indium wire, in continuous lengths of 0.081 cm minimum diameter is available in 0.5, 2.5 or 10 kilogram spools (10). Indium sulfate is available as a dry powder or in solution and is packed in bottles in units of 1 gram. Indium sesquioxide is available as dry powder packed in bottles in 900 gram units (10).

REMOVAL FROM AIR

Conventional techniques for the removal of particulates from air could presumably be used when indium or its compounds are encountered in particulate form. No specific data were found however.

REMOVAL FROM WATER

Indium could presumably be precipitated as the hydroxide or sulfide if it proved a problem in industrial effluents. No specific data were found, however,

SOLID WASTE DISPOSAL

Recovered indium has a recycled value so a minimum quantity is discarded. Closed containers are required for the disposal of indium and its compounds (9).

ECONOMIC IMPACT OF CONTROLS

In view of the limited production volume of the indium industry, sufficient interest has not been focussed on indium pollution to yield any estimates of the economic effects of control of pollution by indium.

REFERENCES

(1) Versar, Inc., *Preliminary Investigation of Effects on the Environment of Boron, Indium, Nickel, Selenium, Tin, Vanadium and Their Compounds, Vol. 2, Indium,* Report EPA-560/2-75-005 B, Washington, D.C., Environmental Protection Agency (Aug. 1975).

(2) White, C.E.T., *American Metal Markets* (Apr. 25, 1973).

(3) Greenspoon, G.N., in *Mineral Facts and Problems,* Wash. D.C., U.S. Bureau of Mines (1970).

(4) Podosinovskii, V.V., *Work Hygiene Problems in the Production and Use of Indium and Gallium and Their Toxicological Characteristics,* Mosk. Med. Inst. (Moscow) pp. 110-12 (1964).

(5) Raiciulescu, N., Niculesco-Zinca, D. and Stoichita-Papillian, M., "Anaphylactoid Reactions Induced by In-113 and Au-198 Radiopharmaceuticals used for Liver Scanning," *Rev. Roum. Med. Interne.,* 9, (1), 55-60, (1972).

(6) Luckey. T.D., Venugoral, B and Hutcheson, D., *Heavy Metal Toxicity and Hormonology,* New York, Academic Press (1975).

(7) Snell, F.D. and Ettre, L.S., *Encyclopedia of Industrial Chemical Analysis,* New York, Interscience (1971).

(8) Sox, I.N., *Dangerous Properties of Industrial Materials,* New York, Reinhold (1968).

(9) Workplace Standards Administration, Bureau of Labor Standards, Material Safety Data Sheet, Wash., D.C., U.S. Dept., of Labor (May 1969).

(10) American Smelting and Refining Co., *Indium,* New York, (Feb. 1965).

LEAD

As has been pointed out by S.K. Hall (1), lead is widely used in industry and in everyday life but does pose serious health hazards. Indeed, Hall ascribes the decline of the Roman empire to extensive lead poisoning in the ruling class.

In excess of a million tons of lead are processed yearly. The total usage of lead has remained relatively stable during recent years, but the consumption by various industries has changed. For example, there has been a decrease of lead usage in the manufacture of home paints and a simultaneous increase in the manufacture of lead storage batteries.

Metal products and miscellaneous categories account for the bulk of lead consumption. The refining and processing necessary to form these products include heating, grinding, and volatilization and these processes produce potentially hazardous industrial atmospheres. The impression should not be left that all workers in these industries are jeopardized, but rather that such uses of lead places them at a risk of lead absorption.

Lead's low melting point (327°C), ductility, malleability, and weathering resistance enabled its use without the need for the more complex equipment that, in modern times, has enabled the use of other metals such as steel that have more desirable properties for many applications.

TOXICITY

As pointed out by Luckey et al (2), lead halides, sulfates, hydroxides and phosphates are very insoluble. This leads to deposition in the bone when lead salts are ingested. Lead is present in all organs and tissues of mammals, although it is not essential to their nutrition. The toxicology of lead has been studied extensively and a number of excellent reviews are available, including those by Passow et al (3), Kehoe (4), and Chisholm (5). A recent guide to the literature on the biological impact of lead is presented by Kinnison (6).

Organic lead poisoning from contact with tetraalkyllead compounds is a special problem in itself. Because of its solubility in lipids, tetraethyllead is readily absorbed from the respiratory tract, gastrointestinal tract and skin. Part of the tetraethyllead is metabolized and the released Pb behaves the way inorganic lead in tissue distribution and excretion does. However, the major portion of volatile organic lead accumulates in the brain due to a special affinity between organic lead and the lipid of nerve tissues. Since lead poisoning is cumulative, experimental lethal and toxic doses of various lead compounds are not very significant.

According to Schroeder (7), tetraethyllead is a hundred or more times as toxic as inorganic lead. Of course it is handled in concentrated form only by industrial personnel equipped with the latest in safety devices and information.

Tucker (8) points out that "the range of (lead) levels found in urban populations, including children who are probably much more sensitive to lead poisoning than adults, actually overlaps the range known to cause accepted classical symptoms of lead poisoning."

A description of effects of lead absorption can be graphic if based on effects seen in industries earlier in this century. Thus, the effects of lead poisoning, from studies of many years ago, are loss of appetite, metallic taste in the mouth, constipation and obstipation, anemia, pallor, malaise, weakness, insomnia, headache, nervous irritability, muscle and joint pains, fine tremors, encephalopathy, and colic.

In lead colic, there may be severe abdominal pain, such that abdominal surgery has occasionally been performed. In workers, who have had repeated attacks of lead colic over many years, there is a tendency towards the occurrence of weakness of extensor muscle groups. This weakness may progress to palsy, often observed as a characteristic "wrist drop" or "foot drop."

The important routes of absorption of lead by man and animals are ingestion and inhalation. Eating of lead-bearing paint by children and drinking of lead-contaminated, illicitly distilled whiskey are important sources of nonindustrial poisoning. Other sources include exposure to burning battery casings, drinking of liquids from improperly fired, lead-glazed containers, and high levels of airborne lead. But man absorbs lead in small amounts not normally leading to poisoning from his food and water, and from the air. These sources lead to the "normal" body burden of lead. Thus, the lead absorbed in the course of occupation exposure is superimposed on lead absorbed from other means.

Descriptions of lead poisoning appear in many texts and reviews, for example *Airborne Lead in Perspective,* a report of the National Academy of Sciences (9), and *The Diseases of Occupations* by Hunter (10). A NIOSH report (11) pertains to the occupational aspects of lead poisoning, with a few notes on effects seen only in children.

Lead can interfere with the synthesis of heme, thereby altering the urinary or blood concentration of enzymes and intermediates in heme synthesis or their derivatives. Thus, lead poisoning can lead to accumulation of nonheme iron and protoporphyrin-IX in red blood cells, an increase in delta-aminolevulinic acid

(ALA) in blood and urine, an increase in urinary coproporphyrin, uroporphyrin, and porphobilinogen, inhibition of blood ALA-dehydrase (ALA-D), and an increased proportion of immature red cells in the blood (reticulocytes and basophilic stippled cells).

Anemia from lead poisoning is associated with a reduced red cell life span and with reticulocytosis and basophilic stippled cells in peripheral blood. Symptoms of this anemia include irritability, fatigue, pallor and sallow complexion. Bone marrow preparations show increased numbers of sideroblasts, and this is useful in differential diagnosis of lead poisoning from iron deficiency anemia.

Gastrointestinal sequelae of lead poisoning include intestinal colic, nausea often without vomiting, and constipation (or, occasionally, diarrhea). Headache usually occurs before or about the time of onset of colic.

Peripheral and central nervous system effects occur in severe poisoning. Peripheral neuropathy of lead poisoning involves considerable loss of motor function but little loss of sensory function. Extensor muscles of the hand and feet are often involved; extensor weakness normally precedes wrist drop or palsy.

Encephalopathy may be either acute or chronic. Acute encephalopathy may follow ingestion or inhalation of large amounts of lead, and may develop quickly to seizures, coma, and death from cardiorespiratory arrest. Chronic encephalopathy usually occurs in children after excessive ingestion of lead, and leads to loss of motor skills and of speech, and to development of behavioral disorders. Lead encephalopathy, often involving psychosis, also occurs from absorption of alkyl lead compounds.

Nephropathy is another effect of lead poisoning. There may be a progressive and irreversible loss of kidney function, with progressive azotemia, and occasionally hyperuricemia with or without gout. Children have developed renal dwarfism, hypertension, marked retention of urea, and low urinary concentration; some children with acute encephalopathy have developed a form of Fanconi syndrome, a kidney disease indicative of severe injury of the proximal renal tubules. Nephritis in adults is not common, but ischemic nephritis may occur after prolonged absorption of lead.

Lane (12) examined the causes of death of storage battery workers, including retired workers, and compared data from this group with data from all English and Welsh males of similar ages during the same period of time. Among the retirees who had been exposed to lead, there were found to be greater numbers of deaths than would have been expected, for their ages, from data on the population as a whole. Most of this excess in expected mortality was accounted for by vascular lesions in the central nervous system. Lead workers who died during employment also showed an excess of deaths from this cause.

Another study of electric storage battery workers was conducted by the Public Health Service over 30 years ago (11). In this study, the incidence of various disease states was studied in relation to lead exposure of 766 workers, most of whom (75%) had worked in a storage battery plant for more than 5 years and some of whom (12%) had worked there for 20 years or more. The incidence of disease (other than plumbism) in men exposed at levels of 0.15 mg/m^3 and higher (high exposure group) was compared to the incidence in men exposed below 0.15 mg/m^3 (low exposure group). Special attention was given to cardiovascular disease because of the common belief that chronic plumbism results in arteriosclerosis; however, the data developed by the PHS team did not show that more severe exposure to lead is associated with a significantly higher incidence of vascular disease.

The incidence of arteriosclerotic-hypertensive disease was not significantly different in the high and low exposure groups. The responses to a standard exercise, in terms of return to preexercise pulse rates and to systolic and diastolic blood pressure, were also compared, and again the two groups were found not to be significantly different from each other. These lead workers were also found not to be significantly different from other, nonlead, workers in terms of blood pressure. From this, it was concluded that exposure to lead in the storage battery industry does not cause cardiovascular effects.

A contrary conclusion was reached by Dingwall-Fordyce and Lane (13) in a study of British battery workers. A significant excess of deaths from cerebrovascular accidents was found in pensioners who had had exposure to lead of sufficient degree to have caused mean urinary lead levels of 0.25 mg/l during many years of lead work.

They compared three groups of workers—those with no occupational lead exposure, those with negligible exposure, and those occupationally exposed to lead—with the general population of English and Welsh males of similar ages. They found a significant excess of death, over that predictable from the population at large, among retirees in the highest exposure group, and this was largely attributable to cerebrovascular incidents.

They also examined records of deaths due to cancer in lead workers, both employed and retired, and concluded that there was no association between malignant disease and lead absorption. While they found an excess of deaths from cancer in the negligible exposure group (in the last decade of the 35-year figures only) there was a slight decrease in deaths, from that expected from statistics on the whole population, among workers absorbing more than negligible amounts of lead, hence their conclusion that malignant disease is not related to lead absorption. As improved working conditions decreased lead exposure, the excess of cerebrovascular deaths diminished.

Malcolm (14) recently conducted similar investigations of past and present employees exposed to lead. Since 1927, airborne lead to which these men had been exposed had been limited to 0.15 mg/m^3, according to Malcolm. He divided the workers into three groups: (a) no exposure, (b) mild exposure, and (c) severe exposure. Average blood lead in group (c) workers, since 1961, has been 0.065 mg/100 g, from which it may be inferred that the 0.15 mg/m^3 air concentration was sometimes exceeded. Urinary leads in subgroups averaged 0.09 to 0.180 mg/l, and averaged 0.119 mg/l for the entire group of workers.

Based on comparison of blood pressures of the two exposed groups (b) and (c) with the control group (a), it was concluded that there was no occupationally induced hypertension (although there might have been lead-induced hypertension before improved hygienic measures were instituted). There was a nonsignificant increase in chest disease among older retired workers, attributed to other causes, since most of these pensioners lived in an urban area with a higher rate of death from chest disease than that in the country as a whole.

Unlike the findings of earlier investigators [Dingwall-Fordyce and Lane (13) and Lane (12)] Malcolm (14) found no evidence of increased frequency of cerebrovascular death in his study, which included deaths occurring between 1963 and 1967, while data from the two earlier reports included deaths from 1926 to 1960. Thus, if all three reports are correct in their conclusions, it would seem that improving hygiene has diminished lead-induced cerebrovascular disease.

For years, chronic nephritis was thought to be a consequence of plumbism, and an analysis of death rates in the U.K. in 1921 (12) and in 1931 (14) shows a considerable excess in plumbers and painters due to nephritis and to cerebrovascular disease. The question of nephropathy from lead has also been raised by Henderson and Inglis, (15) who showed a relationship between chronic nephritis and excessive lead absorption as indicated by elevated lead levels in bone.

Lane (15) described 9 deaths from renal failure in lead workers, men who had been exposed for long periods at lead concentrations around 0.5 mg/m^3. Terminally, they all had evidence of chronic azotemic nephritis. These men, all of whom worked in storage battery industries for over 20 years, died between the ages of 42 and 52 (average age at death was 48.4). Other than two episodes of colic, there had been no previous history of lead intoxication.

In the United States, there have been few reports of renal disease in lead workers, (14) although the PHS survey of storage battery workers discovered an increased incidence of albuminuria in affected workers.

EXTENT OF EXPOSURE

The sources of lead induced into the atmosphere as a consequence of the activities of man have been reviewed by Snyder et al (17). The major stationary sources have been found to be the primary and secondary smelting of lead, smelting of brass, nonferrous foundries, storage battery manufacturing, paint manufacturing and glass and insecticide manufacturing. In addition, lead may be emitted into the air during riveting and paint-scraping operations. Emissions of the metal from waste incineration and open burning have also been observed.

It is well established and widely known that the major anthropogenic source of atmospheric lead is comprised by the variety of lead-based antiknock additives in the gasoline which fuels the internal combustion engine. Arguments regarding the motor vehicle as a source of lead pollution are external to the scope of this report and may be reviewed elsewhere (18). Estimates of lead emissions to the atmosphere by source differ rather widely as shown in Table 27 (19)(20).

TABLE 27: AIRBORNE LEAD EMISSIONS

	Emission Estimate in Tons	
Source	Mitre Corp. (19)	W.E. Davis & Assoc. (20)
Mining and milling	345	60

(continued)

TABLE 27: (continued)

Source	Emission Estimate in Tons: Mitre Corp (19)	W.E. Davis & Assoc. (20)
Primary lead	714	1,700
Primary zinc	250	240
Primary copper	370	1,700
Primary nickel	246	-
Secondary lead	2,020	220
Secondary copper	520	40
Iron and steel	150	1,500
Gray iron foundries	1,400	2,300
Ferroalloys	-	70
Cement	-	500
Battery manufacture	2	480
Lead alkyl additives	810	1,900
Pigments	-	210
Solder	-	110
Cable covering	-	50
Type metal	-	200
Metallic lead products	-	90
Petroleum refining	1,250	--
Incineration	320	5,800
Coal combustion	851	650
Oil combustion	24	90
Total	9,272	17,910

Source: Reference (19)(20)

The emission factors presented in Table 28 are the best currently available. They were determined by W.E. Davis and Assoc. (20) through a combination of methods consisting of: [1] direct observation of emission data and other related plant processing and engineering data; [2] estimates based on information obtained from literature, plant operators, and others knowledgeable in the field; [3] calculations based on experience and personal knowledge of metallurgical processing operations; and, [4] specific analytical results where available.

Table 29 lists some specific occupations and trades where lead exposure occurs and gives examples of general exposure from industrial operations utilizing lead. Examination of this table should give at least a general overview of the extent of occupational exposure to lead.

The diversity of occupations displayed shows why a precise measure of the extent of lead exposure is nonexistent. The National Academy of Sciences' document on lead (9) agrees, stating, "A reliable definition of the extent of risk of occupational lead exposure is unavailable." Because of the changing usage of lead in industry and the widely varied trades where exposure occurs, the United States has no reporting system whereby the prevalence of occupational lead poisoning can be analyzed.

The principal sources of lead wastes of concern have been identified (22) as: [1] waste sludges from petroleum refineries; [2] waste sludges from the manufacture of alkyl lead compounds; [3] waste solvent-based paint sludges from paint manufacture; [4] paint residue left in used paint containers; [5] waste

TABLE 28: LEAD EMISSION FACTORS

Source	Factor	Qualifier
Mining and milling	0.2 lb/ton lead mined (controlled)	Plant visit
Metallurgical industries:		
Primary lead production	5.0 lb/ton of product (controlled)	Questionnaires
Primary copper production	0.6 lb/ton of Cu concentrates (controlled)	Estimate
Primary zinc production	0.3 lb/ton of Zn concentrates (controlled)	Estimate
Secondary lead production	0.7 lb/ton of product (controlled)	Questionnaires
Lead oxide processing	0.7 lb/ton of lead oxide (controlled)	Questionnaires
Consumer product manufacturing:		
Storage batteries	8.0 lb/ton of lead processed (uncontrolled)	Questionnaires
Storage batteries	1.3 lb/ton of lead processed (controlled)	Questionnaires
Gasoline additives	14.0 lb/ton of lead processed (controlled)	Questionnaires
Solder	3.0 lb/ton of lead processed (controlled)	Estimate
Cable covering	2.0 lb/ton of lead processed (controlled)	Questionnaires
Type metal	17.0 lb/ton of lead processed (controlled)	Questionnaires
Brass and bronze	4.0 lb/ton of lead processed (controlled)	Questionnaires
Other emission sources:		
Waste oil combustion	0.04 lb/bbl of oil burned (controlled)	Estimate
Municipal incineration	0.2 lb/ton of charge (controlled)	Stack sampling
Sewage and sludge incineration	0.6 lb/ton of charge (uncontrolled)	Stack sampling
Coal combustion	2.2 lb/1,000 tons of coal burned (controlled)	Estimate
Distillate oil combustion	0.1 lb/1,000 bbl of oil burned (uncontrolled)	Stack sampling
Residual oil combustion	0.04 lb/bbl of oil burned (uncontrolled)	Stack sampling
Steel production:		
Open-hearth	0.14 lb/ton of steel (uncontrolled)	Atomic absorption
Basic oxygen	0.18 lb/ton of steel (uncontrolled)	Atomic absorption
Electric arc	0.18 lb/ton of steel (uncontrolled)	Atomic absorption
Grey iron foundries	0.3 lb/ton of iron (uncontrolled)	Estimate
Cement production	25 lb/1,000 bbl of cement production (uncontrolled)	Stack sampling
Pigments	10 lb/ton of lead processed (uncontrolled)	Questionnaires
Ferroalloys:		
Silicomanganese, electric furnace	0.9 lb/ton of product (uncontrolled)	Estimate
Ferromanganese:		
Electric furnace	0.4 lb/ton of product (uncontrolled)	Estimate
Blast furnace	4 lb/ton of product (uncontrolled)	Estimate

Source: Reference (20)

TABLE 29: GENERAL EXPOSURE FROM OPERATIONS UTILIZING LEAD

Operation	Incidence of Plumbism	Average Lead Concentrations Found: Air (mg/m^3) Average	Air (mg/m^3) Maximum	Urine (mg/l) Average	Urine (mg/l) Maximum
Metalizing	High	-	-	-	-
Paint spraying: red lead	High	1.8	3.5	-	-
Brush painting: red lead	Some	-	-	0.26	0.35
Paint sanding, scraping	High	0.32	-	0.30	0.48
Leaded iron pouring	High	19.5	-	-	-
Bearing bronze pouring	Some	1.86	3.4	0.54	0.82
Bearing bronze grinding	Low	0.84	-	0.33	-
Storage-battery manufacturing:					
Mixing	Some	0.73	3.8	0.70	1.00
Pasting	Some	0.75	2.1	0.26	0.48
Grouping	Some	0.50	4.0	0.22	0.68
Separating	Low	0.15	0.41	0.15	0.27
Casting	Low	0.26	0.65	0.19	0.31
Lead smelting, refining	Some	0.35	1.45	0.35	0.88
Lead burning	Some	0.57	1.5	0.26	0.37
Homogenizing	Some	3.0	-	-	-
Painted-steel burning	Some	-	-	0.41	0.50
Lead powder mixing	Some	2.2	10.2	0.22	0.32
Lead sanding, grinding	Some	4.2	7.4	0.26	-
Paint mixing	Low	1.75	5.8	0.17	0.29
Painting, N.O.C.	Low	-	-	0.09	0.16
Paint spraying: chrome yellow	Low	3.9	-	0.10	-
Wire patenting	Low	0.29	0.60	0.12	0.21
Steel tempering	Low	0.13	0.22	0.10	0.21
Bronze pouring	Low	0.34	1.56	0.20	0.34
Bronze grinding	Low	0.47	1.24	0.17	0.34
Lead casting	Low	0.12	0.35	0.14	0.37
Printing:					
Stereotyping	Low	0.26	0.51	0.15	0.22
Linotyping	None	0.07	0.24	0.08	0.14
Soldering, tinning	Low	0.25	0.62	0.15	0.23
Lead sawing	Low	0.25	-	-	-
Lead glass working	Low	0.01	0.02	0.05	0.10
Gasoline-tank cleaning	Low	-	-	0.07	0.14

Source: Reference (21)

sludges from the manufacture of lead acid batteries; and [6] solvent and water washes from printing ink production.

Lead wastes from petroleum refineries are mainly the sludges from the storage tanks of leaded gasoline. A typical sludge contains a 1% mixture of tetraethyl-lead and lead oxide, along with gasoline hydrocarbons, iron oxide, and silt. Total quantity of tetraethyllead sludges from this source is estimated to be 4,800 tons per year.

Organic lead sludges are also generated from the manufacturing process for alkyl lead compounds. The sludge contains an average of 0.5 to 1.0% tetraethyl and tetramethyllead. Annual production on this waste sludge is estimated to amount to 260 tons.

The major lead-containing pigments include white lead, red lead, leaded zinc oxide, chrome green, chrome yellow, chrome orange, and molybdate orange. Lead is present in these pigments as the oxide, the carbonate, the hydroxide, the chromate, and the molybdate. In paint manufacture, lead is found in the waste solvent-based paint sludges which typically contain 27.5% pigment, 25.0% binders, and 47.5% organic solvents. It is estimated that a total of 640,000 pounds of lead are lost through 37 million pounds of solvent-based paint sludges per year. In addition, an estimated 4.4 million pounds of lead are lost as paint residues left in used paint containers each year.

Lead wastes from the manufacture of lead acid batteries are generated as a result of the mixing operation of the lead acid pastes and the application of these pastes to support grids. It is estimated that over 1.3 million pounds of lead, mostly in the form of lead sulfate in neutralized sulfuric acid, are lost through the waste sludges from battery manufacture each year.

Lead wastes from printing ink production are generated as a result of the operations to clean up the ball mills, mixing tanks, and other equipment. The waste is in the form of a liquid containing 0.5 to 1.5% lead pigment mixed with varying amounts of other metals in organic solvents and water washes. It is estimated that 21,000 pounds of lead are lost through this waste stream each year.

Lead is also a common constituent of plating wastes, although not so frequently encountered as copper, zinc, cadmium and chromium. Fales (23) has reported the lead content of wastewater from one engine parts plating plant as ranging from 2.0 to 140.0 mg/l. The higher concentrations resulted from dumping of spent plating bath solution. The author reported that the quantity of lead discharged per day might amount to 112 pounds from a single lead-plating bath. Pinkerton (24) has reported lead content of plating bath rinse waters which, as expected, contain lower lead concentrations than presented by Fales for spent plating bath solutions. Rinse waters contained 0 to 30 mg/l of lead (24). Table 30 gives the lead levels reported for some wastewaters (25).

TABLE 30: REPORTED LEAD LEVELS IN INDUSTRIAL WASTEWATERS

Industry	Lead, mg/l
Battery manufacture	
Particulate	5-48
Soluble	0.5-25

(continued)

TABLE 30: (continued)

Industry	Lead, mg/l
Plating (23)	2 - 140
Plating (24)	0 - 30
Television tube manufacture	400
Mine drainage	0.02 - 2.5
Mining process water	0.018 - 0.098
Tetraethyllead manufacture	
Organic	126.7 - 144.8
Inorganic	66.1 - 84.9
Tetraethyl lead manufacture	45

Source: Reference (25)

DETECTION

The recommended method of sampling air involves collection of 100 liters of air or more, use of breathing zone samplers with sampling at a rate of 2 liters per minute, and collection on 0.45 μ cellulose membrane filters. Other sampling rates and other collection media (filter paper, nitric acid impinger, electrostatic precipitation) are capable of giving equivalent results (11).

Lead may be monitored in air by the following instrumental techniques (26): colorimetric, emission spectroscopy, x-ray fluorescence and neutron activation. The ASTM standard method for the determination of lead in air is the dithizone colorimetric which is designated ASTM D3112.

For analysis of lead in blood, atomic absorption spectrophotometry and dithizone colorimetry were considered (11). Appreciable consonance can be demonstrated between results obtained with atomic absorption and dithizone methods. Both methods have been used for analysis of air samples, and both are concluded to be capable of giving accurate results.

After a review of the several procedures involving atomic absorption spectrophotometry, it was concluded that not one of these procedures has been sufficiently standardized. Individual laboratories get excellent results with a specific procedure, but these procedures have not been compared in a number of laboratories. Dithizone colorimetry, on the other hand, has been used for a long time and has been thoroughly studied. The procedures, interferences, sensitivity, and replicability have been studied and are described by Keenan, Byers, Saltzman and Hyslop (27).

Dithizone colorimetry is a wet chemical method requiring equipment found in most chemical laboratories, but requires meticulous attention to detail and to the prevention of loss and the exclusion of contamination. Results of lead analysis by this method obtained by well trained technicians are often superior to results obtained by other methods of analysis (11).

A process has been developed by A.O. Walker (28) for determining the presence and quantities of volatile organic lead compounds in air. Air suspected of containing such compounds is passed over iodine crystals to sublime at least one-half thereof. Lead iodine compounds are formed and cooled on a porous substrate. Free iodine is removed from the substrate, thereby developing a yellow color on the substrate if lead was present in the air.

Several methods have been proposed for determining the presence and concentration of volatile lead compounds in the atmosphere. The methods described in U.S. Patent 3,071,446 and the article by Snyder et al appearing in *Analytical Chemistry,* volume 20, pages 772 to 776 (1948) require the performance of several procedural steps before the actual determination of lead in an air sample can be made. These steps are complex and require a trained chemical operator to perform them.

The Walker method (28), on the other hand, makes use of a relatively simple apparatus which can be operated to determine the presence and quantity of volatile lead impurities in the atmosphere which can be operated by a person not skilled in the field of chemistry. After a few hours of training, any person of reasonable intelligence can be taught to operate the apparatus to make determinations of the quantity of volatile lead compounds contained in the atmosphere. A later method developed by Walker (29) uses a tube of light-transmitting material having packed therein iodine crystals, means for driving an air stream through the tube and a photooptical sensor adjacent the tube for sensing the presence of a predetermined amount of organic lead compound by detecting the color of a coating on the inner surface of the tube, and an alarm responsive to the triggering of the sensor.

Lead may be monitored in water by the following instrumental techniques (30): electrochemical, spark source mass spectrometry and x-ray fluorescence. Lead may be determined by flame atomic absorption down to a 10 to 20 ppb detection limit and by flameless atomic absorption down to a 1.0 ppb detection limit.

The determination of lead in water by atomic fluorescence spectroscopy has been discussed by R.F. Browner, R.M. Dagnall and T.S. West (31). The atomic fluorescence characteristics of lead are described in air-acetylene, nitrous oxide-hydrogen, and argon-oxygen-hydrogen flames. An electrodeless discharge tube is used as the source of excitation. A detection limit of 0.01 microgram per milliliter of lead in aqueous solution is obtained by measurement of the direct line fluorescence at 405.8 nm in the argon-oxygen-hydrogen flame.

The effect of 100-fold excesses of 30 cations and anions is examined; only aluminum interferes significantly. Effects of multipass optics and signal collection mirrors are examined and their effect on signal-noise ratios is discussed. The study confirms that lead can be determined more sensitively by atomic fluorescence than by atomic absorption or flame photometry.

A spectropolarimetric titrimetric method for the determination of various heavy metals, including lead, has been reported by R.J. Palma, Sr. and K.H. Pearson of the Department of Chemistry of Texas A & M University (32). Spectropolarimetric methods of analyses of Cd(II), Pb(II) and Bi(III) were developed, based on the stereospecific ligand, D-(-)-1,2-propylenediamine-tetraacetic acid [D(-)PDTA]. The optical rotation of the solution was monitored with a photoelectric polarimeter to determine the end point of the titration.

The effects of pH, wavelength, and dilution were discussed and the optimum conditions for the determination of each metal were established. The range of accurate analyses at 365 nm was 0.1 minus 0.001 M for Cd(II), and 0.1 minus 0.001 M for Pb(II), Hg(II) and Bi(III). The range of the average deviations from visual colorimetry was 0.05 minus 0.29%.

A technique developed by R. Bloch, R. Bauer and B.F. Phillips (33) is one in which which polyvalent metal ions such as lead ions can be determined colorimetrically by using a semipermeable polymeric membrane containing a chelating agent and a solvent therefor, which chelating agent changes color upon contact with the metal ions being determined.

Various methods are known for the determination of such metal ions in solution. The most widely used of these methods are color reactions which are more or less specific for particulate metal ions. The use of most color reactions, however, involves the preparation of reagents and other laboratory procedures which may be cumbersome and difficult, especially for laboratories staffed with relatively unskilled personnel or for use in the field.

It has been found that immobilizing a chelating agent and using it in the solid phase provides a convenient means for the determination of metal ions in solution. To immobilize the chelating agent it is dissolved in a suitable solvent and incorporated into a semipermeable polymeric matrix. The polymeric matrix containing the dissolved chelating agent is then cast in the form of a membrane.

The membrane may be plasticized by means of a plasticizer added to the polymeric matrix or the solvent for the chelating agent may serve as a plasticizer for the membrane. The resulting membrane containing the chelating agent and solvent and/or plasticizer can be used to give a colorimetric response to various polyvalent metal ions in solution including iron, cobalt, nickel, copper, lead and mercury.

The following is one specific example of the operation of the technique. To 60 ml of freshly distilled cyclohexanone were added 5 grams of polyvinyl chloride. The mixture was rapidly slurried and heat was applied until all particles were dissolved. The resulting solution was then cooled to 25°C and brought to a volume of 100 ml by adding cyclohexanone.

A 0.75% solution of dithizone in dipentylphthalate (1 ml) and 2 ml of dipentylphthalate were then added to 20 ml of the above solution of polyvinyl chloride in cyclohexanone. The composite solution resulting was then poured upon glass microscope slides and dried at 60°C for 40 minutes to yield a green colored membrane. The green membrane when dipped into a 10^{-6} M solution of lead nitrate changed to a red color. Higher concentrations of lead salts caused the color to become darker.

A sample technique whereby lead in concentrations in the range 4 to 40 ppm in aqueous solution of pH 1 to 5 is visually detected by adding a solution of a soluble sulfide has been described by Horine (34).

ENVIRONMENTAL STANDARDS

A recommended standard for occupational exposure to inorganic lead has been promulgated by NIOSH (11) as follows: Occupational exposure to inorganic lead shall be controlled so that workers shall not be exposed to inorganic lead at a concentration greater than 0.15 mg Pb/m^3 determined as a time-weighted average (TWA) exposure for an 8-hour workday.

Air quality standards for lead have been set by the states of Montana and Pennsylvania at 0.005 mg/m^3 on a 30-day averaging time basis. The Soviet air quality standard for lead is 0.0007 mg/m^3 on a 24-hour averaging time basis. The Canadian (Ontario) standard is 0.02 mg/m^3 on a 30-minute averaging time basis. Emission standards for lead have been set in the United Kingdom as follows:

115 mg/m^3 for up to 3,000 cfm of exhaust, 100 lb/wk limit
115 mg/m^3 for up to 3,000 to 10,000 cfm of exhaust, 400 lb/wk limit
23 mg/m^3 for up to 10,000 to 140,000 cfm of exhaust, 1,000 lb/wk limit
12 mg/m^3 for up to over 140,000 cfm of exhaust

The permissible concentration of lead in domestic water supplies is below 0.05 ppm and it is desirable that it be totally absent. The permissible lead concentration in irrigation water for continuous use is 5.0 mg/l and for short-term use in fine textured soil is 20.0 mg/l. The maximum allowable concentration of lead in effluents being discharged to a storm sewer or stream is 0.1 ppm.

HANDLING PROCEDURES

NIOSH has set forth procedures (11) for the handling of inorganic lead. Producers of organic lead compounds such as the Ethyl Corporation have issued instructions for handling antiknock compounds containing tetraethyllead (35). The cleaning of petroleum storage tanks where TEL may be present is the subject of a publication by the American Petroleum Institute (36).

NIOSH (11) has suggested the following medical monitoring procedures for the handling of inorganic lead: Medical monitoring (biologic monitoring and medical examinations) shall be made available to workers as outlined below.

A. Biologic Monitoring — Biologic monitoring shall be made available to all workers subject to "exposure to inorganic lead." It consists of sampling and analysis of whole blood, or alternatively, of urine for lead content. Such monitoring shall be performed to ensure that no worker absorbs an unacceptable amount of lead. Unacceptable absorption of lead posing a risk of lead poisoning is demonstrated at levels of 0.080 mg Pb/100 g of whole blood or greater, or at levels of 0.20 mg Pb/liter of urine (with urine specific gravity corrected to 1.024) or greater.

Procedures for sampling and analysis of blood or urine for lead shall be as described (11), or by any method shown to be equivalent in precision and accuracy. In the case of urine, "spot" urine specimens of about 100 ml shall be collected during a workday, and urine specimens with a specific gravity less than 1.010 shall be discarded and another sample obtained.

Half of all workers subject to "exposure to inorganic lead" shall be offered biologic monitoring every 6 months, so that each worker shall have blood sampling and analysis made available to him yearly. If urine sampling and analysis are chosen instead of blood lead sampling and analysis to satisfy the biologic monitoring requirement, every worker shall have urine sampling and analysis made available to him at 6-month intervals.

The schedule of biologic monitoring, above, may be altered if indicated by a professional industrial hygiene survey. If environmental sampling and analysis show that environmental levels are at or greater than the environmental limit, the interval of biologic monitoring shall be halved, i.e., blood analysis shall be conducted quarterly, with each worker sampled semiannually, or urinalysis shall be conducted quarterly on every worker. This increased frequency shall be continued for at least 6 months after the high environmental level has been shown.

If a worker's urine lead level is found to be 0.20 mg/l or greater, calculated to a specific gravity of 1.024, a blood sample shall be obtained and analyzed within 2 weeks. If a blood level of 0.080 mg Pb/100 g or greater is found, and confirmed by a second sample to be taken within 2 weeks, steps to reduce his absorption of lead shall be taken as soon as the high levels are confirmed.

Steps to be considered should include improvement of environmental controls, of personal protection or personal hygiene, and use of administrative controls. A medical examination for possible lead poisoning shall be made available, and the OSHA area industrial hygienist shall be informed of the results of the biologic sampling of those workers with confirmed, high biologic levels of lead.

Biologic monitoring shall also be made available where the OSHA area industrial hygienist has reason to believe operations produce unusual exposure excursions or that environmental samples do not adequately describe worker exposure.

B. Medical Examination – Medical examinations shall be available when a variance has been granted permitting administrative controls or use of respiratory protection, for workers with unacceptable absorption of lead as judged by biologic monitoring, or when environmental levels are at or above the environmental standard.

These examinations should be made available prior to employee placement and annually thereafter. They should include a physical examination, complete blood counts, blood lead determinations, routine urinalysis (specific gravity, sugar and protein determinations, and microscopic examination), and should record any signs or symptoms of plumbism, if present.

Where urine is selected instead of blood for biologic monitoring, the preplacement examination should also include a urinary lead determination. Each employee who absorbs unacceptable amounts of lead as indicated by biologic monitoring shall be examined as soon as practicable after such absorption is demonstrated and confirmed, and at least every 3 months thereafter until his blood or urine levels have returned to normal, i.e., below 0.080 mg/100 g of blood or 0.20 mg/l of urine.

If clinical evidence of plumbism is developed from these medical examinations, the worker shall be kept under a physician's care, in accordance with applicable

Workman's Compensation provisions, until the worker has completely recovered or maximal improvement has occurred.

Medical records shall include information on all biologic determinations and on all required medical examinations. These records shall be available to the medical representatives of the employer, of the Secretary of Labor, of the Secretary of Health, Education and Welfare, and, at the employee's request, to the employee's physician. These records shall be kept for at least 5 years after the last occupational exposure to inorganic lead.

As regards labeling concerning occupational exposure to inorganic lead, NIOSH (11) recommends that areas where exposure to lead at levels over one-half the workroom air standard is likely to occur shall be posted with a sign reading:

LEAD (Pb)
DANGER!

High concentrations of fume or dust
may be hazardous to health.
Provide adequate ventilation.

If environmental levels are at or greater than the environmental limit, or if a variance permitting use of respiratory controls has been granted, add information to the label or placard describing the location of the respirators.

As regards respiratory protection for occupational exposure to inorganic lead, NIOSH (11) recommends the following: Engineering controls shall be used wherever feasible to maintain lead dust and fume concentrations below the prescribed limits. Appropriate respirators shall be provided and used when a variance has been granted to allow respirators as a means of control of exposure to routine operations and while the application is pending. Administrative controls should also be used to reduce exposure.

Respirators shall also be provided and used for nonroutine operations (occasional brief exposures above the TWA of 0.15 mg/m^3 and for emergencies); however, for these instances a variance is not required but the requirements set forth below continue to apply. Appropriate respirators as described in Table 31 shall only be used pursuant to the following requirements:

1. For the purpose of determining the class of respirator to be used, the employer shall measure the atmospheric concentration of inorganic lead in the workplace when the initial application for variance is made and thereafter whenever process, worksite, climate or control changes occur which are likely to affect the lead concentration. The employer shall test for respirator fit and/or make lead measurements within the respiratory inlet covering to ensure that no worker is being exposed to inorganic lead in excess of the standard either because of improper respiratory selection or fit.
2. Employees experiencing breathing difficulty while using respirators shall be referred to a physician for evaluation. This evaluation should investigate if the employee has adequate ventilatory capacity and any evidence of obstructive lung disease. Employees with inadequate ventilatory capacity or evidence of obstructive lung disease shall not wear types A, B and E respirators.

3. A respiratory protective program meeting the general requirements outlined in section 3.5 of American National Standard for Respiratory Protection Z88.2-1969 shall be established and enforced by the employer.
4. The employer shall provide respirators in accordance with Table 31 and shall assure that the employee uses the respirator provided.
5. If both fume and dust are present, the recommended usage is that for fume.
6. Respiratory protective devices described in Table 31 shall be either those approved under the following listed regulation or those approved under 30 CFR 11, published March 25, 1972. The termination date of currently approved respirators described in 30 CFR 11 shall apply.
   a. Reusable or replaceable filter-type air-purifying respirator, 30 CFR 14 (Bureau of Mines Schedule 21 B)
   b. Powered air-purifying positive-pressure respirator, 30 CFR 14 (Bureau of Mines Schedule 21 B)
   c. Type C positive-pressure supplied air respirator, 30 CFR 12 (Bureau of Mines Schedule 19 B)
7. Usage of a respirator specified for use in higher concentrations of lead is permitted in atmospheres of lower concentrations.
8. Employees shall be given instruction on the use of respirators assigned to them, cleaning of the respirators, and how to test for leakage.

TABLE 31: REQUIREMENTS FOR RESPIRATOR USAGE AT CONCENTRATIONS ABOVE THE STANDARD

Exposure	8 Hr TWA mg/m^3	Respirator Type*
Inorganic lead dust	<1.5	A,B
	<15.0	C
	>15.0	D
Inorganic lead fume	<1.5	E
	<15.0	F
	>15.0	D

*A — Reusable or replaceable filter-type air-purifying dust respirator
B — Single-use dust respirator
C — Powered air-purifying positive-pressure dust respirator
D — Type C positive-pressure supplied air respirator
E — Replaceable filter-type air-purifying fume respirator
F — Powered air-purifying positive-pressure fume respirator

Source: Reference (11)

As regards work clothing for occupational exposure to inorganic lead, NIOSH (11) recommends that:

1. Each employee subject to exposure above the environmental standard should wear coveralls or similar full body work clothing and hat, which should be worn during the working hours in areas where there is exposure to lead. Workers subject to "exposure to inorganic lead" at or below the recommended standard should change into working clothes before starting work and should remove clothing before leaving work. This work clothing need not afford full body coverage.
2. Work clothing should be vacuumed before removal. Clothes shall not be cleaned by blowing or shaking.

3. Work clothing should be changed at least twice a week and more frequent changes, especially in high-exposure area, are suggested.
4. Adequate shower facilities should be available and used.
5. When in the judgment of the OSHA area industrial hygienist contamination of clothing or exposed body surfaces can produce significant secondary exposures, items [1], [2], [3] and [4] above shall be mandatory.

As regards apprisal of employees of the hazard from occupational exposure to inorganic lead, NIOSH (11) specifies that:

A. Each employee exposed to lead shall be apprised at the beginning of his employment or assignment to a lead area of all hazards, relevant symptoms, appropriate emergency procedures, and proper conditions and precautions for safe use or exposure and shall be instructed as to availability of such information which shall be kept on file including that prescribed in B. below and shall be accessible to the worker at each place of employment where lead is involved in unit processes and operations.
B. Information shall be recorded on U.S. Department of Labor Form OSHA-20, "Material Safety Data Sheet," or on a similar form approved by the Occupational Safety and Health Administration, U.S. Department of Labor.

As regards work practices for occupational exposures to inorganic lead, NIOSH (11) recommends the following:

1, Emergency procedures
   a. Procedures including fire fighting procedures shall be established and implemented to meet forseeable emergency events.
   b. Respirators shall be available for wearing during evacuation procedures if long distances need to be traversed; supplied air respirators shall be available for employee use where equipment or operations cannot be abandoned.
2. Exhaust systems
   Where a local exhaust ventilation and collection system is used, it shall be designed and maintained to prevent the accumulation of lead dust and fume.
   a. Hazardous types of exposure should not be scattered throughout a plant but, rather, concentrated in a single area where special control procedures can be utilized.
   b. Air from the exhaust ventilation systems shall not be recirculated into the workroom, and should not be discharged outside the plant so as to create an air pollution problem.
3. General housekeeping
   a. Vacuuming shall be used wherever practicable and no dry sweeping or blowing shall be performed.
   b. Emphasis shall be placed upon cleanup of spills, periodic repair of equipment and leaks, proper storage of materials, and collection of lead-containing dust.

As regards sanitation practices for occupation exposure to inorganic lead, NIOSH (11) specifies that: A. Food preparation, dispensing (including vending machines), and eating shall be prohibited in lead work areas. B. Work and street clothing should not be stored in the same locker. As regards monitoring, recordkeeping and reporting requirements, NIOSH (11) suggests that workroom areas where environmental levels do not exceed half the environmental standard shall not be considered to have inorganic lead exposure. But records of surveys, including the basis for concluding that air levels are below half the environmental standards, shall be kept.

1. Employers shall monitor environmental levels of lead at least every 6 months, except as otherwise indicated by a professional industrial hygiene survey. If environmental levels are at or above the standard, environmental levels shall be monitored every 3 months. This increased frequency of monitoring shall be continued at least 6 months (i.e., two more quarterly monitoring periods) after the last sampling that demonstrated levels at or above the environmental limit. Periodic environmental sampling shall be performed to coincide with periodic biologic sampling, i.e., shall be performed within 2 weeks of biologic sampling. Breathing zone samples shall be collected to permit construction of a time-weighted average exposure for every operation. The following number of samples shall be collected and analyzed, as a minimum, based on the number of workers exposed in any given work area:

Number of Employees Exposed	Number of Samples
1 - 20	50% of the number of workers
20 - 100	10 samples plus 25% of the excess over 20 workers
over 100	25 samples plus 5% of the excess over 100 workers

2. When any time-weighted average exposure is at or above the environmental standard, the OSHA area industrial hygienist shall be notified.
3. Records shall be maintained for all sampling schedules to include the sampling methods, analytical methods, type of respiratory protection in use (if applicable), and the concentrations of lead in each work area. Records shall be maintained so that they can be classified by employee. Each employee shall be able to obtain information on his own environmental exposure.
4. Medical records shall include information on all biologic determinations and of all required medical examinations. These records shall be kept for at least five years following the last occupational exposure to inorganic lead.

REMOVAL FROM AIR

Lead emitted from most anthropogenic sources effluxes with the carrier gas stream as small particles. The metallic fumes and dusts emitted from various industrial operations range from about 0.03 to 5.00 μm in diameter (37). Nearly all industrial lead emissions are particulates in this size range.

The efficacy of variations of the baghouse filter has proven to be superior in nearly all instances of applied controls on lead emissions; however, some processes employ other particulate collection devices, principally the electrostatic precipitator and the scrubber.

High temperatures may pose a problem in the application of a baghouse filter. Effluent gases leaving reverberatory furnaces, for example, are typically 2000°F. Exhaust gases may be cooled by direct water spraying with heat exchangers or with cold air dilution. However, these techniques present problems because of the potential for condensation corrosion, and extra costs associated with excess capacity and extra equipment. It has been found successful to employ a water-jacketed cooler followed by one or more air-cooled radiation columns before the gases contact the filter medium. Treated Orlon has found general favor over glass cloth as a high temperature-tolerant medium.

The baghouse functions as a final collector in most controlled metallurgical operations. In certain situations, it is preceded by a settling chamber in addition to gas cooling apparatus. Successful lead collection has been achieved in lead refineries by using a gas cooling scheme and then a baghouse equipped with Dacron bags, using a gas velocity of 2.0 fpm.

The baghouse has also found wide application to reverberatory and blast furnace systems. In one such pulse-jet system, effluent gas has been cooled in a U-shaped venturi scrubber. The gas then enters a bag filter at a temperature of about 90°C. The system operates at a flow rate of 35 to 50 cubic meters per minute with a pressure loss of 80 to 100 inches of water (38).

One battery manufacturer has coped with the emissions of lead oxide and other lead aerosols by prefacing a baghouse filter with 14 cooling towers, which serve to reduce the temperature of the gas from 2400° to less than 450°F. Applications of baghouses provide typical collection efficiencies for lead oxide of 93.7 to 96.2% (39).

Although the electrostatic precipitator is a reasonable efficient collector of very small particles, it has found little use in controlling lead emissions. Besides the high costs of the system, the electrical resistivity of lead oxide dissuades from usage of the apparatus. Although it is possible to precondition the particles to increase the resistivity to a level compatible with effective precipitator application, it has been more desirable to use a baghouse. There are additional problems with temperature, water and SO_3 in the gas stream, particularly at lead smelting plants, which require careful optimization for this type of control system, and afford little opportunity for adjustment during operation (40).

Dynamic scrubbers or mechanical washers have proven to be comparatively less effective than the electrostatic precipitator and the baghouse filter in collecting particles in the submicron size range. This characteristic, along with the disadvantages of high power consumption, mechanical wear and required separation of the collected metallic fumes from the recirculating water often has led to the use of alternative particle collection apparatus. In one application, at a brass smelter, collection efficiencies averaged little better than 50% (39). Improved scrubber designs should provide efficiencies significantly higher than this value, however.

A process developed by W.C. Jaasma (41) involves the recovery of tetraalkyllead compounds from an inert gas stream by contacting the stream at least once with an inert organic scrubbing liquid.

Figure 11 shows the configuration of apparatus for the process as applied to the production of tetramethyllead in the presence of toluene. Referring to the drawing, inert gas vapor from a steam still or other product recovery or production means containing quantities of tetramethyllead and toluene or other volatile condensables such as aluminum compounds enters the process by line **10** and is fed to fume scrubber **11** where it is contacted with aqueous caustic solution entering fume scrubber **11** by line **12** and line **5** to produce a scrubbed gas phase indicated in first sump vessel **13** at **14** and an aqueous liquid phase **15**.

FIGURE 11: APPARATUS FOR RECOVERY OF TETRAALKYLLEAD FROM AIR BY SOLVENT SCRUBBING

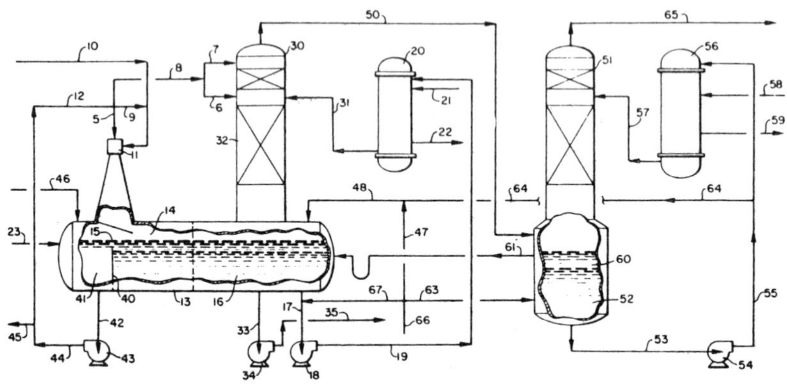

Source: Reference (41)

Should the quantity or quality of volatile aluminum compounds, or other material, in the inert gas stream of line **10** be such that condensation thereof may cause plugging of fume scrubber **11**, additional quantities of aqueous caustic solution may be added to or blended with the entering inert gas stream by line **9** to prevent such plugging.

It is clear also that when the process is used for the recovery of tetraalkyllead product from an inert gas stream containing no or substantially no volatile aluminum compound or like material, fume scrubber **11** may be bypassed or eliminated, and the product containing inert gas fed directly to vapor space **14** without any pretreatment.

A first sump in vessel **13** containing the aqueous liquid phase **15** and a first organic liquid **16** from this first sump is removed by line **17** through pump **18** and line **19** to cooler **20** where the first organic liquid phase may be cooled to a temperature of from -10° to 50°F by brine entering cooler **20** by line **21** and leaving by line **22**. Cooled first organic liquid phase from cooler enters tower **30** by line **31** where it contacts scrubbed gas phase from vessel countercurrently.

Suitable packing section **32** or other contacting means such as distillation plates may be provided in the tower to permit contact of the liquid and gas phases. The first organic liquid phase after contact returns to the vessel and settles to the organic liquid phase with extracted tetramethyllead and toluene contained therein. Tetramethyllead product and toluene recovered from the scrubbed gas phase, along with first organic liquid are removed from the system as product by line **33** to pump **34** and then to process product recovery by line **35**.

By the use of weir **40** located within the vessel, first organic liquid phase is retained within the vessel, except for that removed to product recovery, and the aqueous liquid phase is permitted to overflow into aqueous sump **41** where it may be removed by line **42** through pump **43** and recycled to fume scrubber **11** by line **44** joining line **12**. As necessary, portions of contaminated aqueous caustic solution may be removed from the system to waste by line **45** and fresh aqueous caustic solution added to the system by line **46**.

In operation of the process over long periods it may be necessary or desirable to provide a continuing or intermittent aqueous caustic wash for the tower to prevent buildup of deposits in the contacting section **32** or the demisting section, if any, above the contacting section. When this is necessary or desirable, aqueous caustic solution from line **12** may be supplied in a desired quantity by line **8** through line **6** or **7** or both. As necessary, and generally in a volume approximating that volume removed to product recovery, substantially pure organic scrubbing liquid may be added to the system from a supply not shown by line **66** to line **47** and then to line **48**.

In most process applications, the concentration of desirable products in the second scrubbed gaseous phase exiting the tower by line **50** is reduced to a minimum and the venting gas phase either may be recovered for reuse as inert gas or vented to the atmosphere without pollution.

In operation in this manner as a single tower system, it is desirable to supply substantially pure organic scrubbing liquid in a necessary volume to that portion of the scrubbing liquid being pumped to cooler **20** and this may be accomplished as shown by line **67** from supply line **66**.

Under certain conditions of production and high volumes of vent gas, it is sometimes desirable to remove only a major portion of product from the inert gas system in the tower and to remove the final minor portion of the product in tower **51** which may be substantially identical to tower **30**. When operating the process in this manner, second scrubbed gaseous phase exiting tower **30** by line **50** is fed to the bottom portion of tower **51** where it is contacted countercurrently with organic scrubbing liquid from sump **52** in the bottom portion of tower **51**.

The organic liquid is removed by line **53** and pumped by pump **54** through line **55** to cooler **56** and from there to the top portion of tower **51** by line **57**. As in the case with cooler **20**, chilled brine enters cooler **56** by line **58** and exits by line **59** to cool the organic liquid to a temperature of from $-10°$ to $50°F$. The second contacting organic liquid phase flows down tower **51** to sump **52** which is connected by line **61** to first sump **16** so that product containing organic liquid does not build up within sump **52** beyond the level of connecting line **61** and any excess product containing organic liquid flows to sump **16** within vessel **13** for use in the first contacting in tower **30** and subsequent removal to product recovery. Any aqueous phase which may be present in second sump **52** as indicated at **60** also may flow to the first sump in vessel **13** for joining with the aqueous phase therein as indicated at **15**.

When the system is operated using two contacting towers, substantially pure organic scrubbing liquid may be charged to the system of both sumps; however, it is preferred to charge it by line **63** only in a quantity substantially equal to that being removed to product recovery. In this manner, substantially pure organic contacting liquid is used to remove the remaining trace quantities of product from the gaseous phase entering tower **51** to obtain a more efficient operation.

Connecting line **64** joining line **48** may be provided to pump quantities of organic liquid to first sump **16** to maintain the desired level, if needed, and the substantially product free inert gas exiting tower **51** by line **65** may be vented to the atmosphere without air pollution problems or recovery for reuse in an inert gas system, if desired.

In a typical example of the operation of the process for the recovery of tetramethyllead and toluene from a nitrogen gas stream, a vapor feed composition having a concentration of 57% nitrogen, 40.2% tetramethyllead-toluene mixture and 2.8% water, all percentages by weight, and having a temperature of 100° to 110°F, was fed to a fume scrubber where it was scrubbed with a 4% by weight aqueous solution of sodium hydroxide.

Scrubbed vapors continued to a first tower where they were contacted countercurrently with ethylene dichloride at a temperature of about 35°F in a quantity sufficient to cause the gaseous phase leaving the first tower to have a temperature of about 50°F.

The contacted vapors from the overhead of the first tower were fed to the bottom of a second tower where they were contacted countercurrently with ethylene dichloride entering the tower at a temperature of approximately 32°F. Contacted vapors entered the second tower at a temperature of approximately 50°F and exited at a temperature of approximately 35°F. Both towers were substantially identical and contained 2-inch steel Pall rings in a depth in the tower sufficient to provide about three theoretical distillation plates in each tower and each tower was operated at substantially atmospheric pressure.

The concentration of the sodium hydroxide in the aqueous caustic solution was maintained at 4% by weight by the addition of a 25% by weight aqueous sodium hydroxide solution as necessary and fresh ethylene dichloride was added to the sump of the second tower as necessary to compensate for the volume of ethylene dichloride-tetramethyllead-toluene mixture removed from the sump below the first tower.

Approximately 90% of the tetramethyllead-toluene mixture in the vapor feed stream was recovered in the first tower and a total of 99% of the tetramethyllead-toluene mixture in the original vapor feed stream was recovered in the two-tower system. Based upon total quantity of tetramethyllead produced from a reaction process not utilizing the process, sufficient tetramethyllead-toluene mixture was recovered using the process to increase the overall yield of the tetramethyllead-toluene produced 4 to 5%.

REMOVAL FROM WATER

Treatment of dissolved lead involves for the most part reaction to form a lead precipitate, and sedimentation. Liebig, et al, (42) have reported, however, the treatment of lead-contaminated water by ion exchange. Lead content was reduced from 0.055 to 0.0015 mg/l. However, the process was bench-scale, and very few details were provided. Undoubtedly, lead-contaminated water is susceptible to ion exchange treatment on the same basis and at similar costs as reported in, for example, the hexavalent chromium and the copper treatment feasibility reports. As discussed in those reports, the net cost of ion exchange treatment is reduced if recovery of the removed metal is economically feasible.

In the precipitation processes, lead is normally precipitated as the carbonate, $PbCO_3$ or the hydroxide, $Pb(OH)_2$. The solubilities of both compounds are very low at alkaline pH, and formation of either product is effective in reducing dissolved lead concentrations (43). Lead is commonly reported as being recovered from the sludge formed in the precipitation processes (44).

In forming insoluble lead hydroxide, lime is the treatment chemical of choice. Fales (23) reports the use of lime in precipitating a plating bath water containing 2.0 to 140 mg/l of lead, plus other metals. The author states that after lime treatment and settling for 1 hour the supernatant was clear and contained only a trace of heavy metals. Lead in a municipal drinking water supply has also been reported reduced by lime treatment and settling from an initial value of 0.31 mg/l down to 0.1 or less mg/l of lead (45).

Treatment of a wastewater from a tetraethyllead plant has also been described (44). The effluent was alkaline, and contained inorganic lead salts in solution and suspension, plus organic lead salts. No values of lead concentration, either prior to or after treatment, were included in that report (44). The waste pH was adjusted to 8 to 9, and lead precipitated by addition of lime.

The precipitation was assisted by the addition of ferrous sulfate, which acted as a flocculating agent. The treated effluent flowed to a Dorr-Oliver Clariflocculator, in which the lead sludge was collected at the bottom, dewatered in a filter, and reclaimed by a lead slag refinery. No costs were reported for the process.

Nozaki and Hatotani (46) also have described treatment of wastewaters from a tetraethyllead manufacturing process. The two major categories of waste were inorganic lead wastewaters and organic lead wastewaters. After sedimentation in a holding basin to recover solid lead and lead oxide, the inorganic lead waste fraction was treated by coagulation with ferric and ferrous sulfate.

Influent inorganic lead of 66.1 mg/l was reduced to "trace" levels in the effluent. Treatment of combined organic and inorganic lead waste streams containing respective concentrations of 13.3 mg/l and 84.9 mg/l was not as effective, however. Effluent contained a trace to 1.4 mg/l of inorganic lead, and most of the influent organic lead. A separate bench-scale treatment process was therefore developed for the organic lead waste stream.

The organic lead compounds were removed effectively on a strongly acid cation exchange resin. Reductions of organic lead from 126.7 to 144.8 mg/l down to 0.020 to 0.53 mg/l (as lead) were reported. The organic lead compounds absorbed on the ion exchange resin were eluted by caustic soda. This eluate was collected in a reaction basin into which chlorine gas was injected at 95°C for 45 minutes. The organic lead compounds were thereby almost completely converted to inorganic lead compounds, which could be treated as described above.

Overall lead removals to less than 1 mg/l residual were expected for the final plant effluent, after installation of the organic lead wastewater treatment sequence (46).

A third report on treatment of lead-bearing waste from a tetraethyl plant reported a similar treatment sequence for inorganic lead (47). Lead occurring in the manufacturing process effluent at a concentration of 45 mg/l was reduced to 1.7 mg/l by precipitation at pH 10.4 to 10.8, aided by addition of ferrous sulfate. The lead sludge was recovered, and lead was extracted and reclaimed.

Similar treatment has been reported for lead removal from a municipal water supply (48). Lime and ferrous sulfate are added to the water to precipitate and coagulate lead leached from a nearby abandoned mine. As is common for most water treatment plants, the water is filtered, which undoubtedly acts as polishing treatment by removing nonsettleable lead hydroxide particles carried through the sedimentation basins.

Precipitation of dissolved lead by dolomite ($CaCO_3 \cdot MgCO_3$) to yield lead carbonate has also been reported (49). The process involved filtration of lead wastewater through a calcined dolomite bed, which reacted with and trapped the dissolved lead. The dolomite filters were regenerated by backwashing with water, which removed the lead carbonate formed in the process. Neither treatment efficiencies nor costs are reported for the process. Based upon a comparison of chemical strength and commercial cost, however, dolomite is about equivalent to lime on the basis of net operating cost.

The reaction of lead with dolomite, yielding insoluble lead carbonate, probably explains why very little lead enters receiving streams from the southeastern Missouri lead mines. The lead vein is located in a dolomite deposit, and the mine tailings thus contain crushed dolomite. As the mine tailings are channeled through sedimentation basins prior to release of the waste effluent, dolomite plus reacted lead carbonate are readily removed. As the authors point out, under existing conditions any heavy metals in the wastewater are precipitated rapidly in the settling lagoons at the slightly basic pH (7.5 to 8.2) of the wastewater.

In addition to the standard lead removal processes discussed above, there are several patented commercial processes available such as a Du Pont process (50). All are based upon the recovery and recycle of lead, rather than directed specifically toward waste treatment.

In summary, sparse information is available in the technical literature on the treatment of lead-bearing wastewaters. Precipitation by lime or dolomite, and sedimentation with sludge disposal or lead recovery have been reported. Little data are available on effluent lead values after treatment. However, the extreme insolubilities of both lead hydroxide and lead carbonate, the two most common precipitation products, would indicate that good conversion of dissolved lead to insoluble lead should be achieved. These precipitated lead compounds can then be removed by settling, with or without a coagulant aid or removed by filtration.

A process developed by H.E. Collier, Jr. (51) involves the removal of dissolved alkyllead compounds from the aqueous effluent used in the manufacture of alkyllead compounds. Tetraalkyllead compounds are highly useful antiknock compounds and are manufactured commercially by alkylating a lead-sodium alloy with an alkyl chloride, such as ethyl chloride, methyl chloride or a mixture of methyl and ethyl chlorides.

In these processes, after the excess unreacted alkyl chloride has been distilled off, the rest of the reaction mass is drowned in a large volume of water and the tetraalkyllead steam-distilled therefrom. The still residue is settled, the solids separated from most of the water and further washed with water to remove various lead-containing salts. The water from the settling, separating and washing steps constitutes waste aqueous effluent from the tetraalkyllead manufacturing process.

In order to remove certain sludge-forming impurities from the steam-distilled tetraalkyllead, the impure tetraalkyllead product is further subjected to an aeration or oxygen purification treatment in the presence of water as disclosed in U.S. Patent 2,400,383. The purified tetraalkyllead is then separated from the aqueous phase. This aqueous phase, containing some soluble alkyllead materials formed in the purification, also constitutes part of the aqueous effluent from the tetraalkyllead process.

The aqueous effluent from the tetraalkyllead process may be neutral but generally is strongly alkaline, usually having a pH between 11 and 12, due to the reaction of the water with unreacted sodium-lead alloy to produce sodium hydroxide. It contains large amounts of sodium chloride and small amounts of water-soluble organic lead compounds, such as those formed in the purification process described above.

In addition to the above, a very small amount of water-soluble inorganic lead compounds are sometimes present in the aqueous effluent. The disposal of the aqueous effluent from the tetraalkyllead process is a serious problem, since, in some locations, the maximum amount of soluble lead compounds allowed in aqueous effluent discharge is 5 parts per million parts of water.

The sodium chloride and sodium hydroxide compounds in the aqueous effluent naturally do not present a serious disposal problem. Moreover, the water-soluble inorganic lead compounds in the aqueous effluent do not pose a serious disposal problem since they can be easily removed from the effluent before discharge by simple adjustment of the pH to a range of from 8 to 9.5 in the presence of water-soluble carbonates as described in Canadian Patent 572,192.

However, the soluble alkyllead compounds in the aqueous effluent, such as trialkyllead chlorides, trialkyllead hydroxides, dialkyllead dichlorides and dialkyllead dihydroxides, cannot be discharged into lakes and streams without causing a potential contamination problem. Moreover, the disposal of such soluble alkyllead compounds constitutes a significant loss of lead.

While the soluble inorganic lead compounds may be removed from the effluent by adjustment of the pH to 8 to 9.5 in the presence of water-soluble carbonate as described above, or by precipitation with hydrogen sulfide or other soluble sulfides, the organic lead compounds are not satisfactorily removed by such treatment, and furthermore, any excess soluble sulfide in the effluent would be highly objectionable and would have to be removed from the aqueous effluent before its disposal.

The process constitutes a process for treating an aqueous effluent from the manufacture of alkyllead compounds and containing about 5 to 5,000 ppm of lead as dissolved organic lead not precipitatable by pH adjustment to 8 to 9.5, which process comprises the steps of:

[1] Adjusting the pH of the effluent to between 8.0 to 9.5
[2] Intimately contacting the aqueous effluent with an ozone-containing gas
[3] Precipitating the converted lead compounds
[4] Separating the precipitated lead-containing compounds from the aqueous effluent

This process is based on the surprising discovery that ozone is highly effective in removing water-soluble organic lead compounds from wastewater by producing water-insoluble products. Indeed, the discovery that ozone successfully converted the soluble products to water-insoluble materials was most surprising since air and other oxidizing agents have been found unsatisfactory to effect this removal.

By this process the dissolved alkyllead can be removed from the aqueous effluent to such an extent that the aqueous effluent from tetraalkyllead manufacture can be safely discharged into lakes and streams. The process is easy to operate, and the quantity of materials, the time and intimacy of contact are easily coordinated to reduce the dissolved organic lead content to an acceptable level, below 5 ppm, or substantially nil if desired.

A process developed by E.R. Taylor, Jr. (52) involves the removal of dissolved organic lead from an aqueous effluent produced in the manufacture of alkyllead compounds by contacting a metal more electropositive than lead but essentially nonreactive with water with the effluent until the dissolved organic lead is converted to an insoluble lead-containing product, leaving a reduced dissolved organic lead content in the effluent. The following is one specific example of the conduct of this process.

250 ml of effluent water from the tetraethyllead process with a pH of 11.2 and containing 9 ppm lead [90% as organic lead, tri(mixed C_1 and C_2 alkyl)lead, and 10% inorganic lead as Pb^{++}] were passed at 10 ml/min through a column of No. 40 mesh iron filings 3 cm in diameter and 13 cm deep. The effluent water had a contact time of 9.2 minutes with the column of iron. The effluent water after the iron contact contained no organic phase and a maximum lead content of 0.2 ppm, thereby showing a greater than 97% reduction of lead content.

A process developed by C. Lores and R.B. Moore (53) is a process for treating an aqueous medium containing dissolved organic and inorganic lead compounds, for example, aqueous effluent from the manufacture of tetraalkyllead, with an alkali metal borohydride to substantially reduce the level of dissolved lead in the aqueous medium.

SOLID WASTE DISPOSAL

Because metallic lead is insoluble in water and not attacked by most inorganic acids or bases, lead in small quantities is often disposed of in landfills (54). Though this disposal method is satisfactory under most conditions, it is not recommended because lead is lost to future use. In addition, under certain conditions, lead is solubilized as either the chloride or the sulfate, which are both slightly soluble in water.

Since the lead oxides are insoluble in water, and also are very stable under ordinary temperatures, disposal of lead oxide wastes by means of landfill is acceptable. Mixed wastes of lead oxides such as glass and ceramic products often do not contain any water-soluble toxic components. Hence, the risk of contaminating the ground and surface waters by these waste materials is very minimal.

Disposal of soluble lead compounds such as the acetate, nitrate and nitrite by landfill technique is generally unacceptable. Even the slightly soluble lead chlorite should not be disposed of by landfill. However, lead carbonate is practically insoluble in water and, in fact, occurs naturally as the mineral cerussite. Similarly, lead sulfide is also insoluble in water and occurs naturally as the mineral galena. Therefore, disposal of these two lead compounds by the landfill technique in California Class 1 type disposal sites might be deemed adequate.

Recovery is also a viable option for some lead-containing solid wastes. Thus, in a process described by Fischer (55), used storage batteries are coarsely ground into pieces in an impact crusher and the fine material containing metallic lead is separated and ground again and then classified. The remaining part of the coarsely ground material is subjected to a sink-flotation-separation and the fine material recovered as a sludge is introduced into a tubular vibrating mill filled with grinding bodies and ground therein, whereupon the ground material is screened and classified to recover the lead therein.

ECONOMIC IMPACT OF CONTROLS

General data on equipment for air pollution control which is applicable to the treatment of particulate lead emissions has been presented (56). No waste treatment costs directly bearing on lead wastewaters have been uncovered, despite an extensive search of the industrial and pollution literature, according to Patterson et al (25).

Nozaki et al (48) have reported on the treatment of aqueous wastes from a TEL manufacturing plant; however, the author reports that one-half million dollars (1960 prices) was spent on waste treatment facilities, out of a total plant investment of twelve million dollars. Waste treatment capital costs thus represented 4.2% of the total plant costs. Waste treatment included settleable solids and chloride control, in addition to lead removal.

The treatment costs associated with lead removal by lime and ferrous sulfate treatment are due to the chemical and operating costs of the lime precipitation and sludge disposal processes. Anderson and Iobst (57) have reported costs associated with a mixed metal plating waste treatment by lime precipitation and sludge disposal. The treatment plant had an initial installation cost of $1,108 per 1,000 gallons treated per day.

Operating costs were 80¢ per 1,000 gallons. The waste, while not containing lead, was a typical heavy metal waste which required treatment equivalent to the lead precipitation process. Watson (58) has reported lime treatment costs for a mixed metal plating rinse water at 62¢ per 1,000 gallons, and as $1.62/1,000 gallons for operating plus amortization costs.

REFERENCES

(1) Hall, S.K., "Lead Pollution and Poisoning," *Env. Sci. Tech.,* 6, 30-35 (1972).

(2) Luckey, T.D., Venugopal, B. and Hutcheson, D., *Heavy Metal Toxicity and Hormonology,* New York, Academic Press (1975).

(3) Passow, H., Rothstein, A. and Clarkson, T.W., "The General Pharmacology of Heavy Metals," *Pharmacol. Review,* 13, 185-224 (1961).

(4) Kehoe, R.A., "Industrial Lead Poisoning," *Hygiene and Toxicology,* F.A. Patty, Ed., 2nd Ed., New York, John Wiley (1962).

(5) Chisholm, J.J., Jr., "Lead Poisoning," *Scientific American,* 224, 15-23 (1971).

(6) Kinnison, R.R., "Lead—In Search of the Facts," *Env. Sci. Tech.,* 10, 644-649 (July 1976).

(7) Schroeder, H.A., *The Poisons Around Us,* Bloomington, Indiana University Press (1974).

(8) Tucker, A., *The Toxic Metals,* London, Earth Island Ltd. (1972).

(9) Committee on Biologic Effects of Atmospheric Pollutants, *Airborne Lead in Perspective,* Wash., D.C., Nat. Acad. of Sciences (1971).

(10) Hunter, D., *The Diseases of Occupations,* 4th Ed., Boston, Little Brown & Co. (1969).

(11) Nat. Inst. for Occupational Safety & Health, "Criteria for a Recommended Standard . . . Occupational Exposure to Inorganic Lead," Wash., D.C. (1972).

(12) Lane, R.E., "Health Control in Inorganic Lead Industries—a Followup of Exposed Workers," *Arch. Envir. Health,* 8, 243-50 (1964).

(13) Dingwall-Fordyce, I. and Lane, R.E., "A Followup Study of Lead Workers," *Brit. J. Ind. Medicine,* 20, 313-15 (1963).

(14) Malcolm, D., "Prevention of Long Term Sequelae Following the Absorption of Lead," *Arch. Envir. Health,* 23, 292-98 (1971).

(15) Henderson, D.A. and Inglis, J.A., "The Lead Content of Bone in Chronic Brights Disease," *Australasian Ann. Med.,* 6, 145-54 (1957).

(16) Lane, R.E., "The Care of the Lead Worker," *Brit. J. Ind Medicine,* 6, 125-63 (1949).

(17) Snyder, R.B., Wuebbler, D.J., Pearson, J.E., and Ewing, B.J., "A Study of Environmental Pollution by Lead," Document No. 71-7, Chicago, Illinois, Illinois Inst. of Environmental Quality (Nov. 1971).

(18) Faust, W.J., and Sterba, M.J., *Minimizing Exhaust Emissions—A Realistic Approach,* Special Technical Publication No. 487, Phila., Pa., Am. Soc. for Testing Materials (1971).

(19) Goldberg, A.J., "A Survey of Emissions and Controls for Hazardous and Other Pollutants," Wash., D.C., Environmental Protection Agency (Feb. 1973).

(20) Davis, W.E. and Assoc., "Emission Study of Industrial Sources of Lead Air Pollutants," 1970, Leawood, Kansas (April 1973).

(21) Elkins, H.B., *The Chemistry of Industrial Toxicology,* 2nd Ed., New York, John Wiley & Sons (1959).

(22) Ottinger, R.S., Blumenthal, J.L., Dal Porto, D.F., Gruber, G.I., Santy, M.J. and Shih, C.C., "Recommended Methods of Reduction, Neutralization, Recovery or Disposal of Hazardous Waste," Report PB-224,580, Springfield, Va., Nat. Tech. Information Service (Aug. 1973).

(23) Fales, A.L., "A Plating Waste Disposal Problem," *Sew. Wks. Jour.,* 20, 857-860 (1948).

(24) Pinkerton, H.L., "Waste Disposal," in *Electroplating Engineering Handbook,* 2nd Ed., A. Kenneth Graham, ed, New York, Reinhold Publishing Co. (1962).

(25) Patterson, J.W. and Minear, R.A., "Wastewater Treatment Technology," Report PB 204,521, Springfield, Va., National Technical Information Service (August 1971).

(26) Lawrence Berkeley Laboratory, "Instrumentation for Environmental Monitoring—Air, *Publ. LBL-1, Vol. 1,* Berkeley, Univ. of Calif. (Feb. 1, 1973).

(27) Keenan, R.G., Byers, D.H., Saltzman, B.E. and Hyslop, F.L., "The USPHS Method for Determining Lead in Air and in Biological Materials," *Am. Ind. Hyg. Assoc. J.* 24, 481-91 (1963).

(28) Walker, A.O., U.S. Patent 3,453,081, July 1, 1969, assigned to Nalco Chemical Company.

(29) Walker, A.O., U.S. Patent 3,870,469, March 11, 1975, assigned to Nalco Chemical Co.
(30) Lawrence Berkeley Laboratory, "Instrumentation for Environmental Monitoring—Water," *Publ. LBL-1, Vol. 2,* Berkeley, Univ. of Calif. (Feb. 1, 1973).
(31) Browner, R.F., Dagnall, R.M. and West, T.S., *Anal. Chim. Acta.*, 50, 375-81 (June 1970).
(32) Palma, R.J., Sr. and Pearson, K.H., *Anal. Chim. Acta.,* 49, 497-504 (Mar. 1970).
(33) Bloch, R., Bauer, R. and Phillips, B.F., U.S. Patent 3,635,679, January 18, 1972, assigned to Miles Laboratories, Inc.
(34) Horine, A.A., U.S. Patent 3,809,537, (May 7, 1974).
(35) Ethyl Corporation, "A Guide to the Safe Handling and Mixing of Ethyl Antiknock Compounds," Houston, Texas.
(36) American Petroleum Institute, "A Guide for Controlling the Lead Hazard Associated with Tank Entry and Cleaning," Publication 2015A, Wash., D.C., (May 1975).
(37) Environmental Protection Agency, "Atmospheric Emissions from Wet-Process Phosphoric Acid Manufacture," Publication EH-57, Research Triangle Park, N. Carolina (1970).
(38) Hama, A., "Dust Control of Reverberatory and Blast Furnaces for Battery Scrap," *Journal of Air Pollution Control (Tokyo),* 646-648 (August, 1970).
(39) Danielson, J.A., ed., *Air Pollution Engineering Manual,* Cincinnati, U.S. Public Health Service (1967).
(40) Nilsson, F. and Rudling, B., "Air Pollution Control at the Boliden Copper and Lead Smelting Plant, Ronnskarsverken, Sweden," England, H.M. and Beery, W.T., ed. *Proceedings of the Second International Clean Air Congress,* New York, (1971).
(41) Jaasma, W.C., U.S. Patent 3,403,495, October 1, 1968, assigned to Ethyl Corporation.
(42) Liebig, C.F., Jr., Vanselow, A.P. and Chapman, H.D., "The Suitability of Water Purified by Synthetic Ion-Exchange Resins for Growing Plants in Controlled Nutrient Cultures," *Soil Sci.,* 55, 371-376 (1943).
(43) Sillen, L.G., and Martel, A.E., *Stability Constants of Metal-Ion Complexes,* 2nd Ed., Sp. Pub. No. 17, Chemical Soc., London (1964).
(44) Anonymous, "An Integrated Plant for Tetraethyl Lead," *Indust. Chemist,* 30, 429-436 (1954).
(45) Anonymous, *Annual Report Western Australian Government Chemical Laboratories,* Dept. of Mines, Australia, (1962).
(46) Nozaki, M. and Hatotani, H., "Treatment of Tetraethyl Lead Manufacturing Wastes," *Water Res.,* 1 (2), 167-177 (1967).
(47) Robb, L.A., "Waste Water Treating Facilities at Ethyl Corporation of Canada, Ltd.," *Proc. 8th Ontario Industrial Waste Conf.,* 90-96 (1961).
(48) Anonymous, "East Midland District Meeting," *J. Instit. Municip. Engrs.,* 84 (November): XXV-XXVCC (1957).
(49) Voznesenskic, S.A., Evalonova, A.V. and Suvorova, R.V., "Physico-Chemical Methods of Purification of Industrial Waste Waters," *Wat. Poll. Abst.,* 13, 135 (1940).
(50) E.I. du Pont de Nemours & Co., French Patent 1,361,089, May 15, 1964.
(51) Collier, H.E., Jr., U.S. Patent 3,308,061, March 7, 1967, assigned to E.I. du Pont de Nemours & Co.
(52) Taylor, E.R., Jr., U.S. Patent 3,697,567, October 10, 1972, assigned to E.I. du Pont de Nemours & Co.
(53) Lores, C. and Moore, R.B., U.S. Patent 3,770,423, November 6, 1973, assigned to E.I. du Pont de Nemours & Co.
(54) Powers, P.W., *How to Dispose of Toxic Substances and Industrial Wastes,* Park Ridge, N.J., Noyes Data Corp. (1976).
(55) Fischer, R., U.S. Patent 3,777,994, Dec. 11, 1973, assigned to Klockner-Humboldt-Deutz A.G.
(56) McFarland, A.R., "Control Technology for Toxic and Hazardous Air Pollutants," Chicago, Illinois Institute for Environmental Quality (Dec. 1974).
(57) Anderson, J.S. and Iobst, E.H., Jr., "Case History of Wastewater Treatment in a General Electric Appliance Plant," *Jour. Wat. Poll. Control Fed.,* 10, 1786-1795 (1968).
(58) Watson, K.S., "Treatment of Complex Metal-Finishing Wastes," *Sew. Ind. Wastes, 26,* 182-194 (1954).

MANGANESE

Manganese (Mn), an essential micronutrient, is a respiratory irritant and a systemic poison when inhaled mainly as oxides in excessive amounts. Manganese is not found in a free state in nature, but its minerals are quite abundant. Sea water contains a 0.002 ppm concentration of dissolved manganese, while the average manganese concentration in the earth's crust is calculated to be 1,000 ppm. Manganese functions as an essential element of at least 100 minerals and as an accessory element in many others. Manganese oxide minerals predominate over the manganese carbonate and manganese silicate minerals. The most common manganese minerals are:

	% Mn
Pyrolusite	60-63
Psilomelane	45-60
Hausmannite	72
Rhodochrosite	47
Rhodonite	42
Braunite	63

Relatively little manganese is mined within the United States.

TOXICITY

The toxicity of manganese and its compounds has been reviewed by Luckey et al (2). Inhalation of manganese oxides may cause chronic manganese poisoning or manganic pneumonia. Chronic manganese poisoning is a disease affecting the central nervous system, resulting in total or partial disability if corrective action is not taken. Some people are more susceptible to manganese poisoning than others. Manganic pneumonia is a croupous pneumonia often resulting in death. The effect of long exposure to low concentrations of manganese compounds has not been determined, according to Sullivan (3). Manganese compounds are known to catalyze the oxidation of other pollutants, such as sulfur dioxide, to more undesirable pollutants, for example, sulfur trioxide. Manganese intoxication has acute effects on the lungs, liver, central nervous system and blood system of the body. Epidemiologic appraisals of exposures to manganese have primarily involved industrial exposures. Initial reports of manganese intoxication described a neurologic disorder closely resembling parkinsonism. This disease, known as manganism, is generally considered to result more from inhalation of Mn than ingestion, and its development is dependent on atmospheric manganese concentration, duration of exposure, and the size of the suspended manganese particles (4).

Exposure to manganese in the atmosphere has been suggested as being statistically associated with several types of respiratory diseases, heart diseases, and cancers. One epidemiological study has attempted to correlate the incidence of cancers, bronchitis, and pneumonia (by mortality rates) with atmospheric concentrations of a number of metal particulates throughout Great Britain (5). The investigator concluded that manganese was significantly correlated with lung cancer mortality and bronchitis incidence. However, because of the many variables, a strong cause-and-effect relationship was not established between manganese and these diseases.

Excessive exposure to atmospheric manganese has been associated with the occurrence of pneumococcal pneumonia (acute lobar pneumonia caused by *Diplococcus pneumoniae* bacteria) in Saude, Norway, which had eight times more incidence of it than the general population. The dramatic increase in pneumococcal pneumonia occurred within a fifteen-year period from the start of industrial activity. It was found that a major industry's electric furnace produced ferromanganese containing 80% manganese and silicomanganese, with manganese and silica particles occurring in sizes of less than 5 μ (6). It was suggested that inhalation of manganese dust reduces the resistance of the lungs to pneumococcal invasion, thus predisposing the lungs to the development of pneumococcal pneumonia.

A recent study in Japan conducted in Kanazawa City describes the differences between middle school children attending a school approximately 100 meters from a ferromanganese plant and the children attending a school situated in the central part of the city, about seven kilometers from the ferromanganese plant (7). The students attending the school located near the plant also lived near the plant. The students at the more distant school lived no closer than five kilometers from the ferromanganese plant. Based on a self-administered questionnaire, the children from the school nearer the ferromanganese operation had a significantly higher rate of cough, colds, and upper respiratory infections than did the children attending the "control" school. The mean lung capacity of the students near the ferromanganese plant was found to be lower than that of the control group.

Unfortunately, the manganese concentrations in the air that the children breathed were not reported. However, an estimation of the ambient Mn concentration can be derived. In a one-day sample of down-wind smoke, a manganese concentration of 260 $\mu g/m^3$ was reported. In a five-day sample at a site 330 meters from the ferromanganese plant, the concentration averaged 160 $\mu g/m^3$. From dustfall measurements, it was estimated that there was a manganese-in-air concentration of about 10 to 40 $\mu g/m^3$. Therefore, it was assumed that the actual respiratory exposure to manganese was in the range of approximately 10 to 260 $\mu g/m^3$.

The results of chronic exposure to manganese dioxide fumes or dust have been studied in workers involved in mining, processing, and industrial use of manganese. Summaries of these studies are listed in Table 32 according to exposure level.

TABLE 32: RELATIONSHIP OF HUMAN HEALTH EFFECTS TO CONCENTRATIONS OF Mn IN AIR

Human Health Effects*	Ambient Air Concentrations, $\mu g/m^3$, Mn
Respiratory	
Pneumonia, dyspnea, shallow breathing	30,000-90,000
Psychomotor	
Emotional instability, apathy, hallucinations, compulsory acts	500-46,000
Neurological	
Parkinsonism type	4,800-28,440
Muscular hypertonia	500-46,000
Tremors	20-46,000
Sweating	120,000-872,000
Progressive weakness	100-4,500
Muscular pains	500-872,000
Facial masking and reduced blinking	20-5,900
Miscellaneous	
Muscular fatigue	500-872,000
Headache	4,800-872,000
Somnolence	120,000-872,000
Sexual depression	500-46,000
Synergistic Mechanism	
At this level, the Mn can be expected to catalyze 10 to 30% of the ambient SO_2 to sulfates in the presence of high humidity	0.1
Proposed Standard	
It was calculated that a conversion rate of SO_2 to H_2SO_4 of approximately 0.12% could be achieved at this level.	0.006

*These signs and symptoms appear in most cases. Also, the longer the duration, the greater the accumulation of the toxic material within the body and thus the greater the toxic effects which may reach the level of acute symptoms, i.e., symptoms caused by higher concentrations for shorter periods of exposure.

Source: Reference (1)

Men exposed to MnO_2 and higher oxides of manganese in the air suffered a high incidence of manganese pneumonia, (a form of lobar pneumonia unresponsive to antibiotics) according to a study of pneumonia incidence of workers involved in the manufacture of potassium permanganate. The pneumonia rate was 26/100 among exposed workers while 0.73/100 among unexposed workers (8). Analyses of plant dust samples revealed a 41 to 66% Mn content and nearly all particles to be less than 1 μ in size.

A retrospective study was performed at a manganese mine in Chile in which fifteen cases of manganism were discovered among eighty-three miners. The average time of exposure was eight years and two months (ranged from nine months to sixteen years). Introduction of pneumatic drilling resulted in increased dustiness. The lowest concentrations were 500 $\mu g/m^3$ and the highest concentrations were 46,000 $\mu g/m^3$. The psychomotor symptoms experienced by these workers were emotional instability, apathy, hallucinations, and compulsory acts, while the outstanding neurologic signs were muscular hypertonia, expressionless face, gait changes, and monotonous voice (9). Eleven of 34 workers suffered neurologic symptoms of manganese poisoning in a Pennsylvania manganese ore-crushing and packaging mill in 1939. The lowest concentration at which a case of chronic manganese poisoning was found was 30,000 $\mu g/m^3$; the worker was exposed to manganese dioxide dust. The exposure time was from less than 1 to 15 years. The proportion of workers affected increased with increasing atmospheric manganese concentration and increasing length of employment (10).

Lower concentrations of fumes, however, have produced cases of chronic manganese poisoning. Two cases of chronic neurological disease in manganese steel plant workers were produced by the inhalation of manganese. Neurological examinations revealed inexpressive faces and infrequent blinking among other signs. Air samples analyzed by the membrane filter method reported Mn concentrations of 100 to 4,500 $\mu g/m^3$ (11).

In a dry cell battery plant, 8 workers were found to have neuropsychiatric manifestations and 6 had chronic psychosis after exposure to dust containing 65 to 70% MnO_2. One worker had left hemiparkinsonism, and another had left choreoathetosis. High concentrations of manganese dust ranging from 4,800 to 28,440 $\mu g/m^3$ were found in the working areas. The exposure period ranged from 1 to 16 years. One case of psychosis progressed to parkinsonism. Individual susceptibility seems to be apparent in all of these cases. Headaches were the commonest complaint. In addition, sexual impotence and arrested sleep rhythms were reported (12).

Five cases of manganism were found among workers employed in the production and processing of ferromanganese alloy. Three were exposed to manganese fumes and two were exposed to dust. One of these, a hot blast-man exposed to MnO_2 fumes for 10½ years, showed clinical micrographia, loss of associated arm movements, and reduced blinking reflex among other signs. The manganese fumes in the working area ranged from 20 to 5,900 $\mu g/m^3$, weighted average = 3,300 $\mu g/m^3$ (13).

One hundred fifty cases of manganese poisoning were observed among underground workers in three different mines in Morocco. The majority of these workers were pneumatic-drill operators. About 19 others were underground laborers working in the immediate area of the drills and therefore in an atmosphere of dust from the blasting. Details of exposure time were obtained for 115 of the patients, the majority of whom had been exposed from 3 months to over 10 years.

The disease may be described in three phases. In the prodomal period (first phase), disorders are subjective. The worker complains of somnolence, general asthenia (any weakness, but one especially originating in muscular or cerebellar disease) and anorexia (loss of appetite) as well as muscular pains. In the course of the preceding subjective manifestations, certain objective symptoms (second phase) such as clumsiness of movement, hyperemotionalism, and speech disorders are observed. As the disease progresses, a third phase of complete establishment characterized by serious disorders leading to permanent disability is manifested.

The worker experiences difficulty in walking or in movements, tremors, vasomotor and secretory disorders, and hypertonia (abnormal tension of arteries or muscles). In one case the hypertonia led to permanent closure of mouth (14).

A separate problem is potential exposure to manganese if methylcyclopentadienylmanganese tricarbonyl (MMT) is used as an automotive antiknock compound to supplement or replace tetraethyllead (TEL). In a recent animal study conducted by the EPA, rats and hamsters were exposed for 8 hours a day for 2 months to automotive emissions where gasoline containing ⅛ gram MMT was used as a fuel. No significant pathological lesions were found that could be related to MMT. It has been suggested that additional studies utilizing proposed concentrations of MMT in gasoline are needed to determine lung burden and overall biological effect.

The EPA (15), commenting on a 1968 Ethyl report, noted that in industrial experience typical handling in an open system involved the pouring of MMT from storage tanks into fuel containers. Normal transfer and distribution of the product have not resulted in known cases of illness or reactions in workers. However, several accidental cases of other types of human exposure have been noted. Two individuals, while wearing airline masks, were accidentally exposed during a maintenance operation by having their clothes thoroughly soaked by a spray of MMT, creating a potential skin contact for approximately 1.5 hours.

The immediate symptom was a slight burning sensation of the skin. Urinary manganese levels rose to 46 and 137 μg manganese per liter following this exposure. Urine samples taken a few weeks later showed the Mn levels to be within the normal 2-3 μg per liter range. The men did not exhibit additional significant symptomatology. In another accident, a worker spilled 5-15 ml of MMT on his hands. Within 5 minutes the symptoms included: thick tongue, giddiness, nausea, and headache. A fellow worker, handling a rag soaked with MMT, noted a metallic taste in his mouth.

Again, the EPA (15), commenting on the 1968 Ethyl report, noted that exposures to MMT by inhalation and topically through the skin of the hands and forearms have been reported in 6 individuals; before the skin was cleaned, exposure period was estimated at 5-30 minutes. The symptoms began 5-60 minutes after exposure and in most cases completely subsided within 2 hours, with the exception of two workers in which a vague abdominal distress persisted for two days. No persistent changes on physical or neurological examination were noted in any of the men.

It is suggested that chronic exposure to such chemicals would be an occupational health problem for individuals manufacturing the additive and to gasoline attendants. It may also present a problem to the general public because of the increase in self-service gasoline stations. Manganese carbonyl compounds have also been used as a fuel additive in fuel oil as a smoke inhibitor. The EPA reported the emission of manganese associated with the addition of MCT to a Number 2 oil at a concentration of 1:9,000 to be 1,219 $\mu g/m^3$ or 20 mg/kg of fuel. The Ethyl Corporation is the sole producer of MCT in the U.S. with a production rate of 450,000 kg/year (4).

Presently, there is no information available on the health effects of a very long MMT exposure at low concentrations or on exposures of certain potential high-risk groups such as the aged, pregnant women, children, and those with cardiovascular disorders.

EXTENT OF EXPOSURE

Manganese emission to the atmosphere results primarily from industrial processes and fuel combustion. Industrial emission sources include production of manganese metal from manganese ore, iron and steel industries, blast-furnace and electric-furnace production of ferromanganese (used primarily in steel production), aluminum production, magnesium production, manufacture of dry cell batteries, manufacture and use of welding rods, manufacture of manganese chemicals, and mixing and application of fertilizers (1)(3).

The most likely sources of manganese air pollution are the iron and steel industries producing ferromanganese. Two studies, one in Norway and one in Italy, have shown that the emissions from ferromanganese plants can significantly affect the health of the population of a community. Other possible sources of manganese air pollution are manganese fuel additives, emissions from welding rods, and incineration of manganese-containing products, particularly dry cell batteries. Manganese may be controlled along with the particulates from these sources. Figure 12 and Table 33 detail the sources of airborne manganese (16).

TABLE 33: MANGANESE EMISSIONS

Source	Amount, tons	Pollutant, %
Manganese mining	5	0.03
Primary manganese preparation	325	1.71
Iron and steel		
Blast furnace	1,000	5.27
Open hearth furnace	1,660	8.74
Basic oxygen furnace	1,060	5.58
Electric arc furnace	620	3.26
Gray iron foundry		
Cupola	2,770	14.58
Ferro-alloy preparation		
Blast furnace	1,113	5.86
Electric furnace	3,669	19.32
Nonferrous alloy preparation		
Furnaces	60	0.32
Material handling	Neg	Neg
Silicomanganese preparation		
Electric furnace	4,164	21.92
Manganese chemical preparation	300	1.58
Dry storage battery production	90	0.47
Welding rod production	24	0.13
Sewage and sludge burning	175	0.92
Power plant boilers		
Pulverized coal	1,409	7.42
Stoker coal	162	0.85
Cyclone coal	49	0.26
All oil	2	0.01
Industrial boilers		
Pulverized coal	62	0.33
Stoker coal	218	1.15
Cyclone coal	31	0.16
All oil	2	0.01

(continued)

TABLE 33: (continued)

Source	Amount, tons	Pollutant, %
Residential/commercial boilers		
Coal	20	0.11
Oil	3	0.02
Total	18,993	100.10

Source: Reference (16)

Table 34 shows manganese emission factors from various industrial sources as taken from Anderson (17).

TABLE 34: EMISSION FACTORS FOR MANGANESE FROM INDUSTRIAL SOURCES

Source[a]	Emission factor	Emission factor symbol[b]
Mining	0.1 kg/10^3kg (0.2 lb/ton) of Mn mined	Q
Manganese metal production	13 kg/10^3kg (25 lb/ton) of Mn processed	Q
Production of manganese alloys		
Ferromanganese		
Blast furnace[c]	41 kg/10^3kg (82 lb/ton) of ferromanganese produced	E
Electric furnace	12 kg/10^3kg (24 lb/ton) of ferromanganese produced	E
Siliconmanganese, electric furnace	35 kg/10^3kg (70 lb/ton) of siliconmanganese produced	E
Processing of manganese and and its compounds		
Steel production (carbon and alloy steels)		
Blast furnace	0.4 kg/10^3kg (0.8 lb/ton) of pig iron produced	E
Open hearth furnace[d]		
Oxygen lance	0.06 kg/10^3kg (0.1 lb/ton) of steel produced	OES (1)
No lancing	0.3 kg/10^3kg (0.6 lb/ton) of steel produced	OES (1)
Basic oxygen furnace[e]	0.7 kg/10^3kg (1 lb/ton) of steel produced	E
Electric furnace		
Oxygen lance	0.2 kg/10^3kg (0.3 lb/ton) of steel produced	E
No lancing	0.1 kg/10^3kg (0.2 lb/ton) of steel produced	E

(continued)

TABLE 34: (continued)

Source[a]	Emission factor	Emission factor symbol[b]
Cast iron	0.2 kg/10^3kg (0.3 lb/ton) of metal charged	E
Welding rods	8 kg/10^3kg (16 lb/ton) of Mn processed	Q
Nonferrous alloys	6 kg/10^3kg (12 lb/ton) of Mn processed	Q
Batteries	5 kg/10^3kg (10 lb/ton) of Mn processed	Q
Chemicals	5 kg/10^3kg (10 lb/ton) of Mn processed	Q
Cement plants		
Dry process		
Kiln[f]	0.04 kg/10^6kg (0.07 lb/10^3 tons) of feed	ES (1)
Feed to raw mill[f]	0.005 kg/10^6kg (0.01 lb/10^3 tons) of feed	ES (1)
Air separator after raw mill[f]	0.01 kg/10^6kg (0.02 lb/10^3 tons) of feed	ES (1)
Raw mill[f]	0.004 kg/10^6kg (0.008 lb/10^3 tons) of feed	ES (1)
Feed to finish mill[f]	0.004 kg/10^6kg (0.008 lb/10^3 tons) of feed	ES(1)
Air separator after finish mill[f]	0.005 kg/10^6kg (0.01 lb/10^3 tons) of feed	ES (1)
Wet process		
Kiln[g]	0.005 kg/10^6kg (0.01 lb/10^3 tons) of feed	ES (2)
Clinker cooler[g]	0.02 kg/10^6kg (0.03 lb/10^3 tons) of feed	ES (1)
Clinker cooler[h]	0.1 kg/10^6kg (0.2 lb/10^3 tons) of feed	ES(1)
Clinker cooler[f]	0.0002 kg/10^6kg (0.0004 lb/10^3 tons) of feed	OES, SSMS (1)
Air separator after finish mill[f]	0.00005 kg/10^6kg (0.0001 lb/10^3 tons) of feed	OES, SSMS (1)

[a]Emissions are uncontrolled unless otherwise specified
[b]Numbers in parentheses indicate number of samples analyzed
Q = Questionnaire survey
E = Engineering judgment
OES = Optical emission spectroscopy
ES = Emission spectroscopy
SSMS = Spark source emission spectroscopy
[c]Average particle size of particulate matter is 0.3 micron
[d]Mean particle size of dust is 0.5 micron
[e]Mean particle size of particulate matter is 0.7 micron
[f]Exit from baghouse
[g]Exit from electrostatic precipitator
[h]Exit from two baghouses in parallel

Source: Reference (17)

FIGURE 12: MATERIAL FLOW THROUGH THE ECONOMY SHOWING PRIMARY EMISSION SOURCES–MANGANESE

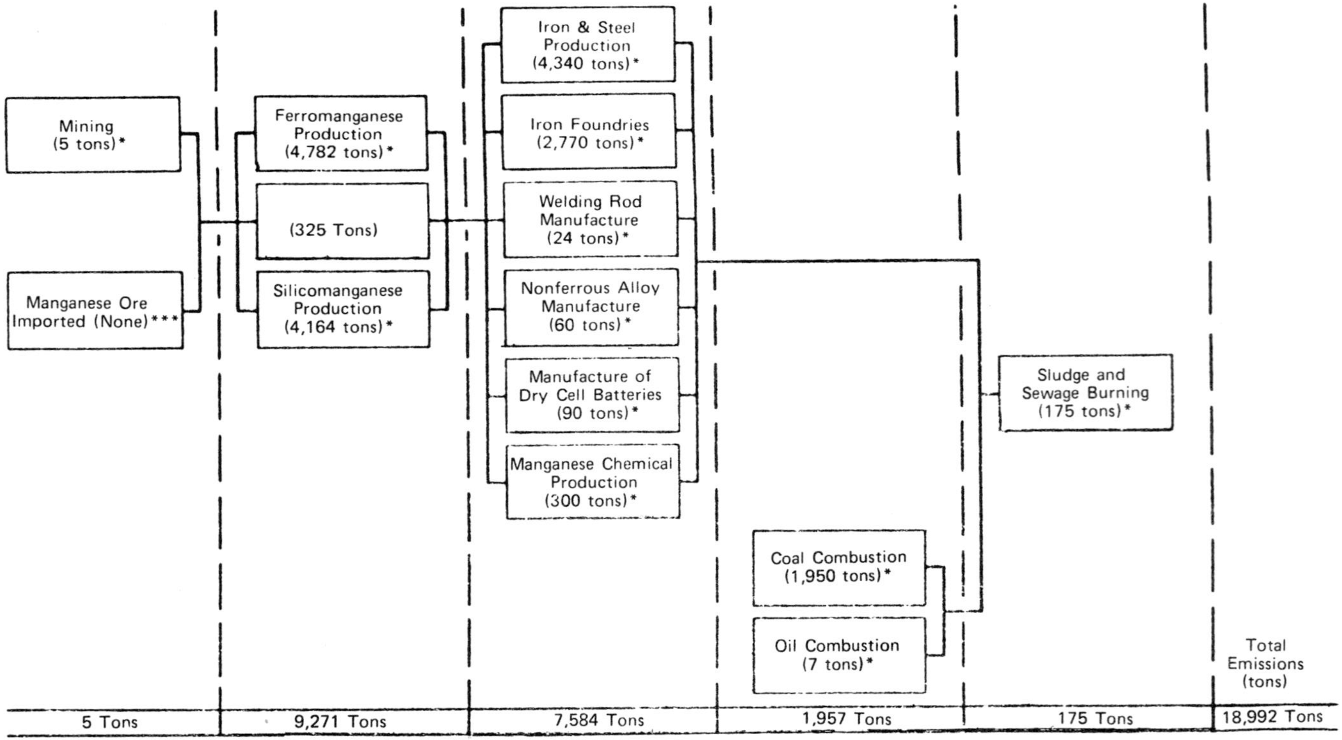

**National Inventory of Sources and Emissions–Manganese,* W.E. Davis and Associates, August 1971
***Preliminary Air Pollution Survey of Manganese and its Compounds,* Litton Systems, Inc., October 1969
***MITRE

Source: Reference (16)

In order to insure proper octane levels and antiknock properties of gasoline in which lead is removed, several gasoline modifications have been suggested. One possible alternative may be the development of other substitute additives including several manganese antiknock additives. Methylcyclopentadienylmanganese tricarbonyl (MMT) and cyclopentadienylmanganese tricarbonyl (CMT) have been mentioned as possible replacements for tetraethyllead (TEL) (18). It is suggested that oxides of manganese, such as MnO_2, Mn_2O_3 and Mn_2O_4 may end up in the exhaust of uncontrolled cars.

In attempting to assess the possible levels of atmospheric manganese due to a complete switch from leaded gasoline to manganese antiknock additives, the EHRC (1) utilized Chicago Ambient Air Quality Data and Ethyl Corporation's projective use level of Mn/gal of gasoline. The calculated increase in atmospheric manganese would be 0.10 $\mu g/m^3$ to yield a total yearly average of approximately 0.15 to 0.40 $\mu g/m^3$ for the Chicago area.

In contrast to the EHRC's predicted increase of 0.1 $\mu g/m^3$ Mn in the levels of ambient manganese contributed by gasoline combustion, Piver (18) predicted a net increase of 0.35 $\mu g/m^3$ Mn while Ethyl predicted only 0.05 $\mu g/m^3$. The reasons for this discrepancy lie in the fact that Piver assumed a concentration of 0.25 gram Mn/gal of gasoline, yet Ethyl affirmed that concentrations of 0.125 gram per gallon Mn were being used; also, Piver assumed lead concentrations in gasoline of 1.5 grams per gallon whereas Ethyl utilized 2.5 grams per gallon based on a national pool of lead content of 2.5 grams per gallon in 1968.

Finally, Piver also assumed an ambient lead concentration of 2.0 $\mu g/m^3$, whereas Ethyl used the median airborne lead concentration (1.0 $\mu g/m^3$) based on quarterly composite samples supplied from the 1969 National Air Sampling Network Survey. The EHRC used the same figures as Ethyl except that the EHRC utilized a 2.0 $\mu g/m^3$ concentration of lead since that was most representative of the Chicago area. It is important to note that such emissions of manganese will greatly exceed the EHRC proposed ambient air standard for manganese which is 0.006 $\mu g/m^3$.

It must be realized that foggy weather without wind could easily increase the atmospheric concentration of manganese 10-20 times the average as well as significantly increase the H_2SO_4 levels. Consequently, individuals caught in traffic tie-ups and inner city dwellers could be exposed to concentrations much higher than normal (18).

McKee and Wolf (19) have reported that manganese and its salts are used in the following industries and appear in their wastewaters; steel alloy; dry cell battery; glass and ceramics; paint and varnish; ink and dye; and match and fireworks. Among the many forms and compounds of manganese, only the manganous salts and the highly oxidized permanganate anion are appreciably soluble. The latter is a strong oxidant which is reduced under normal circumstances to insoluble manganese dioxide. The highly soluble manganous chloride is used in dyeing operations, linseed oil driers and electric batteries, while the equally soluble manganous sulfate is used in porcelain glazing, varnishes and production of specialized fertilizers.

Hill (20) reports manganese concentrations of 1.2 to 8.0 mg/l for mine waters. Kunin and Downing (21) cite a value as high as 50 mg/l for an acid mine water. Since the chemistry of manganese is similar to that of iron, it would be expected that any pickling operation involving manganese steel alloy results in dissolution of manganous ion and the presence of this ion in pickling and rinse solutions. In contrast to iron, the divalent (manganous) form is not readily oxidized to the insoluble manganic form, other than at elevated pH (22).

DETECTION

Dusts and fumes of manganese compounds may be collected by any methods suitable for collection of other dusts and fumes; the impinger, electrostatic precipitator and filters are commonly used (23)(24). Many methods of analysis have been used for the determination of manganese. In several methods, such as atomic absorption, emission spectrography, and neutron activation, the chemical species does not matter. The essential requirement is a representative sample in the activation site. With atomic absorption, all of the manganese analyzed must be in solution or suspension (25)-(29).

With chemical methods, care must be taken to insure that the manganese is not only in solution, but also in a single oxidation state. The most common method of determining manganese in air samples is a colorimetric method. In this procedure, the manganese is oxidized to permanganate by periodate ions. The color of the permanganate ion is very intense and follows Beer's law over a large range. A 0.5 ft^3 sample collected in a midget impinger is sufficient to determine manganese at a 5,000 $\mu g/m^3$ concentration (28). The following sensitivities are listed for the different analytical methods.

Method	Sensitivity, $\mu g/g$
Colorimetric permanganate	0.1
Emission spectrography*	10-1,000
Neutron activation*	0.001-1
Atomic absorption*	0.01-20
Flame photometry*	0.1-500

*Sensitivity depends on interfering elements and these sensitivities were not determined on air samples.

The emission spectrographic method is used to analyze samples collected by the National Air Sampling Network. Working standards are made by diluting a solution containing 20 $\mu g/0.05$ ml by $\frac{1}{2}$, $\frac{1}{4}$, $\frac{1}{8}$...$\frac{1}{1,024}$. The minimum concentration detectable by the emission spectrograph is 0.011 for urban air and 0.0060 for nonurban air. The difference in sensitivity is due to different extraction methods used in preparing the sample for emission spectrography (23)(30). Manganese may be monitored in air by the following instrumental techniques (31): Colorimetric and Emission Spectroscopy. Manganese may be monitored in water by the following instrumental techniques (32): Direct Reading Emission Spectrometry and Spark Source Mass Spectrometry.

As described by W. Baumer et al (33), an indicator for the detection of manganese(II) ions comprises an absorbent support carrier impregnated with a substantially insoluble sulfate salt of an aromatic amine oxidizable by MnO_2 to a colored oxidation product. Atomic absorption and neutron activation may be used for manganese determination down to a detection limit of 0.1 ppb. Sampling and analyses of manganese have also been reviewed in some detail in the publication by the National Academy of Sciences (4).

ENVIRONMENTAL STANDARDS

The threshold limit value (a ceiling value) for manganese recommended by the American Conference of Governmental Industrial Hygienists is 5 mg/m^3, which is generally believed to carry a low margin of safety for those occupationally exposed. According to Public Health Bulletin 247 (10), the lowest average concentration at which a case of chronic manganese poisoning was found was 30 mg/m^3. This was in a manganese mill; the worker was exposed to manganese dioxide dust. Manganese fumes may be more toxic, inasmuch as lower concentrations of fumes have produced cases of chronic manganese poisoning in several steel plants.

As with many other substances, there is a wide range in the amounts of manganese to which workers in various countries are permitted to be exposed. These range from 0.3 mg/m^3 in the U.S.S.R., Poland, and Hungary to a peak concentration of 10 mg/m^3 in Czechoslovakia (4). The United States, with a ceiling value of 5 mg/m^3, stands in the middle, closely followed by East Germany (The German Democratic Republic), West Germany (the Federal Republic of Germany), and Yugoslavia, with 6 mg/m^3.

The Environmental Health Resource Center of the State of Illinois (1) recommends an environmental standard for particulate manganese of 0.006 μg/m^3. As a result of the toxicity of sulfuric acid and particulate sulfates, a standard for ambient manganese was selected which would insure a catalytic rate of sulfur dioxide conversion of approximately 0.12%. This value approaches the rate (0.1%) of noncatalytic conversion of sulfur dioxide to sulfuric acid in air. This standard is based on a 24-hour average sample. The concentration should be reported at 25°C and one atmosphere of pressure.

A distinctive feature of manganese toxicity is the apparent central role of particle size (34). The manganese dioxide aerosols encountered occupationally may be classified either as fumes, which consist of particles less than 1 μm in diameter, or as dust, characterized as suspensions of particles whose diameter generally exceeds 5 μm. Chronic exposure to relatively high concentrations of manganese dioxide dust, e.g., as experienced by pneumatic drillers, is associated with extensive impairment of the central nervous system. In contrast, a much smaller exposure to manganese dioxide fume may produce the same symptoms in those who melt or cut the metal. The dependence of deposition sites and solubility rates on particle size and the resulting difference in etiologic role between manganese dioxide dust and fume imply that a unitary threshold limit value is inherently inadequate. The standards should reflect explicitly the key role of particle size.

The permissible concentration of (filterable) manganese in domestic water supplies is below 0.05 mg/l and it should preferably be absent. The tolerance for manganese in irrigation water on a continuous use basis is 2.0 mg/ml and on a short-term use basis in fine textured soil is 20 mg/l. The maximum allowable concentration of manganese in an effluent discharged to a storm sewer or stream is 1.0 ppm.

HANDLING PROCEDURES

No specific data are available on the handling of manganese and its compounds.

Obviously, inhalation and ingestion of manganese should be avoided, however, in view of its toxic nature.

REMOVAL FROM AIR

The emission of manganese particulates will be controlled at the same time as the particulates from the steel furnaces and incinerators. No special equipment is required to remove manganese. However, the ferromanganese furnace presents special problems (35)(36). Control of manganese from steel furnaces is accomplished by various types of collectors, including electrostatic precipitators, high-efficiency wet scrubbers, and fabric filters (37)(38). Four physical factors make the dust collection economically difficult: the small particle size (as low as 0.03 μ), the large volume of gas, the high gas temperature, and the low value of the recovered material.

Control of emissions from a ferromanganese blast furnace is more difficult than from other furnaces because the waste gas temperature is hotter and the dust is finer. Electrostatic precipitators have been successful in removing approximately 80 to 90% of the dust. No economical way of using the collected dust has been developed (36). Control of emissions from burning fuels containing antiknock and smoke-inhibiting additives may require special systems. The hazards of these organic manganese additives are to be studied as required by the Clean Air Act.

The abatement of manganese emissions from industrial process sources is chiefly a problem of particulate effluent control. Manganese is emitted largely as manganese oxide (MnO), with small quantities of the metal having been observed to efflux from many processes also as manganous oxide (Mn_3O_4). The control of manganese fumes from metallurgical processes is problematic because of the large volume of waste gas typically handled, the high gas temperature, the low value of the recovered material, and the small particle size (as low as 0.03 μm) (38). Reduction of manganese effluent has largely been a consequential by-product of application of standard particulate control systems employed to reduce total particulate material carried in the exhaust gas stream from a particular process.

Control of particle emissions from steel furnaces has been accomplished by use of the electrostatic precipitator, the fabric (baghouse) filter system and the high-energy wet scrubber (37). For coal-fired power plants, a precipitator removed approximately 91% of total manganese, while a cyclone was only about 79% efficient. A precipitator using a cyclone precleaner has a manganese collection efficiency of greater than 96%. Similar studies have shown up to 95% manganese removal from power plant effluent gas through application of an electrostatic precipitator (39).

REMOVAL FROM WATER

Removal of soluble manganese has been practiced for many years in the treatment of municipal water supplies and industrial process waters. Low manganese waters are essential for many industries, including the food and beverage industry. Treatment technology basically involves conversion of the soluble manganous ion to an insoluble precipitant. In water treatment, this generally means simultaneous removal of iron, since conditions under which high iron levels occur in water supplies are essentially the same as those for soluble manganese. Removal is effected by oxidation of the manganous ion, and separation of the resulting insoluble oxides and hydroxides of manganese. Welch (40) in a survey of manganese removal processes has listed those shown in Table 35.

TABLE 35: MANGANESE REMOVAL PROCESSES

Aeration	Slow and ineffective below pH 9.
Chlorine and hypochlorite oxidation	Not particularly effective for organically bound manganese.
Adjustment of pH	Lime soda type treatment gives removal at pH 9.5.
Catalysis	Copper ion enhances air oxidation.
Ion exchange	Effective for small quantities of iron and manganese. Resins quickly fouled by iron and manganese oxides.
Chlorine dioxide	Rapidly oxidizes manganese to the insoluble form, but expensive.
Manganese dioxide or potassium permanganate	Regeneration, green sand filter process. Economic disadvantage of requiring excess permanganate.
Direct potassium permanganate addition	Requires sand filtration.

Source: Reference (40)

Because of the low reactivity of manganous ion to oxygen, simple aeration is not an effective oxidation technique below pH 9, and even above that pH the oxidation reaction is slow. Borgren (41) has reported that even at high pH, organic matter in solution can combine with manganese and prevent its oxidation by simple aeration. Holland (42) reported that at lime and sodium alginate (a coagulant aid) dosage producing optimum iron coagulation, no manganese removal was obtained. The effect of pH on manganese removal by lime has been reported by Owens (43).

A reaction pH above 9.4 is required to achieve significant manganese reduction. The use of chemical oxidants to convert manganous to insoluble manganese, in conjunction with coagulation and filtration, has been recommended, although it may cause filtration problems. Borgren (41) states that the character of manganese flocs produced when strong oxidants are used tends to rapidly clog slow sand filters. Although Welch (40) indicates that manganese oxidation by chlorination may be ineffective, successful manganese removals have been reported.

Owens (43) reports removal of manganese at concentrations as high as 5.7 mg/l with chlorination, followed by lime coagulation and rapid sand filtration. Split flow treatment was employed, to achieve intermediate manganese levels in the final effluent. Chemical costs, adjusted to the treated stream only, are about 2.1¢/1,000 gallons. Sodium silicate was used as a coagulant aid. Harris (44) states that manganese oxidation requires a free chlorine residual of 0.5 mg/l and requires 1.29 mg of chlorine per milligram of manganese oxidized. The successful use of potassium permanganate as a chemical oxidant has been reported. Haney (45) reports that addition of 1.5 to 2.0 mg/l of potassium permanganate reduces 1.0 mg/l of manganese to less than 0.05 mg/l. Addition is 30 minutes upstream from the prechlorination point, which is followed by lime-alum-sodium silicate coagulation (4 to 10 hours detention), and rapid sand filtration. Costs attributed to permanganate addition are less than 0.3¢/1,000 gallons. Since iron is generally present, the quantity of permanganate used must include that required for oxidation of the iron content. The relationship between permanganate required and the quantity of iron and manganese present has been reported by Willey et al (46). The ability to remove manganese at lower pH (an advantage when simultaneous iron removal is desired) by permanganate oxidation is demonstrated by the data of Holland (42). The addition of 0.04 mg/l of sodium alginate reduced 0.35 mg/l of manganese to less than 0.02 mg/l in the filtered supernatant. Borgren (41) cites the absence of residual taste or odor as an additional advantage of permanganate over chlorination.

Permanganate may be used indirectly for manganese removal in the manganese greensand filter method. Willey and Jennings (46) describe the filter as a combination of New Jersey glauconite (greensand), manganese sulfate and potassium sulfate, which results in a filter bed of sand grains containing high oxides of manganese. Upon contact with manganous ion an oxidation-reduction reaction occurs, resulting in the formation of insoluble manganese dioxide which remains in the filter bed. The bed is periodically reoxidized with permanganate, and solids are backwashed from the bed.

Alling (47) reports manganese was reduced from 1.7 to 0.0 mg/l at pH 7.2 using a pressure greensand system. The plant of 500 to 700 gpm capacity cost $26,000 completely installed and operating, compared with an estimated $75,000 for a conventional aeration-lime coagulation-clarifier-filter-recarbonation system. Chemical costs were 3.5¢/1,000 gallons for a permanganate dosage of 13 mg/l. The water had a manganese concentration of 13 mg/l, which required excess permanganate. Treatment costs were roughly equivalent to lime chemical costs required for a similar treatment level. Regeneration for this plant was on a continuous basis with periodic backwashing of the greensand filter. Generalized removal capacities for greensand filters are given as 0.09 pound of manganese per cubic foot of filter volume with regeneration requirements of 0.18 lb/ft^3 of potassium permanganate. Flow rates of 3 $gal/min/ft^2$ are normal.

Alzentzer (48) reports on the use of ion exchange for iron and manganese removal. Many advantages are cited, including smaller capital investment than coagulation-filtration, nearly four times greater flow rates, smaller plant requirements and simple operation. The obvious drawback to ion exchange is the nonselective removal of other ions, which rapidly increases operation costs. Regeneration brine disposal poses an additional treatment problem since the manganese remains in soluble form in the brine. An anaerobic waste stream is an absolute necessity for the ion exchange system, to avoid air oxidation of the manganese with resultant precipitation and clogging of the ion exchange bed.

The application of the above processes to manganese removal from wastewaters should be direct within the constraint that the presence of acids, other oxidizable metals, or organics in the waste stream may substantially increase costs related to manganese removal. The unit operation costs for coagulation-sedimentation-filtration should be similar to those for iron treatment systems. Additional costs would be attributable to chlorine (or chlorine dioxide) and chlorination equipment, plus lime requirements to meet the higher pH requirements. Capital and operation costs for chlorination given by Smith (49) are for treatment of secondary sewage effluents. Chemical costs could be somewhat lower or higher, depending upon the nature of the industrial waste to be treated. A process developed by G.R. Bell (50) relates to the removal of manganese as a contaminant from water sources, and particularly to filter and filtration techniques for effectively achieving such removal. More specifically, this process relates to preconditioning of both ground and surface water to remove soluble manganese by means of a simple filter aid filtration system. Removal of soluble iron compounds will also be achieved by the same process.

This technique involves the incorporation in the liquid to be filtered of a small amount of finely divided particulate material, the filter aid, which functions to form continuously a porous cake upon the filtering surface and to entrap impurities by various mechanisms. The materials most generally used as filter aids are siliceous materials such as diatomaceous silica and perlite, carbon, and fibrous matter such as asbestos and cellulose, and the properties of these materials, e.g., porosity, diversity of shape, incompressibility, etc., make them unique for this purpose. A particularly important feature of filter aid filtration is the fine porosity of the filter aid cake which enables the removal of substantial portions of the suspended particles.

In order to increase the efficiency of the filtration process, a precoat of filter aid particles may be provided on the filter septum in addition to incorporating the particles within the filter feed. This keeps the main filter cake containing the impurities from coming into direct contact with the filter medium and consequently prevents the gummy particles from clogging the medium and lessening the filtration efficiency.

It has been shown that when a strong oxidant such as $KMnO_4$ is used in a filtration process to assist in manganese removal, iron, if present in quantity, is oxidized to ferric hydroxide which deleteriously affects the filtration of both iron and manganese. It has been found that with iron removed or reduced to a low concentration, a filter aid filtration process with suitable preconditioning utilizing filter aid, alkali and a strong oxidant, can effectively reduce manganese to low levels with low capital and operating expenses. The iron concentration may be effectively removed by utilizing concepts of preconditioning with subsequent filter aid filtration.

Briefly, the iron and managanese-containing water is first treated with an appropriate alkali or mixture of alkalis to impart to the water a pH of between 7 and 10. The alkali is so selected as to be a slow reacting material, e.g., Na_2CO_3, CaO, $Ca(OH)_2$, and the addition controlled so as to attain a stable pH with controlled agitation. Filter aid is added along with the alkali and the resulting preconditioned water filtered to remove the iron without prior settling. Once having removed the iron, or at least reducing it to a workable level, the manganese concentration may be reduced to about 0.05 ppm, or lower, by a similar preconditioning technique.

For instance, with a detention time of less than 10 minutes in the preconditioning tank at pH values of about 8.6 obtained by adding Na_2CO_3, filtration preconditioned water containing 1 ppm of $KMnO_4$ and 20 ppm filter aid, reduced an initial manganese content of 1.5 ppm to less than 0.05 ppm with a filter cycle of more than 12 hours and nominal head loss rate. A process developed by G.B. Hatch et al (51) involves oxidizing the manganous ion to higher oxides, hydrous oxides, and hydroxides, to permit its subsequent removal from water. Dissolved manganese is generally present in natural waters in the form of the manganous ion. Oxidation of the ion frequently occurs to some extent on exposure to air. Such oxidation leads to precipitation of dark brown or black hydrous oxides or hydroxides of the higher oxidation states of manganese which are very insoluble.

When these precipitates remain suspended in the water, they cause objectionable discoloration known as black water; when they settle out, black deposits form which can block lines, or act as catalysts causing further manganese deposition. These deposits are very deleterious in textile and laundry operations as they interfere with dyeing processes and leave spots which are difficult to remove. They may increase the corrosion of copper, although this has not been completely proven. They are also troublesome in municipal water distribution systems where their presence makes it extremely difficult to maintain a chlorine residual.

The method usually employed to remove manganese is to oxidize the manganous ion to insoluble higher oxides, hydrous oxides, or hydroxides, which precipitate and may be removed by coagulation and settling, filtration, or both. The oxidation may be effected by raising the pH of the water to 8 or higher where naturrally occurring dissolved oxygen or mechanical aeration brings about oxidation, or by the use of chlorine or permanganate.

However, all of these methods have difficulties which limit their usefulness and effectiveness. The use of a high pH to facilitate oxidation by dissolved oxygen is expensive and tends to cause scale deposition. Chlorine is only slightly more active than dissolved oxygen for oxidation of manganese; chlorine also requires pH elevation and to only a slightly lower level than that required for oxygen. Permanganate is expensive and one may not wish to have the permanganate ion in the water being treated because of its intense color, and because excesses may themselves cause manganese deposits.

In this process, there is added a salt of iron, copper, or cobalt and any compound yielding bisulfite ions in solution to the manganese-containing water. This results in the formation and subsequent precipitation of manganese dioxide or other higher oxidation state oxides, hydrous oxides, or hydroxides which may be removed by filtration. The iron, copper, or cobalt salt already added frequently also serves as a coagulant. Should additional coagulant be desired, it is preferable to add alum or more iron.

In summary then, manganese removal to levels of 0.05 mg/l or less has been accomplished for many years in the treatment of municipal drinking and industrial process waters. Several methods of removal are available, each of which may have distinct advantages depending upon the overall nature of the waste, other required waste treatment operations, and the total treatment goals for a particular waste. The most common general approach seems to involve oxidation of the soluble manganous form to insoluble manganese hydroxide or oxide, with subsequent precipitation. Ion exchange treatment has proven effective however, and has several advantages. No information regarding treatment of wastewaters, industrial or municipal, specifically for manganese removal was found.

SOLID WASTE DISPOSAL

The use for landfill of the insoluble residues from the production of manganese sulfate and manganese chloride by acid leaching is satisfactory (52). These residues consist mostly of silicone dioxide, with some highly insoluble manganese oxides included. There is no air emission or leaching into any ground water. All Mn-containing waste from the production of $MnSO_4$ as a by-product of hydroquinone is moved to a lagoon, where the manganese is precipitated with lime as $Mn(OH)_2$. The overflow from the lagoon contains less than one ppm Mn, which is below the limit set by the State of Illinois. This does, however, exceed the drinking water standard so the water should not be used as a potable source. There is no air emission except for an occasional, accidental release of quinone.

ECONOMIC IMPACT OF CONTROLS

General data on the cost of equipment for particulate removal from air which are applicable to manganese have been published (53). Economics of the removal of manganese from water have been presented for lime coagulation and filtration (43), permanganate oxidation (45), manganese greensand filtration (47), ion exchange (48), and chlorination (49).

REFERENCES

(1) Environmental Health Resource Center, "Airborne Manganese Health Effects and Recommended Standard," Document No. 75-18, Chicago, Illinois Institute for Environmental Quality (July 1975).

(2) T.D. Luckey, B. Venugopal and D. Hutcheson, *Heavy Metal Toxicity and Hormonology,* New York, Academic Press (1975).

(3) R.J. Sullivan, *Air Pollution Aspects of Manganese and Its Compounds,* Report PB 188 079, Springfield, Va., Nat. Tech. Information Service (September 1969).

(4) National Academy of Sciences, *Medical and Biologic Effects of Environmental Pollutants–Manganese,* Washington, D.C. (1973).

(5) P. Stock, "On the Relation Between Atmospheric Pollution in Urban and Rural Localities and Mortality from Cancer, Bronchitis and Pneumonia," *Brit. J. of Cancer* 14, 397 (1960).

(6) D. Elstad, "Manganholding Fabrikervek som Medvirkende Arsak ved Pneumoni–Epidermier i en Industribygd," *Nord. Med.* 3, 2527 (1939).

(7) Z. Yoshipawa et al, "Studies on the Effects on Respiration of Atmospheric Dust, Principally in the Form of Manganese," *Nippon Koshir Eisei Zasshi* 20, 315 (June 1973).

(8) T.A.L. Davies and H.E. Harding, "Manganese Pneumonitis: Further Clinical and Experimental Observations," *Brit. J. of Industrial Med.* 6, 82 (1949).

(9) P. Schuler, H. Oyanguren, V. Matunara, A. Valenzuela, E. Cruz, V. Plaza, E. Schmidt, and R. Haddad, "Manganese Poisoning: Environmental and Medical Study at a Chilean Mine," *Indust. Med. & Surg.* 26, 167 (1957).

(10) R.H. Flinn, P.A. Neal, W.H. Reinhart, J.M. Dallvalle, W.B. Fulton and A.E. Dooley, *Chronic Manganese Poisoning in an Ore-Crushing Mill,* Public Bulletin No. 247, Washington, D.C., Division of Industrial Hygiene, U.S. Public Health Service (1940).

(11) C.M. Whitelock, Jr., S.J. Amuso and J.B. Bittenbender, "Chronic Neurological Disease in Two Manganese Steel Workers," *Am. Ind. Hyg. Assoc. J.* 27, 454 (1966).

(12) A.M. Emara, S.H. El-Ghawabi, O.Z. Madkour and G.H. El-Sumra, "Chronic Manganese Poisoning in the Dry Battery Industry," *Brit. J. of Indust. Med.* 28, 78 (1971).

(13) L.T. Smyth, R.G. Ruhf, N.E. Whitman and T. Dugan, "Clinical Manganism and Exposure to Manganese in the Production and Processing of Ferromanganese Alloy," *J. Occup. Med.* 15, 101 (1973).

(14) J. Rodier, "Manganese Poisoning in Moroccan Miners," *Brit. J. of Ind. Med.* 12, 21 (1955).

(15) J. Starra, W. Moore, D. Hysell, S. Lee, J. Lewkowski, L. Hall, K. Campbell and M. Wiester, *Toxicology of Methylcyclopentadienylmanganese Tricarbonyl (MMT) and Related Manganese Compounds Emitted from Mobile and Stationary Sources,* Report AMPL-Tr-73-123 (1973).

(16) A.J. Goldberg, *A Survey of Emissions and Controls for Hazardous and Other Pollutants,* Washington, D.C., Environmental Protection Agency (February 1973).

(17) D. Anderson, *Emission Factors for Trace Substances,* Report PB 230 894, Springfield, Va., Nat. Tech. Information Service (December 1973).

(18) W.T. Piver, "Potential Dilemma: The Methods of Meeting Automotive Exhaust Emission Standards of the Clean Air Act of 1970," *Envir. Health Perspect.* 8, 165 (1974).

(19) J.E. McKee and H.W. Wolf, "Water Quality Criteria," California State Water Quality Control Board, Publication No. 3-A, (1963).

(20) R.D. Hill, "Control and Prevention of Mine Drainage," presented at the Environmental Resources Conference on Cycling and Control of Metals, Battelle Columbus Laboratories, Columbus, Ohio (1972).

(21) R. Kunin and D.G. Downing, "Ion Exchange System Boasts More Pulling Power," *Chemical Engineering,* 78, 67 to 69, (June 28, 1971).

(22) F.A. Cotton and G. Wilkinson, *Advanced Inorganic Chemistry,* Interscience Publishers, New York (1962).

(23) "Air Pollution Measurements of the National Air Sampling Network, Analyses of Suspended Particulates, 1957 to 1961," U.S. Department of Health, Education and Welfare, Public Health Service Publication No. 978, U.S. Government Printing Office, Washington, D.C. (1962).

(24) H.B. Elkins, *The Chemistry of Industrial Toxicology,* New York, Wiley, (1959).

(25) C.M. Gordon and R.E. Larson, "Activation Analysis of Aerosols," *NRL Quarterly on Nuclear Science and Technology,* Naval Research Lab., Washington, D.C. pages 17 to 22, (January 1, 1964).

(26) W. Horwitz, ed., *Official Methods of Analysis of the Association of Official Agricultural Chemists,* 10th Ed., Association of Official Agricultural Chemists, Wash., D.C. (1965).

(27) T.M. Kolthoff, P.J. Elving and E.B. Sandell, *Treatise on Analytical Chemistry, Part 2–Analytical Chemistry of the Elements,* volume 7, S, Se-Te, F Halogens, Mn, Re New York, Interscience (1961).

(28) *Manual of Analytical Methods,* "Determination of Manganese in Air," American Conference of Governmental Industrial Hygienists, Cincinnati, Ohio (1958).

(29) H.H. Willard, L.L. Marritt, Jr. and J.A. Dean, *Instrumental Methods of Analysis,* 4th Edition, Princeton, New Jersey, Van Nostrand (1965).

(30) *Air Quality Data from the National Air Sampling Networks and Contributing State and Local Networks,* 1966 Edition, U.S. Department of Health, Education and Welfare, National Air Pollution Control Administration Publication No. APTD 68-9, U.S. Government Printing Office, Washington, D.C. (1968).

(31) Lawrence Berkeley Laboratory, *Instrumentation for Environmental Monitoring - Air,* Publication LBL-1, Volume 1, Berkeley, University of California (February 1, 1973).

(32) Lawrence Berkeley Laboratory, *Instrumentation for Environmental Monitoring - Water,* Publication LBL-1, Volume 2, Berkeley, University of California (February 1, 1973).

(33) W. Baumer, D. Schmitt and A. Stein; U.S. Patent 3,770,379; November 6, 1973; assigned to Merck Patent GmbH.

(34) T.F. Hatch and P. Gross, *Pulmonary Deposition and Retention of Inhaled Aerosols,* New York, Academic Press (1964).

(35) S.E.D.A. Nabi and K.S. Kayed, "EMG and Conduction Velocity Studies in Chronic Manganese Poisoning," *Acta Neurol, Scandinav.* 41, 159 (1965).

(36) J. Schueneman, M.D. High and W.E. Bye, *Air Pollution Aspects of Iron and Steel Industry,* U.S. Dept. of Health, Education and Welfare, Public Health Service, Publication No. 999-AP-1 (1963).

(37) D.L. Benz and R. Bird, "Control of Pollution from Electric Arc Furnaces at Oregon Steel Mills Co.," Preprint, Oregon Steel Mills Co., Portland, Oregon (1965).

(38) J.A. Danielson, *Air Pollution Engineering Manual,* U.S. Dept. of Health, Education and Welfare Public Health Service Pub. No. 999-AP-40, U.S. Govt. Printing Office, Wash., D.C. (1967).

(39) S. Cuffe, R. Gerstte, A. Orning and C. Schwartz, "Air Pollutant Emissions from Coal-Fired Power Plants - Report No. 1," *Journal of the Air Pollution Control Association,* 14, 358-359, (September 1964).
(40) A. Welch, "Potassium Permanganate in Water Treatment," *Jour. Am. Wat. Wks. Assoc.,* 55, 735-741 (1963).
(41) G.R. Borgren, "Removal of Iron and Manganese from Ground Waters," *J. New Eng. Wat. Wks Assoc.,* 76, 70-76 (1962).
(42) G.J. Holland, "Removal of Dissolved Aluminum, Iron and Manganese," *Chem. Ind.,* 2016, London (1964).
(43) L.V. Owens, "Iron and Manganese Removal by Split Flow Treatment," *Wat. Sew. Wks.,* 110, R76-R87 (1963).
(44) W.C. Harris, "Iron and Manganese Removal," *Eng. Ex. Sta. Bull. No. 68,* Louisiana State University (1962).
(45) C.G. Haney, "Potassium Permanganate Solves a Problem of Charlottesville, Va.," *Wat. Sew. Works* 111, R353-R354 (1964).
(46) B.F. Willey and H. Jennings, "Iron and Manganese Removal with Potassium Permanganate," *Jour. Am. Wat. Wks. Assoc.,* 55, 729-734 (1963).
(47) S.F. Alling, "Continuously Regenerated Greensand for Iron and Manganese Removal," *Jour. Am. Wat. Wks. Assoc.,* 55, 749-752 (1963).
(48) H.A. Alzentzer, "Iron Exchange in Water Treatment," *Jour. Am. Wat. Wks. Assoc.,* 55, 742-748 (1963).
(49) R. Smith, "Cost of Conventional and Advanced Treatment of Wastewater," *Jour. Wat. Poll. Control Fed.,* 40, 1564-1574 (1968).
(50) G.R. Bell; U.S. Patent 3,340,187; September 5, 1967; assigned to Johns-Manville Corp.
(51) G.B. Hatch and E.A. Guthrie; U.S. Patent 3,349,031; October 24, 1967; assigned to Calgon Corp.
(52) P.W. Powers, *How to Dispose of Toxic Substances and Industrial Wastes,* Park Ridge, N.J., Noyes Data Corp. (1976).
(53) A.R. McFarland, *Control Technology for Toxic and Hazardous Air Pollutants,* Chicago, Illinois Inst. for Environmental Quality (December 1974).

MERCURY

The general aspects of mercury emission, detection and control have been reviewed by H.R. Jones (1). Since 1969, much attention has been directed to mercury in the environment, and consequently, much has been learned. The public recognition of mercury as a potential health hazard followed reports of numerous incidents of human and wildlife illnesses and deaths attributed to mercury poisoning. As a result of this awareness, there arose several excellent reviews of mercury in the environment [see references (2) through (16)] and a valuable bibliography (17).

Reviews dealing specifically with mercury in air appeared as it became evident that mercury exists abundantly in air as well as water and soil (18)(19)(20). The nature of environmental mercury is frequently misunderstood. Authorities differ widely in their attitudes from firm complacency to frenzied alarm. The Lawrence Berkeley Laboratory has published information on instrumentation for monitoring of air pollution (21).

Two huge bibliographies have been prepared by major chemical producers as well. Thus Dow Chemical Co. produced such a compilation in 1970 (22) and Imperial Chemical Industries followed with one in 1973 (23). A review on environmental mercury contamination was published in 1972 (24).

The Organisation of Economic Cooperation and Development published a comprehensive survey on mercury and the environment in 1974 (25) and the Environmental Protection Agency issued a publication on mercury flow in the environment and regulations offered by various bodies on mercury use in 1975 (26).

TOXICITY

As pointed out by Luckey et al (27), the toxicity of mercury varies greatly with its chemical form. Thus, monovalent mercury is quite nontoxic due to the low solubility of its salts. However, tissues and erythrocytes can oxidize monovalent mercury to highly toxic divalent mercury. Organic mercurials are also highly toxic. Elemental mercury, both in liquid and vapor form, is equally toxic due to its lipid solubility, lack of charge, and membrane permeability. Schroeder (28) states that mercury is about five times as toxic as lead, is about as toxic as cadmium and antimony, and is much less toxic than beryllium.

Chloralkali plants and primary mercury-processing plants are known to emit mercury into the atmosphere in sufficient quantities to create a public health problem. The proposed EPA regulations would limit emissions from these two sources to no more than 2.3 kg mercury per day from each facility. Other sources are yet to be identified. However, power plants and incinerators in Illinois and Missouri are suspected of emitting up to 67 kg/day (2,450 kg/yr) of elemental mercury from each installation (20). Measurements for total mercury at a 2,100 Mw plant have indicated a daily discharge of 2.5 kg from each of three stacks (41). Man generally absorbs 75 to 85% inhaled mercury vapor at concentrations between 50 and 350 $\mu g/m^3$ (19).

Experiments, however, have shown that some individuals can have nearly complete absorption at concentrations between 60 and 250 $\mu g/m^3$ (18). Reports regarding the respiratory intake of mercury other than elemental vapor are scarce, but inorganic and organic compounds are known to be absorbed (2)(29). Elemental, and especially organic, mercury are also absorbed through the skin.

Inorganic mercury compounds, phenylmercury, and methoxyethylmercury are converted in the blood to soluble inorganic mercury salts (2)(18). Carried by the blood plasma, the mercury is widely distributed in all tissues, but accumulates primarily in the kidneys and liver prior to excretion (2)(18). Some accumulation of the mercury salts also occurs in the brain, spleen and alimentary tract (18). Nonalkylmercury might, to a limited extent, be transformed into methylmercury within the liver (2)(12). In general, those compounds which are, or readily form, water-soluble inorganic salts are among the least toxic of the mercury compounds.

The more stable organic mercury compounds, notably alkylmercury, are not readily converted to organic salts (2)(18). Methyl- and ethylmercury, which are lipid soluble, are carried and stored for extended periods in the blood cells (2). Because water-soluble salts are not readily formed, little alkylmercury is found in the blood plasma and excretion does not proceed readily through the kidneys. The stability and lipid solubility of the alkylmercury compounds allow them to penetrate into cells more easily and to be retained much longer. Accumulation in the brain and liver is more pronounced than with other mercury forms.

Methylmercury is known to accumulate also in the fetus. Excretion appears to occur mainly via the liver into the feces; over 90%, however, can be reabsorbed as the feces passes through the gastrointestinal tract (2). Some conversion of alkyl to inorganic forms does occur in the kidneys (18), liver (18) and intestine (12). The excretion rate of methylmercury is approximately 1% of the body burden per day (13). Its mobility, retentiveness, and preferential concentration in the brain make alkylmercury among the most toxic of mercury forms.

Elemental mercury has transport and retention properties resembling both inorganic and alkylmercury compounds. Converted to mercury salts, it is partially transported in the blood plasma. While able to penetrate all body tissue, it accumulates in the brain, other nervous tissue, intestinal tract and salivary glands (2).

Having characteristics intermediate between inorganic and alkylmercury, elemental mercury is somewhat more toxic than the former but far less toxic than the latter.

The toxicity of mercury is attributed to its high affinity for sulfur-containing compounds and its lesser affinity for organic ligands (18). Interference by mercury in the synthesis and function of enzymes and other proteins can result in a variety of adverse effects (18).

Acute mercury poisoning can result from inhalation of inorganic mercury at concentrations from 1,200 to 8,500 $\mu g/m^3$ (18). The kidneys and intestinal tract are primarily involved. Symptoms are metallic taste, nausea, abdominal pain, vomiting, diarrhea, headache, salivation, and anuria (18)(29). The stomach, gums and salivary glands may become inflamed. Acute exposure to elemental mercury can also cause pulmonary irritation and neural damage. Chronic symptoms such as muscular tremor may persist in some cases. Extreme cases may lead to hemolysis, insomnia, delirium and ultimate death from exhaustion (2)(18).

Chronic mercury poisoning can result from inhalation of inorganic mercury at concentrations as low as 100 $\mu g/m^3$. Slight anemia, hypothyroidism and increased excitability may result from occupational exposures as low as 10 to 30 $\mu g/m^3$ elemental mercury (2). Among 642 workers in the chloralkali industry, low-level exposures averaging from $<$50 to 270 $\mu g/m^3$ elemental mercury have been strongly correlated with loss of appetite and weight (2). Chronic poisoning by inorganic mercury affects primarily the nervous system.

There may arise anxiety, insomnia, muscular tremor and other psychological disturbances. Other possible symptoms are erethism, inflammation of the gums, gastrointestinal disturbances, weakness and other symptoms similar to those in acute poisoning (2)(18)(29). The effects of phenylmercury and methoxyethylmercury are similar to those of inorganic mercury compounds, since phenylmercury and methoxyethylmercury are readily converted into inorganic forms. In one study, 26 subjects were continually exposed to 250 to 3,200 $\mu g/m^3$ phenylmercuria pyrocatechin for up to 6 years without evidence of injury, suggesting that phenylmercury may be less toxic than elemental mercury (18).

The toxicological effects of mercury on man depend to a large extent on the form of mercury ingested. Elemental, liquid mercury ". . . is not a poison; a person could swallow up to a pound or more of quicksilver with no significant adverse effects . . ." (5). Calomel, primarily mercurous chloride (Hg_2Cl_2), has been used medically as a cathartic, diuretic, antiseptic and antisyphilitic (30). Mercuric chloride ($HgCl_2$) is poisonous when taken orally and in a substantial dose (5)(30). Organomercury compounds including the methyl and ethyl compounds can be lethal to man (5).

Ingestion of 1.0 to 2.0 grams mercuric chloride is frequently fatal to human beings (31). It produces corrosion of the intestinal tract, injury to the kidney and ultimately death due to kidney failure (5). The effects of some other mercury compounds are described (30)(31). The kidney appears to be the concentrator organ for inorganic mercury (32).

The extremely high toxicity of methylmercury to man and its concentration in the environment are of particular concern (25). It has been found that in aquatic ecosystems, other forms of mercury including inorganic mercury can be converted into methylmercury which can be absorbed and concentrated by aquatic species, and further concentrated along the food chain involving higher organisms, man included. Although the efficiency of this process cannot at the present time be specified in advance, it evidently can occur in ways which are potentially harmful to these higher organisms, especially if the uptake of mercury is continued over a long time span. With particular reference to the effects of methylmercury in man, it is recognized that insufficient knowledge exists with which to establish reliable early diagnosis of subacute poisoning following uptake, or adequately to assess the risks of genetic or fetal damage entrained thereby.

Alkylmercury poisoning has somewhat different characteristics than other mercury poisoning. Symptoms may be dormant for weeks or months after acute exposure and if brain damage occurs, effects may be irreversible. Inhalation of alkylmercury can produce dryness and irritation in the nasopharynx and mouth. Alkylmercury poisonings have caused permanent neurological damage resulting in impaired vision and hearing, sensory loss in limbs, ataxia and tremor. Neurological disorders such as mental retardation and convulsive cerebral palsy have occurred in infants whose mothers were exposed to methylmercury during pregnancy (2)(18). Fetal nerve tissue may be especially sensitive to methylmercury (12).

EXTENT OF EXPOSURE

Mercury, usually in trace amounts, is found throughout the ecosystem: in soils, water, air and in living species. Because mercury can undergo a variety of physical and chemical transformations, and because it can be ingested in foods, it is readily transported in and transferred between these media. While much of the mercury involved in these ecological cycles has been introduced by natural processes, for example, by mercury leached from soils or exhausted in volcanic gases, much has been added by the mining and other industrial activities of man (25).

The hazards of mercury are particularly associated with two interrelated factors, the toxicity of mercury, and the localized occurrence of high concentrations of these chemicals in the environment—either as a result of human activity or natural processes—which can augment the likelihood of damage.

There exist many processes by which mercury can become concentrated locally in the environment. The concentration of methylmercury in aquatic species and in the food chain of higher organisms is one such process. The natural processes of weathering of exposed mercury-rich bedrock can also give rise to high mercury concentrations in surrounding soils and streams.

Furthermore, mining and other industrial activities can yield potentially hazardous concentrations of mercury in their emissions and in their products. Some of these emissions have already been the cause of epidemics of poisoning of alkylmercury compounds—such as the epidemics of methylmercury poisoning from industrial wastes from acetaldehyde manufacturing plants at Minamata and Agano in Japan—and have given rise to long-range environmental problems such as exist with the accumulation of mercury discharges in river and estuarine sediments.

Mercury is extremely mobile in the environment. Natural cycles transform and transport mercury in a wide variety of manners throughout air, water and soil. Elemental mercury is constantly outgassed from soil (14). Mercury in the bottom sediment of waterways is converted by anaerobic bacteria into methylmercury (16), which can then evaporate into the atmosphere as dimethylmercury (14). In the atmosphere, many mercury compounds decompose in the presence of sunlight to form elemental mercury. Nearly all forms of airborne mercury can be adsorbed on particulate matter which can then be washed out by precipitation and returned to the surface. Because mercury is readily cycled among air, water and soil, any contribution anywhere is potentially an addition to the mercury pollution problem.

Mercury is removed from the environment when it is lost from the natural cycles. Possible routes of removal may be settling to great ocean depths or drainage to deep underground layers. Such removal does not really reduce the total amount of mercury; instead, it displaces it to regions remote from biological activity and deters its return, thus reducing the mercury found in the biosphere. The processes and rates of mercury removal, which are presently unknown are an important area of future research. Up to a century may be required for the removal of mercury from bottom sediment (2).

In Table 36, an indication is given of the flux of mercury in production, trade, use and emission, in Japan, Sweden, Canada and the U.S.A. at specified dates in the 1960s. The dates 1967 and 1969 have been chosen since they hold the best available figures from these countries at about the time when extensive control measures began to be introduced (Japan, 1966; Sweden, 1966; U.S.A., 1969; Canada, 1970).

TABLE 36: CONSUMPTION AND EMISSION OF MERCURY BY SECTOR[1]

Sector	Japan, 1967	Sweden, 1967	Canada, 1969	U.S.A., 1969
By usage for				
Chloralkali	735	30	110	714
Paints	na	6*	18*	336
Electrical apparatus	60	2	15	643
Control instruments	-	-	14	240
Dental applications	6	3	12	105
Catalysis	490	0.3	-	102
Agriculture	91	3	12	90
General lab use	na	na	9	70
Pharmaceuticals	8	0	1	25
Pulp and paper[2]	0	4	1	19
Other	385[3]	-	62[4]	330[5]
Some additional emissions[6]				
Mercury production	-	-	27	50
Sulfide ore processes	-	100	-	-
Coal combustion	-	-	11	232
Oil combustion	-	-	4	5

[1] It is important to note that the usage figures given for Japan and the U.S.A. refer to measured or estimated consumption of mercury in each sector; those for Sweden and Canada refer to measured or estimated emission [excepting those figures asterisked (*) which refer to consumption]. Measured consumption may include exports. Units are metric tons; (na) indicates figure is not available.
[2] Includes only slimicides and preservatives; excludes mercury contaminated caustic soda.
[3] Includes chemicals used elsewhere in commerce, e.g., mercury in batteries, paints, lab uses.
[4] Includes catalysts.
[5] Includes particularly startup requirements for mercury cell chloralkali plants.
[6] These represent preliminary estimates only.

Source: Reference (25)

Table 37 is a schematic representation of the nature and relative magnitude of the environmental problems posed by the occurrence and use of mercury. For each industrial sector, it records the chemical nature of the mercury emitted and directs attention to those effluents (in manufacturing industries) and those media (in use of manufactured products), where at least one of the countries concerned has experienced mercury pollution problems. The distinction between manufacturing and product use is itself an important one from the point of view of mercury pollution control, since the former are concentrated (point) sources more amenable to control by in-process changes of an engineering nature, whereas the latter are dispersed sources where control may depend more on prohibitions on use.

TABLE 37: NATURE AND RELATIVE MAGNITUDE OF MERCURY EMISSION PROBLEMS[1]

Sector[2]	---- Chemical Nature of Mercury ---- Metal	Inorganic	Alkyl	Other Organic	Production-Manufacturing Uses	Manufactured Products
Chloralkali industry	xx	x	–	–	GLS	aW-
Paints	–	x	–	xx	-ls	aWt
Electrical apparatus	xx	x	–	–	gls	AWt
Control instruments	xxx	–	–	–	gls	AWt
Dental applications	xxx	–	–	–	-ls[3]	- - -
Catalysis	–	xx	x	–	gLS	- - -
Agriculture	–	x	x	x	-Ls	-wT
General lab use	xx	x	–	–	- - -	-w-
Pharmaceuticals	–	xx	–	x	-Ls	-w-
Pulp and paper	–	–	–	xxx	-L-	aWt
Production inorganic Hg compounds	x	x	–	x	-Ls	awt
Mercury primary production	xx	x	–	–	GLS	
Mercury secondary production	xxx	–	–	–	GlS	
Sulfide mining (other)	xx	x	–	–	GLS	aWt
Metal extraction[4] (chemical)	xx	x	–	–	gL-	
Fossil fuel combination	xxx	–	–	–	Gl-	
Phosphate rock[5]	–	xxx	–	–	gLs	-wt

[1] Key: Each industrial sector in which mercury occurs or is used is characterized as follows:

a. by chemical nature of the mercury emitted. Three x's (representing 100% of the emission), are distributed among the chemical forms of mercury usually emitted, in proportion to their relative importance;

b. in production and manufacturing uses of mercury, by the type of effluent carrying the mercury burden. g = gaseous, l = liquid, s = solid effluent respectively. Upper case letters (G, L, S) indicate effluent problems of major importance.

c. in manufactured products, by the media which may be polluted by the use of these products, a = air, w = water, t = land (terrestrial) media, respectively. Upper case letters (A, W, T) indicate the media in which major problems may arise.

It should be noted however, that industrial sectors (read vertically) should not be compared without reference to tables similar to Table 36 of this report, for the country concerned; and entries in the table under uses and products indicate problem areas identified by at least one of the four countries (Japan, Sweden, Canada, U.S.A.). These entries do not indicate that all these countries have experienced each problem indicated.

[2] Some sectors where mercury occurs, e.g., municipal sewage, have been omitted.

[3] Use at dental surgeries.

[4] Chemical extraction of metals, e.g., gold, silver, by amalgamation.

[5] For fertilizer production.

Source: Reference (25)

Taking Tables 36 and 37 together, attention is directed in particular to the following matters:

A. The mercury cell chloralkali industry is the single most important user of mercury, and appears to be the most important of the point dischargers of mercury to waters. Being a point source, this discharge is controllable in principle, and the technology for so doing exists in fact. (Alternative processes which do not use mercury are also available for the production of chlorine and caustic soda.)

B. Mercury used in agriculture and catalysis is of major importance, however small the quantity used, since mercury emissions may involve alkylmercury compounds—one of the most toxic groups of mercury compounds. Agricultural usage of alkylmercury is a dispersed use, and it is therefore difficult to control by technical means any environmental impacts it may have, and in addition such usage of alkylmercury is subject to misuse which can result in direct poisoning of animals and humans.

C. While measures to control some mercury sources have been taken in Canada, Japan, Sweden and the U.S., attention is also focused upon the need for an assessment and possible control of additional sources. These include particularly mercury vapor from coal combustion. Whereas the mercury content of coal is quite low in general (preliminary measurements indicate a concentration range of 0.01 to 4.0 mg/kg on average), the very large quantities consumed in many countries indicate a source of mercury output into the environment of large magnitude. Mercury emissions in mining and refining operations (particularly those involving sulfide minerals), and the fate of mercury contained in the widely dispersed products of industry (such as electrical apparatus) also warrant urgent attention at this time.

For the U.S. as a whole, the percent distribution of mercury discharges from man-related sources to the various environmental receptors is: 31% to air, 6% to water, and 63% to land.

Less than 5% of the man-related discharges to air are from regulated industries (i.e., chloralkali and mercury mining). Natural emissions (from degassing of the earth's crust) are estimated to be three times greater than the man-related discharges. The fate of mercury and its compounds in air is not clear (although they do not appear to present a direct hazard to man because of the low ambient concentrations), but the mercury eventually returns to the land and to the water.

About 13% of the man-related discharges to water are from regulated industries (currently chloralkali and mercurials manufacturing). Natural sources (primarily runoff) contribute about two times as much mercury to water as do man-related discharges. Some mercury is removed from water in sewage treatment processes as sludge and is either returned to land or incinerated and returned to air. Most mercury discharged to water appears to be incorporated rapidly into sediments and may be released by biological and mechanical action over a period of time.

Some 25% of the man-related discharges to land (as solid waste) are from regulated industries. Although mercury and its compounds appear in trace amounts in virtually all soils, these cannot be compared with man-related discharges. Solid wastes incorporated in regulated landfills (which currently receive about one-third of all solid wastes) do not discharge appreciable amounts of mercury to water; the discharge of mercury vapors to air from landfills is a possibility which needs further study. Also needed is additional study of mercury losses from improperly operated or unregulated landfills. Solid wastes which are incinerated do release mercury to air and are of major concern with respect to pollution.

TABLE 38: TOTAL MERCURY LOSSES IN 1973 FOR THE COTERMINOUS UNITED STATES, BY SECTOR AND SIC CATEGORY (THOUSANDS OF KILOGRAMS)

Sector	SIC No.	Description	Total Losses to: Air	Water	Land	Total Mercury Lost	Total Recycled
I		Mercury mining and smelting:					
	1092M	Mercury mining	0.01	0.00	0.01	0.02	0.00
	1092P	Mercury processing (including secondary)	7.84	0.00	0.41	8.25	0.00
		Subtotals	7.85 (1.7%)	0.00 (0.0%)	0.42 (0.0%)	8.27 (0.5%)	0.00 (0.0%)
II		Other mining:					
	1021	Copper mining	0.02	0.01	0.08	0.11	0.00
	1031	Zinc and lead mining	0.00	0.00	0.01	0.01	0.00
	3331	Copper smelting	40.77	2.26	2.26	45.29	0.00
	3333	Zinc smelting	4.59	0.25	0.25	5.09	0.00
	3241	Cement processing	0.50	0.25	1.76	2.51	0.00
	3274	Lime processing	0.08	0.04	0.29	0.41	0.00
	3332	Lead smelting	4.75	0.26	0.26	5.27	0.00
		Subtotals	50.71 (10.8%)	3.07 (3.5%)	4.91 (0.5%)	58.69 (3.8%)	0.00 (0.0%)
III		Unregulated sources:					
	021	Livestock	0.00	0.00	17.70	17.70	0.00
	2911C	Fuel oil—residential, commercial, industrial	16.94	0.00	0.02	16.96	0.00
	2911M	Refineries	1.15	0.00	1.15	2.30	0.00
	2951C	Tars and asphalt	1.10	1.67	14.99	17.76	0.00
	3312M	Coke ovens	7.16	0.51	2.56	10.23	0.00
	4091C	Coal—residential, commercial, industrial	9.97	0.00	1.11	11.08	0.00
	49110	Utilities—oil and natural gas	11.99	0.00	0.01	12.00	0.00
	4924C	Natural gas—residential, commercial, industrial	15.46	0.00	0.01	15.47	0.00
	4911C	Utilities—coal	40.71	0.00	4.52	45.23	0.00
		Subtotals	104.48 (22.2%)	2.18 (2.5%)	42.07 (4.4%)	148.73 (9.8%)	0.00 (0.00%)
V		Manufacturing and processing:					
	2.5B	Caustic	0.00	7.61	1.90	9.51	0.00
	2819M	Catalyst manufacture	0.00	0.02	0.00	0.02	0.00
	2851M	Paint manufacture	0.01	0.20	0.05	0.26	0.00
	2879M	Pesticide manufacture	0.00	0.06	0.00	0.06	0.00
	2833M	Pharmaceuticals manufacture	0.00	0.02	0.00	0.02	0.00

(continued)

TABLE 38: (continued)

Sector	SIC No.	Description	Total Losses to: Air	Water	Land	Total Mercury Lost	Total Recycled
	2812	Chlor-alkali	14.84	2.93	226.83	244.60	0.00
	2261	Textiles	0.00	0.15	7.63	7.78	0.00
	2851F	Paint formulation	0.29	0.35	0.00	0.64	0.00
	38291M	Control instrument manufacture	0.00	0.00	1.97	1.97	0.00
	3079P	Catalyst usage	0.05	0.10	18.85	19.00	0.00
	3.0	Other	10.28	10.23	8.70	29.21	20.66
	36292M	Tubes/switches manufacture	0.00	0.00	1.57	1.57	0.00
	36410M	Lamp manufacture	0.40	0.00	1.34	1.74	0.00
	36420M	Battery manufacture	0.13	0.05	2.49	2.67	0.00
		Subtotals	26.00 (5.5%)	21.72 (24.8%)	271.33 (28.1%)	319.05 (21.0%)	20.66 (14.1%)
VIIIA		Final consumption: Commercial & industrial:					
	2.5A	Urethane and miscellaneous	0.12	0.00	2.22	2.34	0.00
	2879N	Nonagricultural pesticide use	4.39	17.56	21.95	43.90	0.00
	2879A	Agricultural pesticide use	0.00	2.83	16.02	18.85	0.00
	38291C	Control instrument consumption	16.54	0.00	107.50	124.04	82.69
	36292C	Tubes/switches consumption	7.53	0.00	46.26	53.79	0.00
	26410C	Lamp consumption	6.07	0.00	37.31	43.38	0.00
	7391	Laboratory usage	2.28	5.92	1.59	9.79	12.98
		Subtotals	36.93 (7.8%)	26.31 (30.0%)	232.85 (24.1%)	296.09 (19.4%)	95.67 (65.2%)
VIIIB		Final consumption: Consumer goods:					
	2834C	Pharmaceuticals consumption	1.04	17.77	2.09	20.90	0.00
	2851C	Paint consumption	173.61	0.00	9.14	182.75	0.00
	36420C	Battery consumption	69.71	0.00	403.30	473.01	24.89
	8021	Dental applications	0.93	16.65	0.00	17.58	5.55
		Subtotals	245.29 (52.0%)	34.42 (39.2%)	414.53 (42.9%)	694.24 (45.5%)	30.44 (20.7%)
		Total	471.26	87.70	966.11	1,525.07	146.77
		Final disposal:					
		Sewage	4.01	19.92	22.88	46.80	-
		Urban runoff	0.00	11.70	0.00	11.70	-
		Natural sources:					
		Degassing	1,018.70	0.00	0.00	1,018.70	-
		Runoff and groundwater	0.00	188.30	0.00	188.30	-

Source: Reference (26)

Table 38 delineates in some detail the discharges of mercury to air, water and land in the U.S. as of 1973.

The atmospheric emissions estimated by the EPA (26) are only in order of magnitude agreement with earlier EPA estimates (33) which are shown in Table 39 and Figure 13. There are some major points of agreement in that paint application is estimated to be the largest single source of airborne mercury, and mercury mining and smelting are insignificant sources.

TABLE 39: MERCURY EMISSIONS

Source	Amount in Tons	Percent
Mercury mining	3	0.33
Chlorine fluxing, nonferrous metals	55	6.94
Secondary mercury	11	1.39
Pulp and paper industry	Neg	Neg
Organic chlorine chemical preparation	70	8.86
Paint, varnish, lacquer production	1	0.13
Instrument manufacture	3	0.33
Electrical apparatus manufacture	3	0.38
Dental preparations manufacture	1	0.15
Use of pesticides, herbicides, fungicides	19	2.40
Use of pharmaceuticals	3	0.33
Laboratory use of mercury	51	6.44
Consumption of paint	215	27.15
Incinerators	135	17.04
Sewage and sludge burning	11	1.39
Power plant boilers		
Pulverized coal	150	18.94
Stoker coal	17	2.15
Cyclone coal	6	0.76
All oil	1	0.13
Industrial boilers		
Pulverized coal	6	0.76
Stoker coal	23	2.90
Cyclone coal	3	0.38
All oil	1	0.13
Residential/commercial boilers		
Coal	2	0.25
Oil	3	0.38
TOTAL	793	

Source: Reference (33)

FIGURE 13: MATERIAL FLOW THROUGH THE ECONOMY SHOWING PRIMARY EMISSION SOURCES OF MERCURY

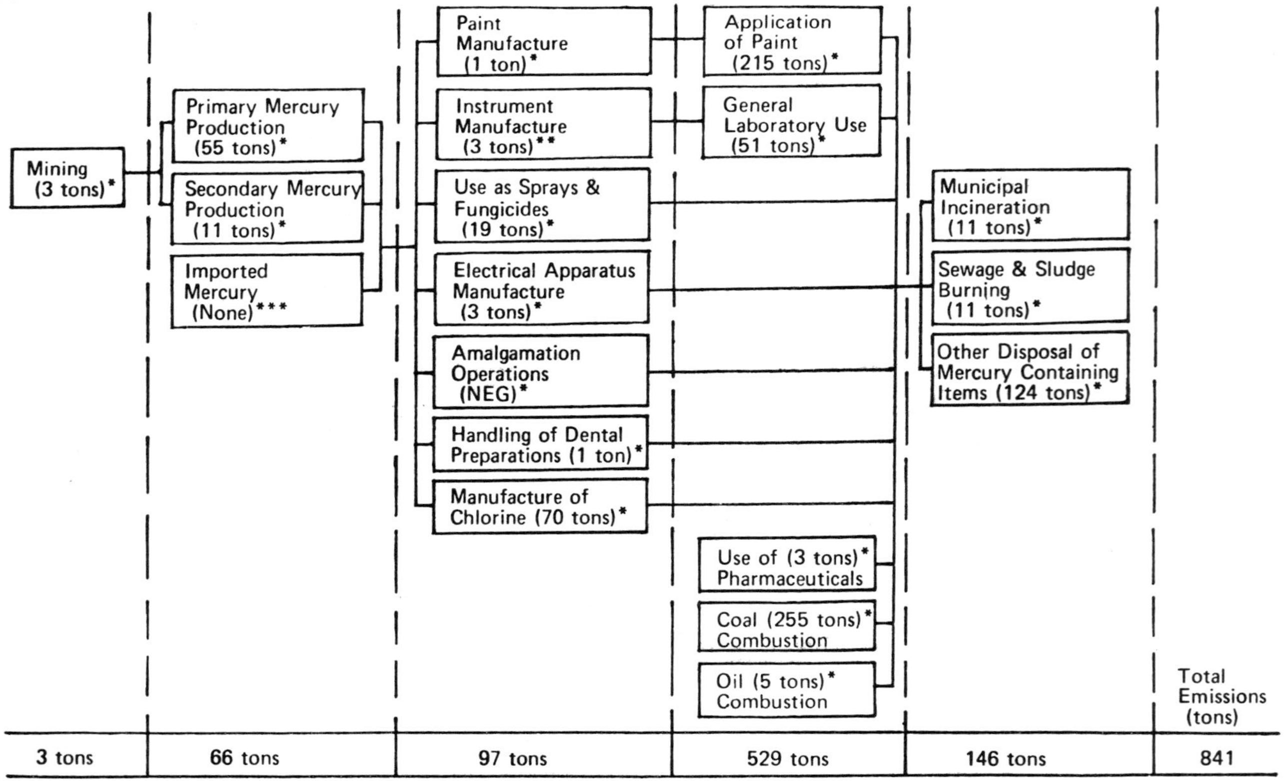

**National Inventory of Sources and Emissions–Mercury*, W.E. Davis and Associates, September 1971.
***Preliminary Air Pollution Survey of Mercury and Its Compounds*, Litton Systems, Inc., October 1969. ***MITRE.

Source: Reference (33)

However there are also some major differences in that copper smelting, a significant source in the later survey, was omitted in the earlier one; organic chlorine chemical preparation was a significant source in the earlier survey and was not included in the later tabulation; incinerators as a source of airborne mercury emissions are not included in the later tabulation; batteries are listed as the second largest emission source in the more recent survey but were not listed as such in the earlier survey; and fuel oil and natural gas are listed as significant sources in the more recent data whereas fuel oil was estimated at much lower levels and natural gas was not even considered in the earlier studies.

Turning now to an analysis of some of the major controllable emission sources in some detail (26), Figure 14 is a flow diagram of a typical mercury recovery unit in operation in 1973. The ore may be obtained by either surface or underground mining and in recent years has contained about 2.3 kg (5 lb) of mercury per ton of ore. Dust from the mining operation is generally a very minor problem and since the tonnages are low, emissions from this source are trivial (estimated to be less than 0.02 ton annually). Generally, the ore is crushed and fed from a storage bin into a rotary kiln (in a few locations, screen or table concentration is employed). Hot gases are fed into the lower end of the kiln; the ore is thus heated to well over 1200°F and the mercuric sulfide is decomposed with an efficiency of about 96%. Mercury vapor, sulfur dioxide, combustion products, and dust are then passed off.

FIGURE 14: FLOW DIAGRAM OF MERCURY RECOVERY USING 1973 TECHNOLOGY

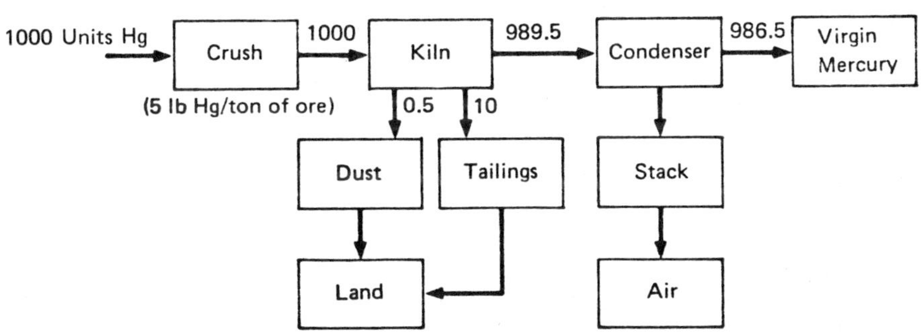

Source: Reference (26)

The dust is separated in a cyclone unit and the vapors are passed through a simple water-cooled iron or steel condenser. Liquid mercury is collected as shown on Figure 14 and is bottled on-site in 34.4 kg (76 lb) flasks. Exhaust gases contain sulfur dioxide and traces of mercury vapor (which are not allowed to exceed 2.3 kg per day total discharge, by EPA standards) and are then discharged through a low stack. The calcined tailings are collected and dumped in a nearby waste pile; they typically contain about 5 ppm of residual mercury. The tailings do not appear to present any leaching problems. The water used in the condensers does not come in contact with the mercury and is discharged or reused. The mercury exhausted from the stack can be assumed to remain primarily in the air, although there is doubtless some local deposition on the land surface; however, no harmful effects from this have been reported.

Figure 15 is a flow diagram of a mercury recovery unit which can be considered the prototype for 1983 technology. This flow chart is for the new McDermitt Mine, which has several unique features. The ore, which has an unusually high mercury content (about 4.6 kg/ton), is obtained from an open pit. After being fed through crushers and a mill, it is concentrated by flotation processes to about 60% mercury content.

FIGURE 15: MERCURY RECOVERY USING 1983 TECHNOLOGY

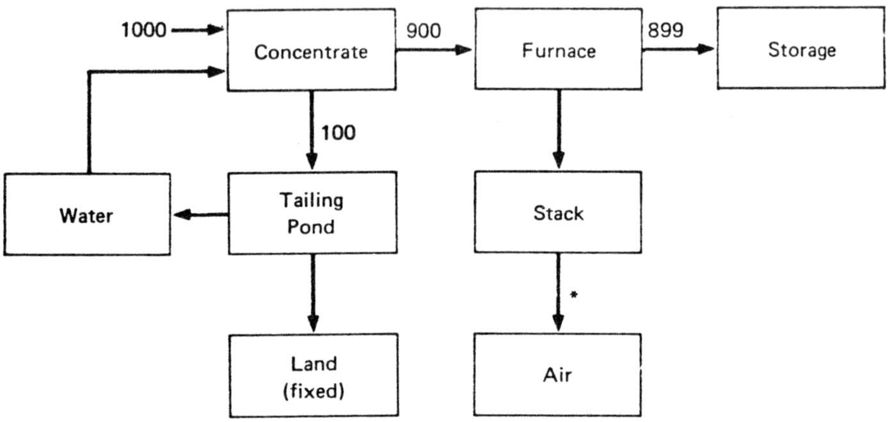

*Maximum discharge of 2.3 kg (5.1 lb)/day. For a production rate of 2,000 kg/day of mercury (equivalent to 20,000 flasks/yr), the discharge to air represents a loss of 0.1%.

Source: Reference (26)

It is then fed at a rate of some 2,200 kg (1,000 lb) per hour through a six-hearth furnace and the calcined tailings are discharged, through a seal, to a waste solution system and carried to a tailing pond for permanent storage. The reclaimed water is returned to the mill for further use. (All makeup water comes from underground wells.) The vapor from the furnace goes through tube condensers that continuously collect most of the mercury, which is then fed through a cleaning process to bulk storage (a unique feature of this mine is that it will use 1,000 kg shipping containers).

The vapor is further cleansed of mercury and the sulfur dioxide is removed; final traces of mercury are removed in a refrigerated gas cooler (utilizing the principle that mercury is very insoluble in air at temperatures of less than 100°F) and are discharged to the atmosphere. The discharge is designed to meet the EPA standard of 2.3 kg per day maximum. Although the technology used in this new mine is not yet proven, it appears to meet high standards in minimizing discharges to the environment. The cost of the mine and facilities construction was approximately $9 million, according to the developers (Placer-Amex Corp.) (26).

Copper mining and smelting is a very significant source of mercury emissions in those areas where the copper industry operates on a large scale. Thus, a study (26) indicated that the copper industry (mining and smelting) in Arizona accounted for almost all of the mercury emissions to air in the state and contributed appreciably to those to land and water. Of major concern are the pyrometallurgical (smelting) methods, since they represent the most commonly used technology, utilize ores with higher average mercury content, and in the course of heating and smelting emit mercury as a vapor to air.

The production of copper is a mercury emission source with widely varing regional impact. A simplified flow diagram showing estimated losses of mercury in the various major processes in the mining and production of copper is presented in Figure 16. The input value of 100 could represent the 100 metric tons of mercury estimated to be present in the copper ores mined in the U.S. in 1973, or it could represent the flow in a typical facility.

Typically, copper ore is obtained from large, open-pit mines and trucked to the mill. A small amount of mercury is no doubt lost in the dust generated from this operation. (An additional small amount of mercury might be released directly to the air as a result of the increased degassing due to the exposure of the mineral deposit; however, as discussed elsewhere, this increase over background is small.) At the mill the ore is crushed, ground, and mixed with water and selective reagents; selective flotation concentrates the copper content in the froth (the contentrate) which is then recovered. The worthless residue (gangue), which contains some mercury, is discharged to a tailing pile. The concentrate, which has about a 20% copper content, can then be economically moved to the smelter even if it is not located nearby.

The copper concentrate is fed into a reverberatory furnace (smelter) which is heated with hot combustion gases and which is also charged with slag from the copper converter and with limestone and siliceous fluxes. (The use of electrical furnaces, another form of smelting, to replace the reverberatory furnace is a rather recent innovation.) The operating temperature of about 1480° Celsius drives off some of the sulfur, which is oxidized to sulfur dioxide and escapes with the flue gases.

Excess iron is eliminated in the slag. The copper matte, which is periodically withdrawn from the furnace, consists of a mixture of copper sulfide and iron sulfide. In Figure 16 it is indicated that most of the mercury goes up the stack along with the sulfur dioxide, but that a small amount remains in the slag and some is carried forward in the copper matte. Considering the operating temperature of the reverberatory furnace, it is more likely that almost all of the mercury is emitted from the stack at this point, but for want of definitive information, the division shown in the figure was used (26).

The stack gases from the reverberatory furnace are approximately 1% sulfur dioxide (SO_2); this is too dilute for economical production of sulfuric acid. On the other hand, the stack gases from electric furnaces usually contain 3 to 4% SO_2, from which sulfuric acid can be economically produced. The gases from the furnace are passed through cyclonic dust collectors and electrostatic precipitators. Thus, an estimated two-thirds of the mercury content is released to the air (through the typically high stacks used by copper smelters) and the remaining one-third is assumed to be caught in the fly ash and deposited on land.

FIGURE 16: 1973 TECHNOLOGY FOR COPPER MINING AND SMELTING WITH MERCURY INPUT AND OUTPUT

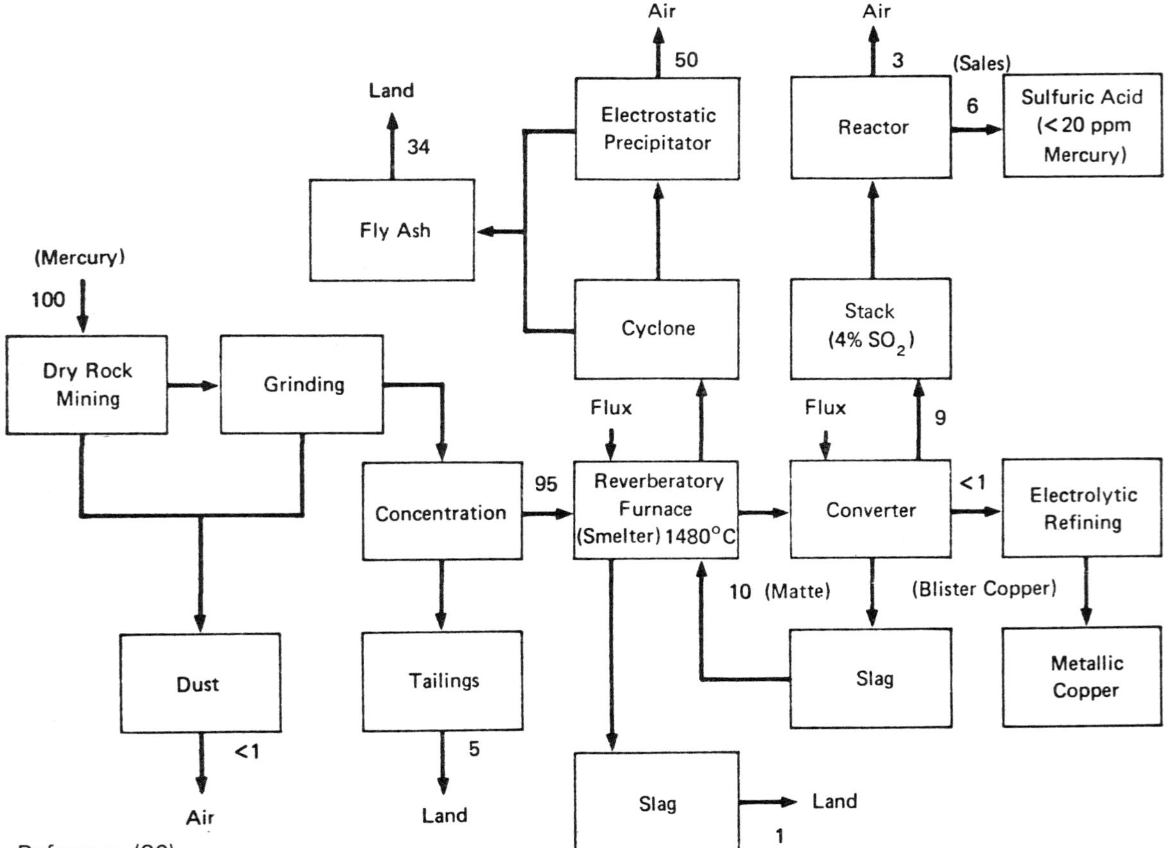

Source: Reference (26)

From the smelter the molten matte is transported to the converter, where it is treated with air or oxygen which reacts with the copper sulfide to discharge SO_2 (in concentrations of about 4% out the stack). As part of the conversion process, iron sulfide is oxidized, creating additional SO_2, and the iron combines with the siliceous flux to form a slag which contains copper and other precious metals and is returned to the reverberatory furnace.

The resultant blister copper is 99.5% pure and is normally further purified, usually by electrolytic refining. It can be assumed that at this point the mercury content of the copper is virtually nil, since most of the mercury was exhausted to the stack during the smelting stage in the reverberatory or electric furnace. Any remaining mercury in the gases would find its way into the sulfuric acid, probably in concentrations of less than 20 ppm. A minute quantity of mercury would go out the stack.

It should be noted that approximately 80 to 90% of the mercury exhausted remains airborne for at least 10 km downwind and may become part of the global burden. However, the remaining 10 to 20% is assumed to be deposited evenly on either land or water surfaces. At least half of this mercury eventually reaches the watercourse, either directly or by runoff from land surfaces.

Figure 17 is a flow diagram of the technology which will probably be in use in 1983 for the pyrometallurgical processing of copper. Again, an input of 100 units of mercury is assumed, since better forecasts are not available. The major changes for 1983 will occur in the pollution control technology associated with the treatment of the stack gases and in the use of electric furnaces which produce a higher SO_2 gas concentration.

In this flow diagram it is assumed that the SO_2-laden gases from the smelter and converter are combined and that a considerably larger fraction of these gases is then reacted to form sulfuric acid. The remainder of the stack gas stream undergoes further treatment for recovery of mercury, and discharges to air and land are greatly reduced. Mercury is also recovered from the sulfuric acid produced, thus adding to the revenues of the facility.

While it is impossible to describe in detail the technology likely to be used in 1983 for control of mercury emissions and for recovery of mercury itself, the current literature suggests various possibilities. For example, the Gortdrum mines in County Tipperary, Ireland, now recover mercury from copper concentrates. However, more interest has been displayed in a technology involving recovery of mercury from flue gases. For example, the use of a countercurrent sulfuric acid wash to recover up to 90% of the mercury content of the stack gases has been proposed. The Outokumpu Oy zinc plant at Kokkolla, Finland, is reportedly recovering 600 flasks per year of mercury as a by-product of the roaster gases in a process which should be adaptable for copper smelting.

As indicated above, the mercury content of the stack gases may be removed prior to the sulfuric acid unit. However, if such technology is not employed, mercury will appear in the concentrated sulfuric acid. This problem has apparently been successfully attacked by the Bunker Hill Company, which reports that it has reduced the mercury content of by-product sulfuric acid to less than 1 ppm at its Coeur d'Alene (Idaho) smelter. Other processes are reported to be under development for removal (and recovery) of mercury from by-product sulfuric acid.

FIGURE 17: 1983 TECHNOLOGY FOR COPPER MINING AND SMELTING WITH MERCURY INPUT AND OUTPUT

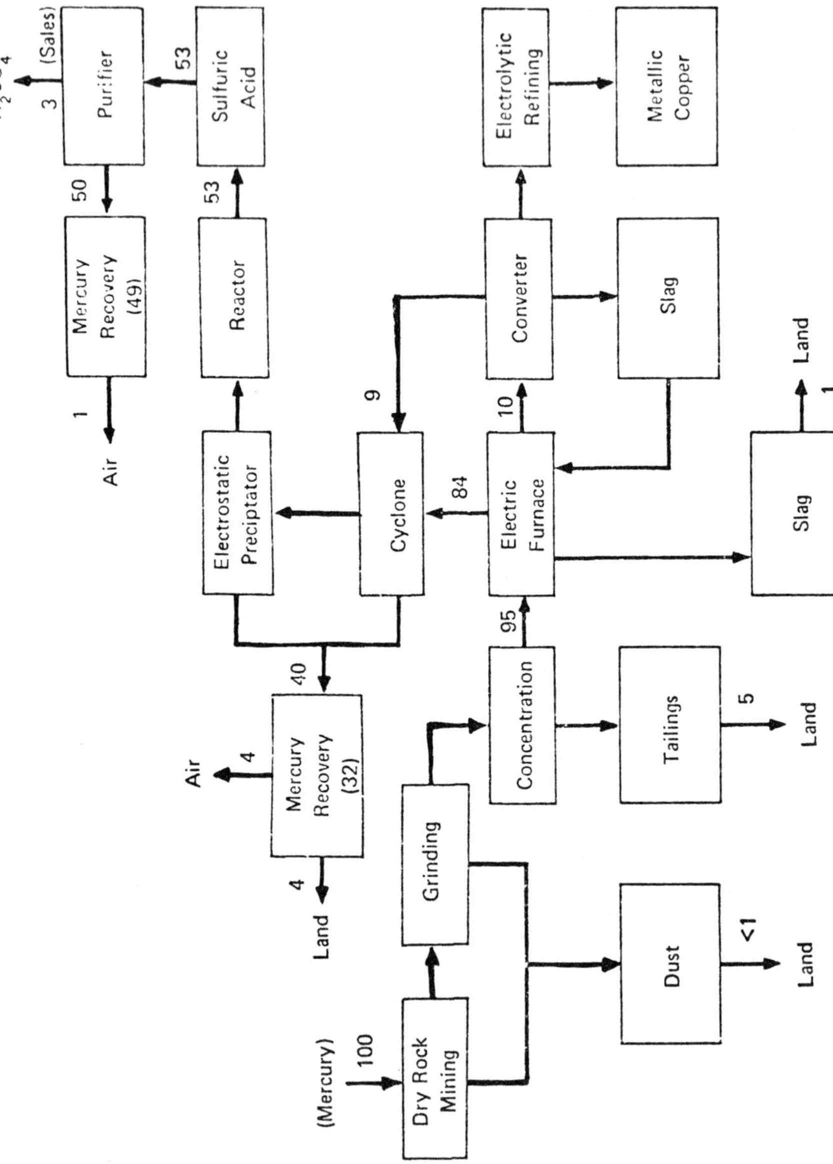

Source: Reference (26)

Two flow diagrams for typical 1973-vintage power plants are shown in Figure 18. The only difference between the two diagrams is in the mercury emission factors of the fuel consumed for electricity generation. The emission factors for coal are 90% to air and 10% to land. Natural gas and petroleum, because of their cleaner burning characteristics, emit 99.9% to air and 0.1% to land.

FIGURE 18: MERCURY LOSSES FROM TYPICAL POWER PLANTS, 1973

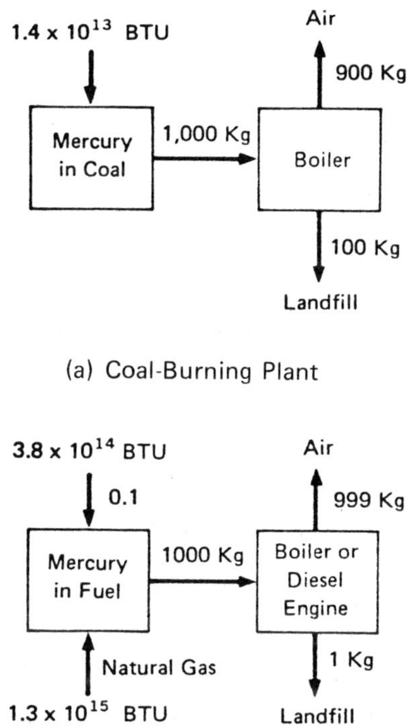

(b) Petroleum- or Natural Gas-Burning Plant

Source: Reference (26)

The principal reason for the high air emissions is the low boiling temperature of mercury (685°F). The air emissions are primarily in the metallic form, with minor losses in the oxide form. Losses to land come from bottom ash, which ranges between 4 and 20% by weight of coal, and fly ash (stack emissions). To date very little effort has been expended by the government or the utilities to reduce mercury losses from fossil fuel power plants.

Figure 19 is a flow diagram for the partial recovery of mercury from power plant operations. These diagrams, which represent prototypes for 1983, are patterned after a sulfur dioxide (SO_2) scrubber pilot plant operated by the Tennessee Valley Authority (TVA). The Colbert plant, which burns high sulfur coal, was equipped with a SO_2 scrubber to remove the sulfur gases from stack discharges, but preliminary tests conducted by the EPA found that it also removed about 33% of the stack mercury discharges. The reasons for the high mercury recovery are that the scrubber removes the fly ash (total emission 0.7 kg mercury per 10^6 kg ash) and that heavier mercury vapor particles are found in the flue gas (total emission 0.08 kg mercury per 10^6 kg gas).

FIGURE 19: MERCURY LOSSES FROM TYPICAL POWER PLANTS, 1983

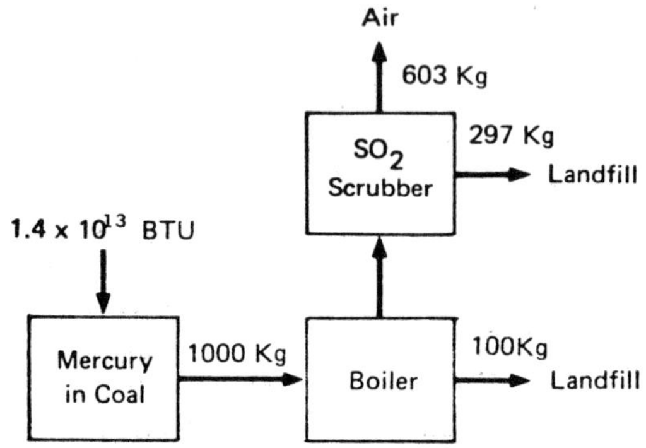

(a) Coal-Burning Plant

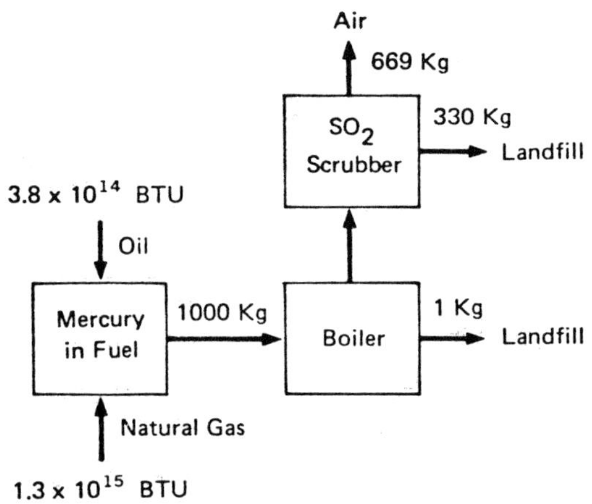

(b) Petroleum- or Natural Gas-Burning Plant

Source: Reference (26)

Because of increased constraints placed on SO_2 emissions to meet clean air standards, scrubbers probably will be installed on most fossil fuel power plants by 1983. This should result in about a one-third decrease in uncontrolled mercury emissions to the atmosphere. This quantity of mercury will then accrete in controlled landfills.

Fossil fuel consumption is expected to increase about 50% between 1973 and 1983 (at a 3.4% annual rate compounded), if current demands and usage continue, with coal increasing 55% and natural gas and oil decreasing 5%. Mercury emissions, however, should increase only 15% for the low prototype technology and should decrease by 14% for highly developed technology. Mercury emissions are not expected to increase at the same rate as fuel consumption because mandated control of SO_2 emissions by 1983 will coincidentally lead to removal of part of the mercury (26).

Figure 20 is a flow diagram of a typical chlorine and caustic soda plant which manufactures 500 tons of chlorine gas and 570 tons of caustic per day. The basic unit in the plant is the mercury cell, which theoretically recirculates the same mercury continuously. However, mercury losses do occur from this cell, and makeup mercury must be added occasionally. The Uhde cell, which may contain 2,651 kg of mercury, is typically 2.0 x 17.6 meters and is constructed of heavy gauge steel to which hard rubber-covered channels are bolted to form the sides.

FIGURE 20: MERCURY LOSSES FROM A TYPICAL CHLORALKALI MANUFACTURING PLANT, 1973

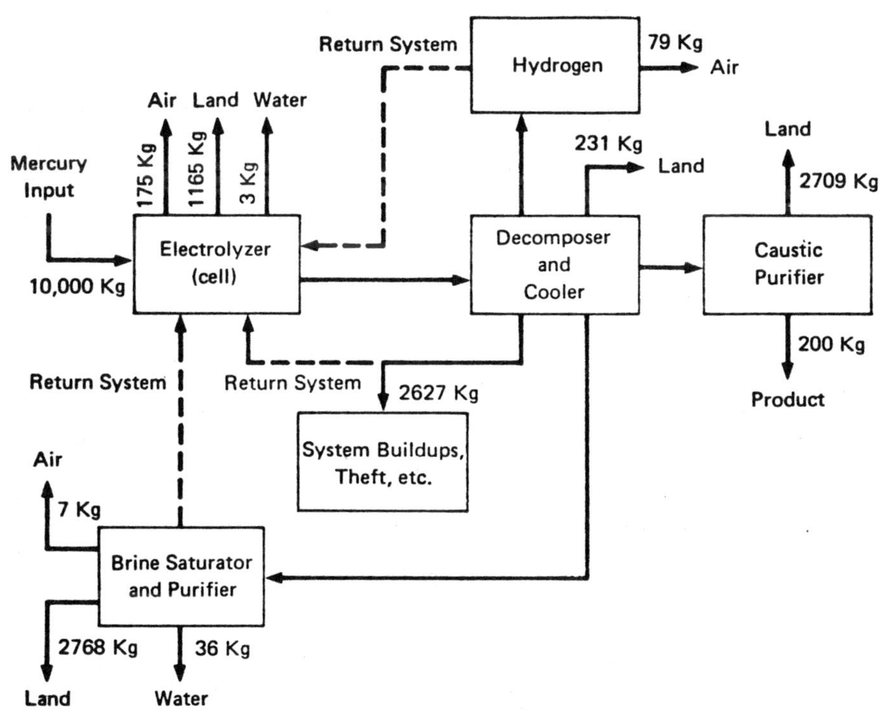

Source: Reference (26)

The bottom has a slight slope (4 to 5 mm/m) so that mercury, which is introduced at the top of the cell through an end box seal, flows uniformly down the flat bottom of the cell to a depth of about 0.003 meter and is removed (with an amalgam content of less than 0.5% sodium) through an outlet end box. Together with the mercury, purified brine is introduced at the top of the cell, covering the flowing mercury to a depth ranging from less than 0.03 meter at the top of the cell to almost 0.12 meter at the bottom of the cell.

Multiple anodes extend down from the cell cover (which may be attached to the body of the cell with a flexible gasket or with a firm connection). A high-density, low-voltage current (less than 0.5 amp/cm^2), which is carried by massive bus bars from cell to cell in series connection, activates these anodes. The anodes then electrolyze the brine to form chlorine, which is collected at the top of the cell or at slots in the anode. The sodium amalgam is deposited in the moving mercury cathode. After leaving the end box, this amalgam enters a decomposer (also called a denuder) which is typically affixed to the end of the cell.

The decomposer is a small reactor that is normally packed with pieces of graphite which, in the presence of trickling water, cause the sodium amalgam to decompose, forming 40 to 60% caustic and releasing the mercury for return to the inlet of the cell. Hydrogen is also formed (99.9% pure). Its temperature of over 80°C allows mercury vapor to be incorporated in the gas. To minimize losses, the hydrogen is cooled at the decomposer to about 55°C; this removes over 98% of the mercury, which is then returned to the cell. The remainder of the hydrogen stream is further purified so that 99% or more of the mercury is ultimately removed. The gas is then normally flared.

The brine which leaves the cell is depleted of its sodium chloride content and it carries some mercury with it. This depleted brine is returned to the brine saturator and is treated to remove undesirable chemicals (primarily calcium and magnesium salts) which are introduced in that operation. This purification step results in the formation of sludges which incorporate significant quantities of mercury. These sludges must be segregated for further treatment and disposal. The purified and saturated brine is recycled to the cell. With the 1973 technology, mercury consumption is assumed to be 0.4 lb of mercury per ton of chlorine produced, or 31,780 kg/yr for the 500 tpd plant.

In this materials balance certain assumptions have been made to account for all mercury losses. For example, a buildup of the mercury inventory within each cell occurs over time as a result of changing operating conditions; these may include the formation of a mercury butter which effectively reduces the mercury available for reaction temporarily, or the need to operate at higher current densities to increase production, which requires a higher mercury level, etc. Mercury also accumulates in piping and other equipment and is an apparent loss. Other poorly defined losses, for example theft, can occur. Estimates have been made to cover all known or suspected losses so that the amount of mercury introduced (i.e., 0.4 lb/ton of chlorine produced) is accounted for.

Actually, such accounting is not possible in a typical mercury cell chloralkali plant. Instead, for the industry as a whole, only about 50% of its annual mercury consumption can be accounted for. This does not imply that mercury is indiscriminately lost to the environment; rather, it is most difficult to estimate where mercury may accumulate in the system and to what extent. For example, it is estimated that each year, in this typical plant, 523 kg of mercury enter structural members and can only be recovered when the plant is disassembled after its active lifetime. The industry is concerned with this lack of accountability and is continuing to seek better results. Many plants now run radioactive inventories of the mercury content of each cell at least once a year, but this expensive technique is only accurate to about ±1% (26).

Losses to air from a typical chloralkali plant occur chiefly at the end box, the electrolyzer cell, from hydrogen and waste air flaring, and from the brine recycling system. Except for the brine recycling system, air losses are assumed to be within EPA requirements of 2,300 g/day. The small recycling system losses, however, occur at a variety of uncontrolled locations within the plant (brine purifier and saturator, leaks at fittings and seals, and settling ponds).

As a part of good housekeeping practice, the inevitable mercury leaks which occur under the cell house are regularly collected, by hosing with water, into sumps. The collected mercury is returned to the system and the mercury-containing water is sent to the water purification plant where it is joined by the supernatant liquid from the sludge settling pond. The collected waters are usually treated with sodium sulfide, forming the very insoluble mercuric sulfide which is easily removed by filtration prior to discharge to a local receiving water body.

The amount of mercury that can be discharged to the aquatic environment depends on the type and size of the water body. Proposed EPA regulations limit the maximum mercury discharge to 0.735 kg/day for streams, 0.612 kg/day for lakes, 12.2 kg/day for estuaries, and 14.7 kg/day for coastal waters. The mercury discharge for the typical plant was conservatively assumed to be 0.79 kg/day (276 kg/yr).

Mercury losses to land are the most significant waste problem for chloralkali plants. Most of this waste is discharged in sludge (approximately 200 ppm of mercury) generated by the brine and caustic purifiers. The sludge is concentrated in settling ponds and deposited in local industrial landfills. Other solids with higher mercury levels include spent graphite (anodes and decomposer), filter cakes (sludge and water treatment residues), and miscellaneous solids (discarded wood and plastic cell room flooring, machinery, pipes, fittings, etc.). These solids may be incorporated in the sludge or shipped to high level industrial disposal sites (26).

Chlorine gas contains, at most, trace amounts of mercury (assumed to be 0.005 ppm). However, caustic, even after treatment, contains approximately 3.7 ppm of mercury. Most of this caustic is shipped out of the plant to other manufacturers and is consumed in their products.

Theft of mercury from traps, flasks, sumps, and other sources is a problem at most chloralkali plants. It is believed that most of this mercury is sold to local salvage firms for resale. Complete elimination of this problem is impossible because of the quantity used at the plants and the accessibility of mercury.

Small amounts of mercury will remain in empty mercury flasks, because they are discarded rather than recycled. It is estimated that this loss is approximately 0.0454 kg/flask (3 to 4 drops). The other loss in this category, long-term structural, occurs as mercury vapor or metal is absorbed into a building's superstructure. This loss remains within the plant until the structure is discarded.

Figure 21 is a flow diagram for the technology which will probably be used in 1983 by the typical chloralkali plant (500 tons of chlorine per day). Mercury consumption in chloralkali plants by 1983 is estimated to be reduced to 0.10 lb of mercury per ton of chlorine produced, for a total of 7,945 kg/yr. This reduction is expected to be brought about by the industry's efforts to control mercury emissions to conform with federal, state, and local regulations. Individual plants are expected to have almost complete accountability of their mercury consumption, which will enable them to control internal plant discharges. Also, technological changes by 1983 are expected to help chloralkali plants in the recovery and final disposition of mercury waste.

FIGURE 21: MERCURY LOSSES FROM A TYPICAL CHLORALKALI MANUFACTURING PLANT, 1983

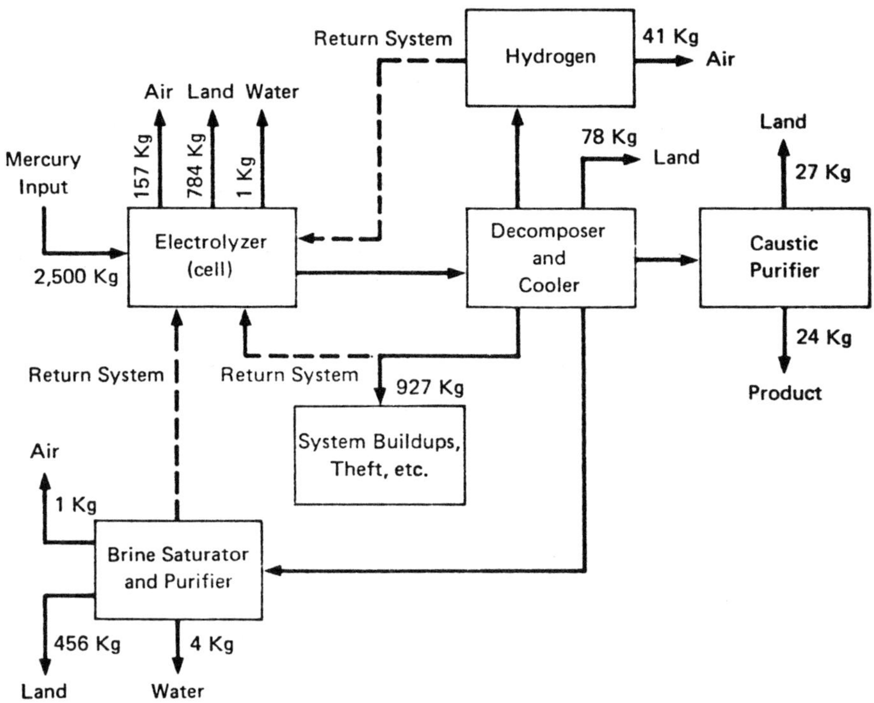

Source: Reference (26)

Mercury losses to air may be reduced 40%, or to 492 kg, for the typical plant by 1983. Most of this reduction will occur at the end boxes and hydrogen streams when the waste gases are chilled, demisted, and put through microsieve filters. Also, with tighter internal plant controls on mercury consumption, losses in the recycling system will be reduced.

Losses to water will be reduced substantially because of wastewater purification technology advances. This advancement will be brought about by government regulations requiring industries to meet best practical waste treatment standards by 1977 and best available waste treatment standards by 1983. Therefore, the mercury discharge for the typical plant should be approximately 0.05 kg/day, or 16 kg/yr.

The major changes in the 1983 technology occur in pollution control technology associated with the treatment of sludges. Two attractive processes are now under development for recovery of mercury from sludges. One, developed and patented by Georgia-Pacific, involves retorting the sludge with condensation of the released mercury. The other, which is reported in the literature, involves the dissolution of the mercury in the sludge in a solution of sodium hypochlorite, with possible further purification steps and subsequent return to the system. It is forecast that some high level wastes will continue to be generated; these will continue to be segregated in special industrial repositories. It is expected that sludge treatment will reduce the mercury entering the environment to 403 kg in 1983 for the typical plant.

Other significant land loss reductions will occur in decomposer graphite emissions. (Anode losses are expected to be eliminated by 1983 as the industry completes the switch from the graphite anode to the dimensionally stable anode.) Also, reductions in decomposer graphite discharges to the environment are likely to result from adoption of good housekeeping practices such as treating the waste graphite with hypochlorite solution. Mercury losses to product (almost entirely caustic soda) will decrease significantly. This reduction will take place at the caustic purifier where the mercury enters the sludge system for treatment. Technological changes are expected to reduce the residual mercury content of caustic alkali to 0.4 ppm by 1983.

Losses due to theft, systems buildups, and miscellaneous escapage will decrease by 1983 because more attention will be concentrated on locating these sources and eliminating them. Also, leaks will be fixed promptly, traps will be collected regularly, and plant structures will be sealed properly. Generally, the attitude will change from moderate concern to major concern with controlling internal plant mercury consumption. Estimated mercury losses from a typical chlor-alkali plant's operation in 1973 and 1983 are compared in Table 40. The 1983 technology values are based on a mercury consumption rate of 0.1 lb/ton of chlorine produced and no expansion or decrease in mercury cell chlorine production. Mercury loss to the environment is expected to decrease by 78% during the decade.

TABLE 40: ESTIMATED MERCURY LOSSES FROM TYPICAL CHLORALKALI PLANT OPERATIONS, 1973 AND 1983 (Kilograms)

Mercury Losses to	1973	1983
Air	15,869	12,527
Water	1,951	113
Land	244,875	48,310
Product	8,661	1,035
Totals	271,356	61,985

Source: Reference (26)

Numerous studies can be cited to identify environmental levels of mercury found in the work environment of various mercury-using industries (34). A review of these studies shows there have been wide ranges of mercury exposures encountered by workers at their places of employment.

Benning (35) reported levels of total mercury, ranging from 0.20 to 0.75 mg Hg per cubic meter, in the workroom atmosphere of a company using copper amalgam compound in manufacturing carbon brushes for electric motors. No industrial hygiene practices, from either an engineering or sanitation standpoint, were in effect at this plant, so that there were most likely multiple exposures of workers to mercury at the plant through inhalation, skin absorption, and ingestion. In addition, mercury-contaminated work clothing was worn home, permitting a certain amount of mercury to be carried into the home.

The installation of ventilation control measures reduced the air concentrations to a range of 0.05 to 0.07 mg Hg per cubic meter of total mercury. Even with this reduction of airborne concentration, high levels of mercury in worker urine samples continued. They were reduced, however, upon institution of strict sanitation requirements for plant housekeeping, handwashing, eating arrangements, and initiation of health orientation programs for employees.

For example, one worker showed a reduction from 1,810 to 330 μg Hg/l of urine after these measures were introduced. This experience indicates the need for evaluation of the total environment for effective control of the hazards associated with exposure to mercury. Reliance on the control of the atmospheric levels of mercury will not, by itself, necessarily assure that absorption of mercury by the worker will be sufficiently reduced if workers have poor personal hygiene or work practices which permit them to be exposed through routes other than inhalation.

Copplestone and McArthur (36) reported effective reductions of airborne mercury levels after installation of ventilation control measures in a company manufacturing jewelry. The peak reading in the general air of this plant reached a high of 0.35 mg Hg/m^3 during the summer months. Almost immediately after installation of an improved ventilation system, these levels dropped to 0.03 mg Hg/m^3.

An investigation of the environment of workships repairing direct current electric meters by Bidstrup et al (37) showed the significance an enclosed environment may have upon the concentration of mercury in air. Air levels of mercury vapor were measured during the summer months when the workshops were open to outside ventilation through windows and again during winter months when the windows were closed.

In two workshops the concentration of mercury in the general atmosphere during the summer reached 0.223 mg Hg/m^3 and 0.23 mg Hg/m^3, respectively. At work stations, levels as high as 1.6 mg Hg/m^3 were recorded. In the other workshops studied the general atmosphere concentrations during the summer ranged from 0.005 to 0.067 mg Hg/m^3. Sampling in these other workshops during the winter months with the windows closed showed that the general atmosphere in most of the shops exceeded levels of 0.2 mg Hg/m^3, while at work stations, levels significantly above this were frequently recorded.

In the one shop which had mercury vapor concentrations of 0.223 mg Hg/m^3

in the general atmosphere (range for all locations sampled was 0.08 to 1.6 mg Hg/m^3) during the summer months, ventilation equipment was installed prior to winter sampling. The winter samples in that shop ranged from 0.03 to 0.1 mg Hg/m^3. The environmental conditions observed in this study emphasize the effectiveness of ventilation for reducing airborne concentrations of mercury. In addition, it illustrates the impact which changes brought about by seasonal conditions may have upon concentrations of mercury in the atmosphere of the workplace.

One of the largest users of metallic mercury is the chloralkali industry in which brine is electrolyzed in large cells with mercury as the cathode. Although the mercury is totally enclosed most of the time, many tons of the metal are present and, inevitably, mercury vapor enters the ambient air during overhaul or cleaning of equipment and from accidental leaks; thus, exposure to elemental mercury vapor is potentially a major hazard to chloralkali workers. Because chlorine gas is frequently present in the atmosphere and reacts with mercury vapor to produce chlorides of mercury, the potential exposure to mercury chlorides also exists.

Smith et al (38) reported on the results of numerous mercury in air determinations in 21 chloralkali plants in the U.S. and Canada. The actual range of the time-weighted average of samples collected was 0.001 to 2.64 mg Hg/m^3, with the highest reading in the cell bed grinding operations. The average air concentration was 0.065 mg Hg/m^3 with more than half (59%) having exposure at or below 0.05 mg Hg/m^3. These results would indicate that engineering controls can limit airborne concentrations of mercury in chloralkali plants to the standard recommended.

The above studies illustrate that effective control of the work environment to limit airborne concentrations of mercury to 0.05 mg Hg/m^3 is feasible. In those instances where ventilation systems were installed or improved, the reduction of airborne levels of mercury to or below a level of 0.05 mg Hg/m^3 of air was prompt. In addition, the study of Benning (35) is significant for showing that the worker plays an important role in controlling his own exposure to mercury by being aware of the hazards inherent in mercury and having good work and personal hygiene practices.

Limited data is available regarding actual wastewater mercury levels. Wastewater emanating from an acetaldehyde production unit was reported to contain 20 mg/l of mercury and the combined wastewater, prior to treatment, contained mercury at 18 mg/l (39). An unidentified wastewater was reported to contain 400 to 600 mg/l of mercury at discharge, and mercury levels of 20 mg/l were observed 1 mile downstream (40). Effluent mercury from one chloralkali plant is estimated to have been 1.4 to 2.8 mg/l, based on 600 gpm waste discharge and 10 to 20 lb/day mercury loss (41).

Cranston and Buckley (42) report the mercury effluent data shown in Table 41. They also reported river water mercury concentrations of 0.134 to 0.380 mg/l near a municipal outfall. In conjunction with these values it was mentioned that commercial bleach preparations alone contain up to 0.200 mg/l of mercury. In contrast, D'Itri (43) reports normal background levels of mercury in natural waters to range from 0.02 to 0.7 μg/l. Minamata Bay water concentrations have been reported to be 0.36 to 1.6 μg/l, and it has been estimated that 600 tons of mercury have entered the bay (43)

TABLE 41: LEVELS OF MERCURY IN INDUSTRIAL EFFLUENTS

Waste	Dissolved Mercury, mg/l	Suspended Solids Mercury, mg/kg dry weight of solid
Paper mill settling pond	0.00008	10
Paper mill effluent	0.002-0.0034	5.6
Fertilizer mill	0.00026-0.004	32.0
Smelting plant	0.002-0.004	-
Chloralkali plant	0.080-2.0	14.0

Source: Reference (42)

The principal sources of mercury wastes in solid, semisolid, or concentrated liquid forms have been identified as: (A) brine sludges from mercury cell chloralkali plants; (B) waste sludges from paint manufacturers; (C) paint residue left in used paint containers; (D) mercury used in electrical apparatus, industrial and control instruments, etc. that are not currently recycled. Mercury wastes from pharmaceutical and battery manufacture have been found to be insignificant (44).

The constituents of the brine sludges from mercury cell chloralkali plants include barium sulfate, calcium carbonate, calcium sulfate, magnesium oxide, magnesium hydroxide, graphite, some iron, aluminum, mud, rocks, and typically 100 ppm mercury in the form of $HgCl_4^=$. An estimated 16,500 lb of mercury are lost through 57,000 tons of brine sludges per year.

Phenyl mercury compounds are still used as mildewcides in water-based paints. The waste sludges from the latex washing system in paint manufacture typically contain 15% pigments, 20% binders, 65% water, and 100 to 150 ppm mercury. It is estimated that 1,800 lb of mercury are lost through 26 million lb of water-based paint sludges per year. The paint residues left in used containers discarded in municipal dumps often contain 0.02 to 0.10% mercury. It is estimated that as much as 32,700 lb of mercury are lost through these paint residues per year.

Of the mercury used for other potentially recyclable uses, such as electrical equipment, measurement and control apparatus, and general laboratory uses, approximately 1 million lb/yr (in batteries, fluorescent tubes, switches, etc.) are disposed of in landfills, dumps and incinerators.

DETECTION

Many analytical methods have been used for the determination of mercury and a review of the literature on this subject has been completed by Smith. Sampling for mercury in air is much more difficult than the subsequent analysis. For best results, sampling and analysis must be considered simultaneously, covering the whole range of types and concentrations of mercury hazards which may be present in the occupational environment (34).

The choice of method of analysis is largely dependent upon what method of sampling is employed. In addition, the degree to which the sampling method collects the different forms of mercury provides the basis for the usefulness of the entire procedure.

The recommended methods for sampling air involve the use of scrubbers to remove mercury vapor and mercury compounds (34). More specifically, the recommended method incorporates two bubblers, each containing a solution of sulfuric acid and potassium permanganate as the collecting medium. Although there is some inconvenience involved in the use of bubblers, their use is justified because of uncertainties in sampling and analysis of nonelemental mercury by other methods as discussed below.

The sampling and analysis of mercury are particularly complex because of the numerous forms in which mercury may exist in the air. The interconvertibility of the various forms of mercury further complicates sampling and analysis. Mercury metal is not highly reactive but does form numerous compounds of varying thermal and chemical stabilities. Mercury reacts to form inorganic and organic compounds. Common oxidation states of mercury are Hg(I) and Hg(II), although only Hg(II) forms organic mercurials.

The compounds of mercury generally have very high volatility compared to those of the alkaline earth metals, and collection of mercury compounds is complicated by their high vapor pressures. Accordingly, it is questionable whether particulates of certain mercurials will remain unvolatilized if collected on simple filters because of the large volumes of unsaturated air which are drawn over the filters.

The chemical stabilities of mercury compounds after collection are important considerations because the relatively easy conversion of one mercurial to another may have a significant effect on the volatility and method of analysis.

The volume of literature regarding the sampling and analysis of mercury attests to the difficulties which are encountered in sampling for mercury. Only in recent years has it been possible to determine, with reasonable accuracy, the form in which mercury exists in the environment. This determination is possible, in part, because of such developments as chromatography and mass spectrometry. These forms of analysis may soon make it possible to clearly and routinely differentiate among the different forms of mercury.

The methods which have been reported for sampling of mercury in air are many but are separable into two basic categories, (A) those methods which remove mercury and mercury compounds from air by scrubbing, and (B) those methods which collect an air sample.

The scrubber type of sampling utilizes bubblers, filters, adsorbants, and amalgamable collectors. The category of bubblers may include impingers, bead-packed towers, and a wide variety of scrubbing solutions. These scrubbing solutions are used to collect the mercury either discriminately or indiscriminately, and frequently to convert the mercury to an easily determinable form such as Hg(II).

The most popular scrubbing solutions are acidic permanganate, iodine-potassium iodide, and iodine-HCl with acidic permanganate being used the most frequently in industrial hygiene surveys. Collection efficiencies for all mercury contaminants have not been reported for all of these scrubbing solutions. However, for mercury vapor and inorganic compounds of mercury, efficiencies greater than 90% have been reported. For organic mercurials, except short chain alkyl mercurials, collection efficiencies appear to be greater than 80% (34).

Uncertainties which still exist in the collection efficiencies for organic and inorganic compounds stem primarily from inadequate methods of standardization and uncertainties in methods of analysis. With a very few exceptions, experimenters have not prepared standard dust chambers of mercury compounds and tested collection efficiencies of scrubbing solutions with these chambers (34).

Filters have been used with scrubbers to collect various mercury contaminants in air. Elemental mercury vapor has been determined by filtering through papers impregnated with selenium, selenium sulfide and potassium iodide. Particulates have been filtered by cellulose filter papers, fibrous glass filters, asbestos wool and quartz wool. However, collection efficiencies of all of these media have not been determined and may be highly variable (34).

Adsorbents are among the most popular collectors and range from charcoal to sea sand. Evaluation of the collection efficiencies of these devices for compounds of mercury are also lacking. Also lacking are evaluations of collection efficiencies for particulates or dust which contain mercury.

Materials developed by Campbell et al (45) are adsorbing media for sampling metallic mercury vapor from an air stream which comprise silver metal coated on finely divided particulate siliceous substrates. Such substrates may range from sea sand to calcined diatomaceous earths. Particularly useful for this purpose is Chromosorb P, a calcined diatomaceous earth. The amount of mercury adsorbed on the silvered substrate is readily quantitatively determined by flameless atomic absorption spectrophotometry. Heating the silvered substrate to 700°C volatilizes all adsorbed mercury and renders the substrate once again useful for mercury sampling.

A method developed by P.E. Trujillo et al (46) involves simultaneously sampling particulate mercury, organic mercurial vapors, and metallic mercury vapor in the working and occupational environment and determining the amount of mercury derived from each such source in the sampled air. A known volume of air is passed through a sampling tube containing a filter for particulate mercury collection, a first adsorber for the selective adsorption of organic mercurial vapors, and a second adsorber for the adsorption of metallic mercury vapor. Carbon black molecular sieves are particularly useful as the selective adsorber for organic mercurial vapors. The amount of mercury adsorbed or collected in each section of the sampling tube is readily quantitatively determined by flameless atomic absorption spectrophotometry.

Amalgamable collectors are also very popular but have been demonstrated to be efficient only for the collection of elemental mercury vapor. If compounds and dust are collected in a separate pyrolysis tube, presumably the mercury in compounds and dust can also be determined. Some work has been done on this method, but a thorough documentation is lacking at this time. Amalgamable collectors, which have been reported to collect mercury, include gold and silver in various forms. The efficiencies of these methods of collection are not well documented.

The second category of sampling involves collection of a direct air sample. There are two general methods of direct air sampling. The first involves a static sample, commonly known as a grab sample, which may be collected in a plastic bag or syringe. This method has not been used or documented extensively for collection of mercury and mercury compounds. The second method is a dynamic monitoring method in which air containing mercury is drawn directly through monitoring instruments called mercury vapor meters. Both hand-held portable units and remote units have been used. To date, these instruments have been designed to monitor only elemental mercury vapor.

Of all the methods for monitoring of mercury, hand-held mercury vapor meters have been used the most extensively. Of all methods used, these are probably the least foolproof (34). There are a number of major difficulties which may be encountered with mercury vapor meters: (a) standardization must be done prior to monitoring mercury with mercury vapor meters, and this is easily overlooked; (b) they respond to many other substances in the air such as dust, cigarette smoke, humidity, ozone, and sulfur dioxide and common organic solvents such as acetone, the presence of which may not be known; (c) some types of mercury vapor meters reverse in response to high concentrations of mercury vapor, indicating meter readings much lower than true mercury vapor concentrations; (d) they may be affected by high magnetic fields that may exist in chloralkali plants; (e) the volume of air sampled by a hand-held mercury vapor meter is small and usually gives a poorly representative sample of the environment. Typically, the volume of air sampled may be 100 ml or less.

The recommended method for analysis of the sample is flameless atomic absorption spectrophotometry. Both colorimetric dithizone methods and atomic absorption methods give similar results in parts added/parts found studies and comparison studies (34). However, the speed in analysis which atomic absorption spectrophotometry affords, without any loss in accuracy and precision, makes it the method of choice as a reference method for analysis of mercury in bubbler solutions when properly standardized and calibrated.

It is imperative that calibration curves be constructed using conditions as close as possible to the actual samples. For example, when flameless atomic absorption is selected, the acid permanganate solutions must be used in the bubblers used for calibration. Calibration curves that are constructed for the dithizone method must also be obtained from acid-permanganate solutions spiked with appropriate mercury compounds.

No single detailed method of analyzing for mercury compounds has been collaboratively tested by numerous labroatories, but the atomic absorption method described by NIOSH (34) represents a culmination of several flameless atomic absorption methods developed and used by various investigators.

Mercury vapor monitors with remote sensors have been used and have problems similar to those of the hand-held type except that (a) problems due to magnetic fields are alleviated, and (b) a large volume of air is sampled and thus provides more representative sampling. With these remote monitors, samples of air are conveyed via Teflon or PVC tubes from sampling ports in the contaminated environment to the monitor. Generally, the sampling points are located in the breathing zone of workers on a grid system, whereby the various parts of the work environment are sampled sequentially.

However, the influence that contamination which may enter and contaminate tubing may have upon subsequent instrument reading does not appear to have been evaluated, and the reliability of these sampling systems has not been documented. Tubing which would be required in a remote sampling system could introduce unknown contaminants and make such a system impractical. In summary, there are many problems associated with the use of mercury vapor meters for determining air concentrations of elemental mercury vapor. In addition, mercury vapor meters do not monitor for compounds of mercury which may be present in the environment.

From the above, it can be seen that for accurately determining concentrations of mercury in air, one must consider not only the form in which the mercury may be present but a multitude of other factors which can have major influence upon the results.

An isotope-shift, Zeeman effect atomic absorption spectrometer has been described by T. Hadieshi et al (47) as having sufficient sensitivity to monitor the total mercury content of ambient air. The minimum total mercury concentration detectable with this device is 0.2 $\mu g/m^3$ of air. The authors point out that this is the first technique which responds to both mercury vapor and particulates which is available for continuous monitoring at this concentration.

Mercury may be measured in water by a variety of techniques, including: atomic absorption spectrometry (AA), emission spectroscopy, and neutron activation analysis.

Atomic Absorption Methods (AA): Mercury concentrations in water are often too low for direct analysis using flame AA. To compensate, a large volume of sample may be taken or mercury may be concentrated by solvent extraction, amalgamation, absorption or ion-exchange.

In the widely used Hatch and Ott method, potassium permanganate is added to a water sample contained in a bottle to convert mercury to the inorganic mercuric form. Excess permanganate is removed by addition of hydroxylamine hydrochloride and mercury reduced to the elemental form by addition of stannous chloride. An air stream is directed through the sample solution to serve as a carrier gas for the mercury which is swept out as a vapor. The vapor is then passed into an absorption cell where the 253.7 nm Hg line emitted by a ultraviolet lamp is absorbed by the vapor. The mercury in the sample is totally converted to the atomic form; the sensitivity is increased substantially over that of the conventional flame technique where only a small fraction of the mercury is converted to the atomic form.

The Hatch and Ott method was modified in the EPA method to include oxidation with potassium persulfate to ensure complete oxidation of some organomercury compounds (48). Other modifications utilize addition of sulfuric and nitric acids to destroy residual organics (49), and use of helium as the carrier gas (50).

In the method by Moffitt and Kupel, biological and aquatic samples are decomposed with nitric acid and mercury ions reduced to mercury metal using stannous chloride (32). The mercury is swept out of solution by an air stream and absorbed into charcoal. The charcoal is then introduced into a small boat-shaped tantalum vessel which is inserted directly into an oxidizing air-acetylene flame. The minimum detectable quantity of mercury by this method is 20 ng.

Chau and Saitoh used a combination of solvent extractions and flameless atomic absorption (51). Mercury in water samples was extracted with dithizone dissolved in chloroform. Dithizone has been found to react with some organomercury compounds as well as inorganic forms. The dithizone extraction also serves to concentrate the mercury in the smaller chloroform volume.

Following dithizone extraction/concentration, the mercury was extracted back into an aqueous solution; stannous chloride was added to reduce mercuric mercury to the metal, and measurement made using flameless AA. A sensitivity of 0.008 ppb was achieved using this technique. Other methods for analyzing mercury in water utilize amalgamation of mercury onto copper or silver collectors. In one fell swoop, these methods can be used to simultaneously separate, concentrate and convert various mercury forms from water samples (52)-(56).

In these methods, the copper or silver wire is electrically heated to drive off collected mercury as a vapor. The mercury vapor is then passed into the optical cell where the change in intensity of the 253.7 nm Hg line is measured. Collection of mercury on the wires is done either as a heterogeneous chemical reaction or as electrodeposition under an applied potential of 3 volts. Less time may be needed for electrochemical collection; 20 minutes versus 2 hours in one case (53). However, if other metals are present in the water, they may also plate out and interfere with the mercury analysis. The method can be applied to measuring concentrations at the nanogram level in laboratory prepared samples.

Other flameless atomic absorption methods of collecting and simultaneously concentrating mercury from aqueous samples onto solids include absorption of mercury vapor into charcoal (32), and evaporation on a tantalum boat (57). The latter procedure involves extraction of mercury from urine samples into an organic layer, followed by evaporation of the organic layer (methyl isobutyl ketone) onto a tantalum boat. The boat is heated in the flame and mercury measured as the vapor.

Laboratory automation of atomic absorption spectrophotometers was achieved by using a Technicon Auto-Analyzer (58)(59)(60). One such system automates the Hatch and Ott method for mercury. The rate of flow of the air stream through the sample and to the absorption cell is critical; the sensitivity increases with increasing flow rate.

It is possible to analyze 22 to 50 samples per hour (58)(60). Comparative throughput for nonautomated operation is not given, but automation should free the laboratory operator to do other work.

A total mercury monitor has been developed by I.A. Capuano (61). It permits analyzing liquid streams for their content of mercury including elemental mercury, ionic mercury and undissociated mercury compounds such as phenylmercury and methylmercury. The sample is automatically measured and first treated with an acidic metal salt reagent containing an oxidant for oxidizing elementary mercury in the sample to ionic mercury and then with hydrazine to form a suitable catalytic colloid. Elementary mercury is then formed by the action of hydrazine and the catalytic colloid on the ionic and combined mercury and is transferred to a photometric detector. The results are indicated and/or recorded automatically, accurately and rapidly. The apparatus and method are especially useful in the control of mercury contamination in the environment and also for monitoring chemical process streams.

An apparatus developed by D.L. Hammitt (62) is one in which means for determining the amount of mercury in a liquid or gas sample is provided comprising means for sequentially oxidizing and reducing a nonorganic fluid containing mercury combined with organic molecules to form mercury metal and means for passing an inert gas to carry the mercury metal to a UV spectrophotometer wherein the amount of mercury is recorded.

An analyzer developed by W.M. Grengg (63) utilizes a vessel, such as a test tube, having a predetermined interior side surface area coated with gold. The closed end of the tube is uncoated. Mercury from a fluid sample is adsorbed by the gold coating, such as by passing a gas through the tube or by placing a liquid in the tube and shaking the tube for a short period of time. After mercury from a given amount of sample has been adsorbed by the gold coating, the test tube is placed in an inverted generally upright position with the closed end of the tube above the open end and with the open end vented. Then the tube is rapidly heated to a temperature in excess of 400°C to vaporize the mercury from the gold coating.

The mercury vapor will rise in the test tube to the closed end thereof and form a bubble of mercury vapor mixed with hot air. Ultraviolet light in the 2537 A wave length band is passed endwise through the test tube. Some of the ultraviolet light is absorbed by the mercury vapor in the test tube and the transmitted light is detected and measured. Next, the mercury vapor laden air is flushed from the test tube with mercury-free air. Ultraviolet light is again passed through the test tube, this time through mercury-free air and the amount of light transmitted through the mercury-free air is detected and measured. The difference between the light transmission measurements is indicative of the concentration of mercury in the fluid sample.

Emission Spectroscopy: Conventional emission spectroscopy is not sufficiently sensitive for direct measurement of mercury in water samples (64). Methods involving plasma excitation sources do have the required sensitivity, and have been applied. Another method utilizes a DC discharge on samples of mercury vapor which were generated from water samples by reduction with sodium borohydride (50). In this last method, volatile mercury compounds can be separated and analyzed using a gas stream and rubber diaphragm. Only voltailized mercury diffuses through the diaphragm, so that both free and total mercury can be analyzed in the same sample. The above methods can have sensitivities in the parts per trillion (10^{-3} μg/l) range.

In the April and Hume procedure (64), 10 ml of sample solution is constantly aerated with helium. After the base line is stable, 0.5 ml of 2% titanous chloride in 1 M H_2SO_4 is added to reduce mercury to elemental form. The purging of elemental mercury by the helium stream into the helium plasma causes an emission maximum in about 20 seconds. The emission, monitored by a conventional monochromator-photomultiplier tube, is linearly proportional to the mercury concentration from 0.001 to 1 μg/ml. The detection limit might be as low as 0.0002 μg/ml in the 10 ml sample, lower than the 0.001 μg/ml limit in the Hatch and Ott flameless atomic absorption procedure (65).

Ag, Pd and Au interfere if present in concentrations 100 times greater than mercury. Pb, Cd, Zn, Cu do not interfere in concentrations 1,000 times greater than mercury. Concentrations of Cl^- over 0.1 M cause a slight depression. Phenyl-

and methylmercury can be analyzed by the same procedure. Elemental and dimethylmercury are immediately aerated by the helium stream before addition of reductant unless oxidized to less volatile forms. The plasma torch unit is not commercially available but could be constructed for under $1,000.

SpectraMetrics, Inc. has introduced argon plasma emission spectrometers which should be applicable to the April and Hume procedure for mercury. Detection limits have not been indicated but might be comparable to helium plasma instruments.

In a method by Braman for measuring mercury in solution (50), the mercury is reduced to elemental form with $NaBH_4$ solution and allowed to diffuse through a latex or Saran Wrap membrane into a helium carrier stream; the mercury is excited as it is carried through a 900 volt, 31.5 watt/inch DC discharge. A monochromator and photomultiplier tube measure the 253.7 nm emission intensity.

With 200 ml of sample solution in an open beaker and a latex membrane area of 12 cm^2, the limit of detection is 4×10^{-6} μg/ml Hg. Analyses of natural water samples gave measurements of 1 to 16×10^{-5} μg/ml. Dimethylmercury could be determined without reduction, with the same sensitivity as elemental Hg. Oxidation during pretreatment has been successfully done with $KMnO_4$-H_2SO_4. The method can be used to determine ng quantities of As and Sb (66).

Emission spectroscopy is less susceptible than atomic absorption to interference from aromatics, since aromatics do not usually emit in the 253.7 nm region. But as in atomic absorption, cobalt is a potential interference, as it has an emission line at 253.649 nm, close to the mercury line at 253.652 nm (67). The cobalt problem is avoided if the mercury is selectively removed such as in the Braman or the cold vapor method.

Neutron Activation Analysis: This technique was used to measure the total mercury content in fish (68)(69), some common foods (70) and samples of fresh water (71).

In each of these analyses, mercury was separated using anion exchange or precipitation (69). For example, the mercury in fresh water (71) was converted to $HgCl_2$ by adding sufficient HCl to a 4 liter sample contained in a polyethylene bottle to make the final solution 0.1 N in HCl. The acidified water sample is passed through anion-exchange resin loaded filter paper to concentrate the Hg. The filter paper is allowed to air dry, sealed in Mylar bags and irradiated. Irradiation time was 2 hours at a thermal neutron flux of 1.3×10^{13} n/cm^2-sec. The samples were cooled three to four days, transferred to a polyethylene bag and counted. Mercury in samples of fresh waters was found to range between 0.03 to 6.6 ppb. The detection limit of the instrument is calculated to be 0.005 μg Hg.

A sample of methylmercuriguanidine was analyzed by first treating for 30 minutes with chlorine gas to degrade the compound and convert it to an analyzable form. Anion exchange followed by irradiation showed that recovery was about 100% for the methylmercuriguanidine sample. Only 0.36% was recovered for samples which were not chlorine treated. Other organomercury compounds may possibly be analyzable using this procedure (71). The potential of neutron activation/anion exchange technique to distinguish between inorganically bound and organically bound mercury is indicated by the low recovery of the latter when chlorine pretreatment is not used.

ENVIRONMENTAL STANDARDS

The problem of environmental standards with respect to background concentrations and emissions is very complex in the case of mercury as pointed out earlier by Sittig (72).

Firstly, even background (natural) concentrations exceed standards in some cases. Tuna and swordfish caught in unpolluted waters near Malaya and Africa had mercury concentrations exceeding the Food and Drug Administration tentative standard of 0.5 ppm (11). The air above ore deposits may contain as much as 20 $\mu g/m^3$ of mercury (6), higher than the Environmental Protection Agency's recent suggestion of 1 $\mu g/m^3$ maximum for public, long-term average exposures (19).

Secondly, as a result of man's activities, mercury concentrations are elevated above background levels, natural hazards are enhanced, unsafe conditions are created and human sickness and deaths have occurred. In industrial and research work areas where mercury is used, air in poorly ventilated spaces can easily contain over 50 $\mu g/m^3$ of organic mercury, the maximum safe concentration in occupational air as recommended by the American Conference of Government Industrial Hygienists (73).

Uncontrolled plant emissions into air and water can cause the surrounding ambient air to contain average mercury concentrations over 1 $\mu g/m^3$ (19). Similarly, liquid wastes can induce high levels of mercury in local waters and wildlife. Perch caught upstream from a pulp and paper mill outfall contained from 0.18 to 0.70 ppm mercury; those caught downstream from it contained from 1.90 to 3.02 ppm mercury (2).

Thirdly, because of the mobility of mercury, discharged mercury does not remain localized. Through natural cycles, it slowly disperses and extends itself into much broader regions to remain almost indefinitely.

Figure 22 shows the relation of various levels of occurrence of mercury in the environment to various environmental standards (26).

Assuming a maximum allowable average daily intake of 30 μg mercury per day (12)(19), and an average dietary intake of 10 μg/day (19), one finds that the average intake of mercury from air should be restricted to 20 μg/day. If the inhalation of air is 20 m^3/day, the average mercury concentration should be no more than 1 $\mu g/m^3$ (19). Occupational health standards based on the 7 or 8 hour workday and 40 hour workweek have been adopted by the American Conference of Governmental Industrial Hygienists. These ACGIH standards suggest maximum, time weighted average values (threshold limit values) of 10 $\mu g/m^3$ for alkylmercury and 50 $\mu g/m^3$ for other mercury (73).

The ACGIH standards are commonly adopted by the U.S. Bureau of Mines, U.S. Department of Labor (73)(74), and other regulatory agencies. Short discussions of health standards for airborne mercury may be found in the references (2)(12)(18). Included are standards of other nations.

FIGURE 22: VARIATIONS IN MERCURY CONTENT OF THE ENVIRONMENT IN RELATION TO EXISTING STANDARDS

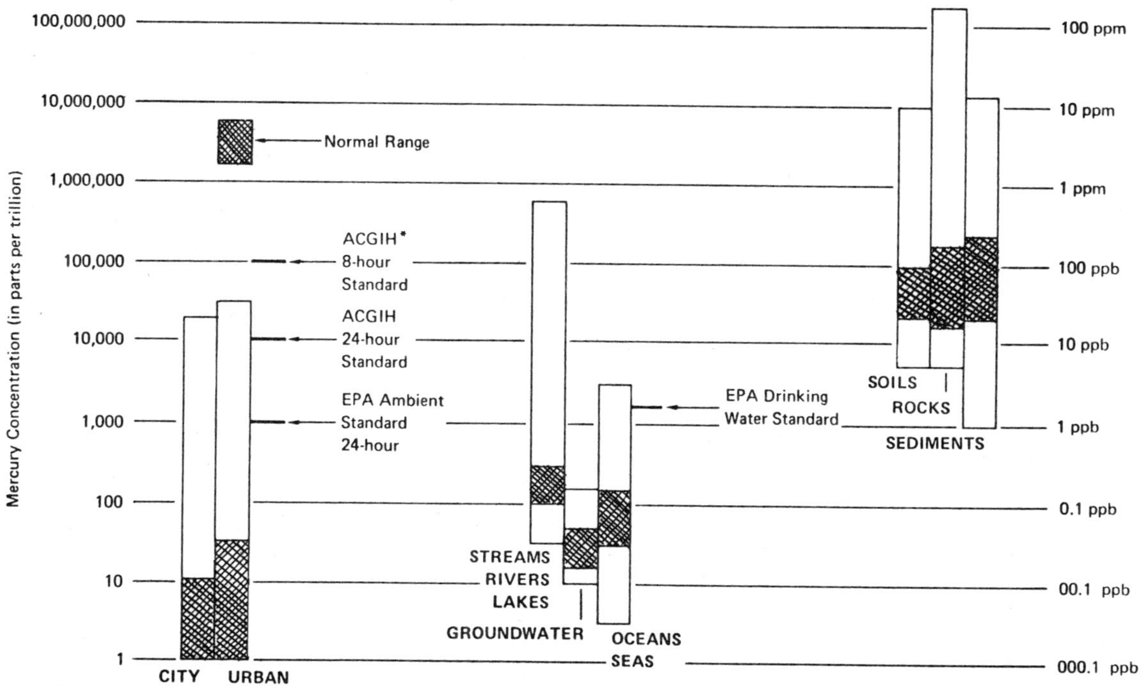

* American Congress of Governmental Industrial Hygienists.

Source: Reference (26)

According to a recommended standard for occupational exposure to inorganic mercury promulgated by NIOSH (34), occupational exposure to mercury shall be controlled so that workers are not exposed to inorganic mercury at a concentration greater than 0.05 mg Hg/m^3 determined as a time-weighted average (TWA) exposure for an 8 hour workday.

The Clean Air Act of 1970 authorized EPA to establish national standards for emissions of hazardous air pollutants. The law did not specifically denote mercury as such a pollutant. However, on the basis of evidence that mercury is emitted by coal-burning power plants, municipal incinerators, and industrial plants, in March 1971 the EPA designated mercury as a hazardous air pollutant (36 FR 5931). Standards for mercury were proposed in December 1971 (36 FR 23239) and were promulgated as regulations in April 1973 for mercury ore processing facilities and chloralkali plants (see Table 42).

There is no direct comparison possible between the recommended criteria for occupational exposure and the proposed national emission standard for mercury that the levels of exposure to the general public of varying health status and age on a 24 hour day, 7 day week, basis should be substantially lower than occupational standards based on an 8 hour day, 40 hour week. However, the amount of mercury which an individual absorbs from the general atmosphere will be superimposed on that which he would receive from his occupational exposure. This additional amount is not expected to adversely affect workers when occupational levels are not above the 0.05 mg/m^3 as recommended by NIOSH (34).

A limit of 0.5 ppm (500 ppb) of mercury in fish to be consumed by humans was set by the Food and Drug Administration (9). Departments of industrial hygiene usually consider 50 ppb of mercury in urine as indicative of excessive exposure (75). A tentative standard for mercury in drinking water was set at 5 ppb by the U.S. Public Health Service in 1970 (2). A Swedish toxicological committee recommended a maximum acceptable intake level of 0.030 mg/day of methylmercury for a 70 kg (154 lb) male (13).

Under the Drinking Water Act of 1974, EPA was given the authority to propose and promulgate standards for mercury in drinking water. The proposed standard is 0.002 ppm, which is much more strict than the previous standard of 0.005 ppm. The most widely known action taken by EPA in this early period involved application of an old law to a new situation. In 1970, the agency conducted a nationwide survey to determine mercury levels in surface waters. Based on these results and on the Rivers and Harbors Act of 1899, EPA proposed standards for mercury discharge into surface waters, at rates dependent upon the type of receiving water body. Since then, several states have enacted their own limitations on mercury discharges to water from a given emitter or class of emitters.

The Federal Water Pollution Control Act amendments of 1972 gave EPA the basis for enforcing the regulation of mercury discharges into the water environment. In September 1973, EPA designated mercury as a toxic pollutant (38 FR 24342), and in December 1973 the agency proposed toxic pollutant effluent standards, which were amended in March 1974 (Table 42). In October 1973, EPA issued regulations prohibiting dumping any material containing mercury into the ocean.

TABLE 42: EPA MERCURY DISCHARGE REGULATIONS

Federal Register	Date	Applicable to	Standard
38 FR 8820*	4/06/73	Mercury ore processing facilities and chlor-alkali plants	2,300 gm/24-hour period
38 FR 35388 (proposed) (39 FR 10603 amendment)	12/27/73	Paper and allied products; oil and gas extraction; industrial organic or inorganic chemical; alkalis and chlorine; ferrous metal production; nonferrous metal smelting and rerefining; lumber and wood products; bituminous coal and lignite mining; storage or primary battery manufacturing; or metal mining facility discharging into navigable water	(1) Into streams, lakes, or estuaries with flow less than 10 cfs or lakes less than 500 acres --no discharges; (2) other streams and lakes--20 μg/ℓ per discharge or 1/10th this concentration when low flow is less than 10 times the waste flow; (3) other estuaries and all coastal waters-- 100 μg/ℓ per discharge or 1/10th this concentration where low flow is less than 10 times the waste flow; (4) stream--not to exceed 0.000162 times flow in cfs (or) 1.62 lbs/day; (5) lake--not to exceed 0.000135 times flow in cfs (or) 1.35 lbs/day; (6) estuary--not to exceed 0.00027 times flow in cfs (or) 2.70 lbs/day; (7) coastal water--not to exceed 0.000324 times flow in cfs (or) 3.24 lbs/day
38 FR 28610	10/15/73	Ocean dumping	No mercury except as trace contaminants
39 FR 38064	10/25/74	Wastewater treatment plant sludge incinerators	3,200 gm/24-hour period

*38 FR 8820 = Vol. 38, *Federal Register*, page 8820.

Note: This table does not include regulations dealing with pesticides. There have been and continue to be many such regulations, all involving either cancellation or suspension of pesticide use; however, limitations of time and space have precluded listing them here.

Source: Reference (26)

The agency has also designated mercury compounds (acetate, ammonium chloride, bromide, chloride, cyanide, iodide, nitrate, oxide, sulfate, and thiocyanate) as hazardous substances, and any person or entity spilling such substances is liable for cleanup and/or fine (38 FR 30466, August 1974). In October 1974, standards for mercury emissions from wastewater treatment facilities were proposed (26).

Beginning in 1970, EPA took a series of actions to cancel pesticides that contained mercury, under authority obtained from the Department of Agriculture. The Federal Environmental Pesticide Control Act of 1972 strengthened EPA's authority to regulate, control, and cancel harmful pesticides, including those containing mercury.

One form of governmental regulation not mentioned previously concerns the interstate transportation of hazardous substances. Many mercury compounds are designated as Group B poisons by the U.S. Department of Transportation, and their shipment is regulated in accordance with applicable regulations. Also, because metallic mercury readily amalgamates with aluminum, the shipment of metallic mercury by air is restricted by the Civil Aeronautics Board (26).

HANDLING PROCEDURES

Only those persons having a need to be there should be permitted in mercury work areas, and each mercury processing area should be separate from other areas where possible.

According to NIOSH (34), medical precautions relating to the handling of inorganic mercury should be as follows. Comprehensive medical examinations (which should include complete urinalysis) shall be made available to all workers subject to exposure to inorganic mercury prior to employee placement and annually thereafter. These examinations should place emphasis on any symptoms or signs of unacceptable mercury absorption such as loss of weight, sleeplessness, tremors, personality change, or other evidence of central nervous system involvement.

Medical records shall be available to the medical representatives of the employer, of the Secretary of Labor, of the Secretary of Health, Education, and Welfare, and of the employee at his request. These records shall be kept for at least five years after the employee's last occupational exposure to inorganic mercury.

As regards labeling in the case of handling of inorganic mercury, NIOSH (34) suggests that: The following warning shall be posted to be readily visible at or near entrances or accessways to work areas where there is potential exposure to inorganic mercury.

WARNING!

MERCURY WORK AREA

Unauthorized Persons Not Permitted

The following warning shall be posted in readily visible locations in any work area where there is potential exposure to inorganic mercury.

WARNING!
MERCURY
High Concentrations
Are Hazardous to Health
Maintain Adequate Ventilation

If environmental levels are at or greater than the recommended standard, add information to the warning describing the location of the respirators. These warnings shall be printed in English and in the predominant primary language of non-English-speaking workers, if any.

NIOSH (34) suggests the following regarding the use of personal protective equipment such as respirators, when handling inorganic mercury. Engineering controls shall be used wherever feasible to maintain inorganic mercury concentrations in the workplace air at or below the prescribed limits. Appropriate respirators, as prescribed in Table 43, shall be provided and used when a variance has been granted to allow respirators as a means of control of routine operations and while the application is pending.

TABLE 43: REQUIREMENTS FOR RESPIRATOR USAGE AT CONCENTRATIONS ABOVE THE STANDARD

Mercury, mg/m^3	Respirator Type*
Less than 5.0	I, II, III
Greater than 5.0	II, III

*Type I: Full facepiece gas mask equipped with a high efficiency filter plus canister containing iodine-impregnated charcoal.
Type II: Type C (positive pressure) supplied air respirator.
Type III: (Positive pressure) self-contained breathing apparatus.

Source: Reference (34)

Administrative controls can also be used to reduce exposure. Respirators shall also be provided and used for nonroutine operations (occasional brief exposures above the environmental standard and for emergencies); however, for these instances, a variance is not required but the requirements set forth below continue to apply. Respirators shall only be used pursuant to the following requirements:

A. For the purpose of determining the class of respirator to be used, the employer shall measure the atmospheric concentration of inorganic mercury in the workplace when the initial application for variance is made and thereafter whenever process, worksite, climate or control changes occur which are likely to affect the mercury concentration. The employer shall ensure that no worker is exposed to inorganic mercury in excess of the standard because of improper respirator selection or fit.

B. Employees experiencing breathing difficulty while using respirators shall be evaluated by a physician to determine the ability of the worker to wear a respirator.

C. A respiratory protective program meeting the general requirements outlined in section 3.5 of American National Standard for Respiratory Protection Z88.2-1969 shall be established and enforced by the employer.

D. The employer shall provide respirators in accordance with Table 43 and shall ensure that the employee uses the appropriate respirator.

E. Respiratory protective devices described in Table 43 shall be either those approved under 30 CFR 11, published March 25, 1972, or under the following regulations.
   1. Gas masks—30 CFR 13 (Bureau of Mines Schedule 14 E)
   2. Self-contained breathing apparatus—30 CFR 11 (Bureau of Mines Schedule 13 E)
   3. Supplied air respirator—30 CFR 12 (Bureau of Mines Schedule 19 B)

F. Usage of a respirator specified for use in higher concentrations of inorganic mercury is permitted in atmospheres of lower concentrations.

As regards work clothing for use in the handling of inorganic mercury, NIOSH (34) suggests that:

A. Each employee subject to exposure to inorganic mercury shall be provided coveralls or similar full body work clothing, shoes or shoe covers, and hat, which shall be worn during the working hours in areas where there is exposure to inorganic mercury. A daily change of clean work clothing shall be supplied by the employer.

B. Adequate shower facilities provided with hot and cold or tempered water shall be available for use and used by workers.

C. Work and street clothing shall not be stored in the same locker.

D. Work clothing should be vacuumed before removal. Clothes shall not be cleaned by blowing or shaking.

Work clothing should take the form of coveralls, as opposed to shirts and trousers, and have a minimum of seams, with no cuffs or pleats. Clothing should also be of a nonwoven or tightly woven fabric, which exhibits a minimum tendency to absorb mercury. Shoe covers, rubber boots or shoes which can be washed should be provided where floor contamination is a problem. Contaminated clothing should be stored in covered containers or vaporproof bags pending laundering. Laundering of such work clothes should be provided by the employer, and precautions taken to minimize exposure of laundry workers to mercury.

As regards the apprisal of employees of hazards which may result from the handling of inorganic mercury, NIOSH (34) suggests that:

A. Each employee exposed to inorganic mercury shall be apprised at the beginning of his employment or assignment to an inorganic mercury work area of hazards, relevant symptoms, appropriate emergency procedures, and proper conditions and precautions for safe use or exposure. He shall be instructed as to availability of such information including that prescribed in (B) below. Such information shall be kept on file and shall be accessible to the worker at each place of employment where inorganic mercury is used.

B. Information as specified in Appendix III (34) shall be recorded on U.S. Department of Labor Form OSHA-20, "Material Safety Data Sheet" (see Appendix III) (34) or on a similar form approved by the Occupational Safety and Health Administration, U.S. Department of Labor.

As regards work practices for the handling of inorganic mercury, NIOSH (34) specifies the following.

A. Emergency procedures
   1. Procedures, including fire fighting procedures, shall be established and implemented to meet foreseeable emergency events.
   2. Respirators shall be available for wearing during emergencies. Self-contained respirators shall be available for employee use in the event of fire or other emergencies where equipment or operations cannot be abandoned because of an emergency.

B. Exhaust systems
   Where a local exhaust ventilation system is used, it shall be designed and maintained to prevent the accumulation or recirculation of mercury vapor, dust, and fumes into the workroom.

C. General housekeeping
   1. Floors, work surfaces, and equipment shall be so contructed and maintained as not to have cracks, crevices, or other areas which may retain mercury.
   2. Spills and leaks of mercury shall be promptly cleaned up either mechanically or chemically, or by other appropriate means. No blowing or dry sweeping shall be permitted. When vacuum cleaners are used, they shall be equipped with mercury vapor absorbing filters to prevent dispersal of mercury vapors into the workplace air and shall be maintained so they will not disperse mercury-laden dust into the workplace.
   3. Waste mercury or materials contaminated with mercury shall be kept in vaporproof containers, under water, or in chemically treated solutions, pending removal for disposal or processing for reuse.

D. General procedures
   1. Containers of mercury shall be kept covered when it is not necessary to have them open for process operations.
   2. Open containers of mercury, to the greatest extent possible, shall have the surface of the mercury covered with an aqueous layer maintained at a temperature below its boiling point to prevent vaporization of the mercury.

As regards sanitation practices for the occupational handling of inorganic mercury, NIOSH (34) specifies the following:

A. Food preparation, dispensing (including vending machines), and eating shall be prohibited in mercury work areas.

B. Smoking materials shall not be permitted in mercury work areas.

C. Handwashing facilities, including hot and cold running water, soap, and towels, shall be made available adjacent to mercury work areas. Employees shall be instructed in the importance of thoroughly washing their hands before eating or smoking.

D. Soiled clothing shall be stored in vaporproof containers pending removal for laundering.

E. Laundering of work clothing shall be provided by the employer. Persons responsible for laundering mercury contaminated clothing shall be informed of the hazards involved.

As regards monitoring and recordkeeping requirements for the occupational handling of inorganic mercury, NIOSH (34) specifies that workroom areas where it has been determined, on the basis of an industrial hygiene survey or the judgment of a compliance officer, that environmental levels do not exceed 40% of the environmental standard shall not be considered to involve worker exposure to inorganic mercury. An additional survey shall be made if there is a change in process or engineering controls. Records of these surveys, including the basis for concluding that air levels are below 40% of the environmental standard, shall be kept.

Requirements set forth below apply to inorganic mercury exposures.

A. Employers shall monitor environmental levels of inorganic mercury at least every 6 months. Breathing zone samples shall be collected to permit calculation of a time-weighted average exposure for every operation.

B. When any time-weighted average exposure is at or above the environmental standard, immediate steps shall be taken to reduce environmental levels. Samples shall be taken every 30 days until the environmental level has been reduced below the standard.

C. Records shall be maintained for all sampling schedules to include the sampling and analytical methods, type of respiratory protection in use (if applicable), and the air concentrations of mercury in each work area. Records shall be maintained so that each employee shall be able to obtain information on his own environmental exposure.

REMOVAL FROM AIR

Control techniques for mercury emissions from extraction and chlor-alkali plants have been detailed by the EPA (76). Other general techniques for mercury removal from air and gases have been reviewed by Sittig (77). The condenser gas stream is the major source of mercury emissions from a primary mercury processing plant. The amount of emissions can be reduced by converting to processes that inherently produce fewer emissions or by treating effluent gases to remove mercury. Process changes that inherently produce fewer emissions include beneficiation of ore, retort processing, and hydrometallurgical processing. Appropriate control techniques include cooling and mist elimination, wet scrubbing, or adsorption beds (76).

In one type of commercially available mist eliminator, the mercury-particulate-laden gases pass horizontally through a fiber bed enclosed in screens. Clean gases exit from the fiber bed, whereas the separated mercury drains to the bottom of the mist eliminator. Inertial impaction, Brownian movement, and direct interception are the three basic mechanisms utilized for mercury mist separation. In order to increase the rate of drainage of the collected mercury from the fiber bed, water or other liquid is intermittently or continuously sprayed onto both sides of the bed. The lower section of the eliminator serves as a coalescer for the separation of the collected mercury from the water. Figure 23 illustrates this device.

FIGURE 23: MERCURY MIST ELIMINATOR

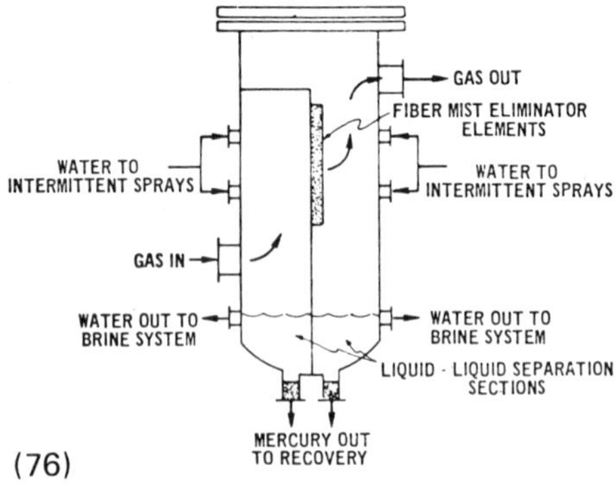

Source: Reference (76)

Since mercury mist eliminators are currently applied in the treatment of relatively clean gas streams (that is, low particulate loading of contaminants other than mercury), the effectiveness of this device for extracting particulate mercury from a stream containing high amounts of particulate matter is not known. Additional particulate loading may cause the efficiency of this technique to be lower than normal. Use of additional flushing water within the mist eliminator or installation of air precleaning equipment prior to the mist elimination device may effectively eliminate the problem.

The particulate mercury removal efficiency of a mercury mist eliminator has been estimated at 86%. Wet scrubbing devices employ a variety of mechanisms to collect particulate matter. Interception of particulate matter by liquid droplets resulting in a heavier dust-liquid agglomerate is the most important of these mechanisms. A particle that has collided with a liquid droplet resists separation because of van der Waals forces. Particulate matter collected on a liquid droplet in this manner can be efficiently removed from the gas stream by a centrifugal collector. The particulate matter collection efficiencies of these devices vary with energy input and can extend over a wide range depending on the design.

The liquid-solid slurry formed can be sent to a settling tank either directly from the device or from a subsequent centrifugal collector. The settled scrubbing liquid may be pumped back to the scrubber and reused. A bleed stream must be taken off the settling tank to maintain a stable level in the tank since particulate matter and moisture are removed from the gas stream. The bleed stream usually represents 3 to 5% of the liquid-solid slurry and is a potential source of water pollution. This stream can be discharged into a secondary settling tank where the mercury solids are separated, and the liquid either discharged or given additional treatment.

Scrubbing systems have several advantages when compared to other particulate matter collection devices:

1. The required capital expenditure is normally lower for scrubbers than for other types of gas-cleaning equipment.
2. The collection efficiency of the scrubber is flexible, depending on the power input.
3. A large range of particulate sizes can be collected.
4. There are no secondary dust problems since the disposal of collected contaminants is a wet operation.
5. The cost of maintenance is low since there are few or no moving parts.
6. Scrubbers are compact in size.
7. Since the collecting operation is wet, there is virtually no limitation on the gas stream humidity and temperature.
8. Wet scrubbers allow the simultaneous collection of pollutants that are in the gaseous, liquid, or solid form.

There are numerous types of wet scrubbers in use that remove particulates from a gas stream, examples of which are the packed-bed scrubber and the venturi scrubber. Several variations of low-energy, low-efficiency, packed-bed scrubbers have been employed in the primary mercury industry to remove mercury from gaseous streams. In one domestic primary extraction facility, a system has been used that consists of water sprays in a packed redwood tank. Another extraction operation scrubs its condenser gases in a similar packed-bed water scrubber. In most situations, the low-energy scrubbing system will prove inadequate for mercury removal because of its low particulate removal efficiency.

The venturi scrubber is a high-energy, high-efficiency scrubber that has been used to remove particulate matter from gas streams similar to the condenser stack gas streams of primary mercury extraction facilities. The water necessary to scrub a 1,600-standard cfm gas stream is estimated to be between 5 and 9 gpm, depending on the gas loading and size distribution of the particulate matter. It is expected that a particulate mercury collection efficiency of 95% could be achieved with a pressure drop of 20 to 30 inches of water through the venturi scrubber and cyclone. A suction fan located after the cyclone is necessary to overcome this pressure drop.

A venturi scrubber that produces sufficient turbulence within the venturi to allow cooling and condensing to occur is commercially available. The scrubbing water required for this situation would necessitate cooling prior to injection. This type of venturi scrubber would not require precooling of the gas stream in order to achieve good collection efficiencies.

The use of either sulfur- or iodine-impregnated activated carbon as a control technique for the removal of mercury vapor from condenser gases of a retort operation should prove adequate. Some problems could arise if substantial amounts of particulate mercury are present in the gas stream; however, this situation can be corrected either by preheating the gas stream or by using a mist elimination device prior to the carbon bed. Although activated carbon has not been used specifically for the treatment of the condenser stream of an indirectly heated primary extraction facility, this technique has been successfully used to treat similar process streams. The mercury vapor is adsorbed by the carbon and reacts with the impregnated sulfur or iodine to form mercury compounds.

In order for treated activated carbon to perform efficiently, the gas stream velocity through the bed should be in the range of 20 to 40 fpm. This low gas velocity is required to allow sufficient contact time between the mercury and the treated carbon. As previously mentioned, the particulate loading of the gas stream, including both particulate mercury and other solid and liquid contaminants, must be low or the carbon bed will plug and lose its efficiency. Destructive distillation of the spent carbon appears practical for recovering the adsorbed mercury.

Because of the low condenser gas flow rates and the low particulate loading in a retort operation, it is estimated that the mercury vapor collection efficiency of the treated activated carbon could approach 99%. Major emissions of mercury from a chlor-alkali plant using mercury cells are from the hydrogen gas stream, the end-box ventilation stream, and the cell room ventilation air. The emissions from all sources can be eliminated by converting to the diaphragm-cell process. The cost of converting a 100 tpd plant is estimated to range from $3,700,000 to $8,000,000.

Mercury emissions can also be reduced by the installation of control systems and the use of good housekeeping practices. The hydrogen gas and the end-box ventilation air streams can be treated by cooling and mist elimination, chemical scrubbing, or adsorption beds. No techniques are presently available to treat the cell room ventilation air; therefore, the control of mercury emissions from this source is dependent on good housekeeping practices. Various techniques have been developed for the control of mercury emissions from chlor-alkali operations. These techniques apply to the by-product hydrogen stream and the end-box ventilation system.

Hydrogen leaves the decomposer at 200° to 260°F and passes into the primary cooler, where ambient water is normally used in a shell-and-tube heat exchanger to cool this stream to 90° to 110°F (Figure 24). The primary cooler usually has a device, known as a mercury knockout drum, for separating condensed mercury from the hydrogen stream. After passing through this initial drum, the hydrogen stream may be subjected to additional cooling in a secondary cooler that uses chilled water or brine as the cooling medium. The condensed mercury from this secondary cooling step is partially removed in a second knockout drum. The temperature of the hydrogen stream discharged from the secondary cooler ranges from 37° to 55°F.

Because of the rapid variation of mercury saturation concentration with temperature, the hydrogen stream temperature should be decreased as much as possible in order to condense the largest amount of mercury vapor.

A large percentage of the condensed mercury may remain in the gas stream as a mercury mist that is difficult to efficiently separate from the hydrogen stream. One approach to resolving this problem is the use of a direct-contact cooler with chilled water or brine instead of a shell-and-tube heat exchanger. The chilled aqueous medium is often in a closed loop with a shell-and-tube heat exchanger that uses mechanical refrigeration. The bleed-off liquor from the direct-contact cooling system is sometimes treated with chemicals such as sodium hydrosulfide to remove the mercury; the clean liquor is then recycled back into the closed loop. An alternate method of treatment is to pump the bleed-off stream into a waste-water settling pond. The use of a direct-contact cooler has the disadvantage of requiring a water treatment system.

The temperature of the end-box ventilation air ranges from 160° to 180°F, and the air may be 10 to 50% saturated with mercury and nearly saturated with water. For calculation purposes, however, it is assumed that both the end-box stream and the hydrogen stream are saturated with mercury vapor. The maximum mercury content of the process streams is therefore considered. Some end-box ventilation gases contain enough chlorine to form mercuric chloride, which remains in the gas stream as particulate matter.

The techniques used for cooling the end-box ventilation air are similar to the methods used to cool the hydrogen stream. However, the presence of mercuric chloride in this gas stream has led to a greater usage of direct-contact coolers for cooling and particulate matter removal. The high water content of this system has limited the temperature to which this gas stream can be cooled due to the formation of ice crystals. The presence of chlorine and mercury salt contaminants has also required the use of more corrosion-resistant construction materials such as titanium and titanium-clad steel.

A technique used in industry to treat end-box ventilation air is to cool the air stream with a direct-contact packed tower employing water as the coolant. This tower is followed by an indirect cooler, which uses 40° to 68°F cooling water.

The resulting entrained particulate mercury is removed by a mist eliminator. A mercury emission rate of 1 ppd from the end-box ventilation system for each 100 tpd of chlorine capacity has been estimated for such a system.

FIGURE 24: COOLING AND CONDENSING OF HYDROGEN STREAM

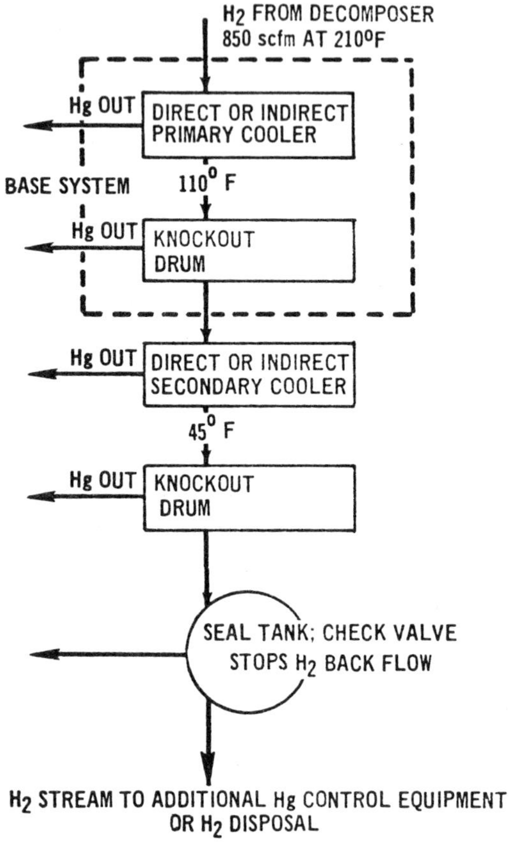

Source: Reference (76)

A considerable reduction in the capital and operating costs of the end-box system can be obtained by the reduction of the end-box ventilation flow rate. This reduction can be accomplished by the installation of leak-tight covers on all cell end-boxes and by the replacement of submerged mercury pumps with in-line pumps.

There are two basic types of mist eliminators that are commercially available to remove mercury mist from gaseous streams. One of these consists of fiber pads to remove entrained mist by mechanisms of impaction, interception, and Brownian movement. When this type of eliminator is used for mercury control in an end-box ventilation gas stream, backwashing with water or a strong reducing agent may become necessary to prevent plugging.

Another type of mist eliminator which is being used in at least one application utilizes a converging-diverging nozzle arranged so that the gases being cleaned follow a curved path and are acted upon by high centrifugal forces in the throat area (Figure 25). These forces are reported to cause the coalescence of mercury mist and entrapment of submicron mercury mist upon the upper wall of the divergent section. Separated particles are washed away from the walls by sprayed liquid. This mist eliminator is reported to have an efficiency comparable to that of the fiber pad type of mist eliminator. The entire unit can be made by plastic or special alloys which resist amalgamation. Figure 26 presents a typical control system utilizing a mist elimination device which is applicable to either the hydrogen or the end-box streams.

FIGURE 25: CONVERGING-DIVERGING NOZZLE MIST ELIMINATOR

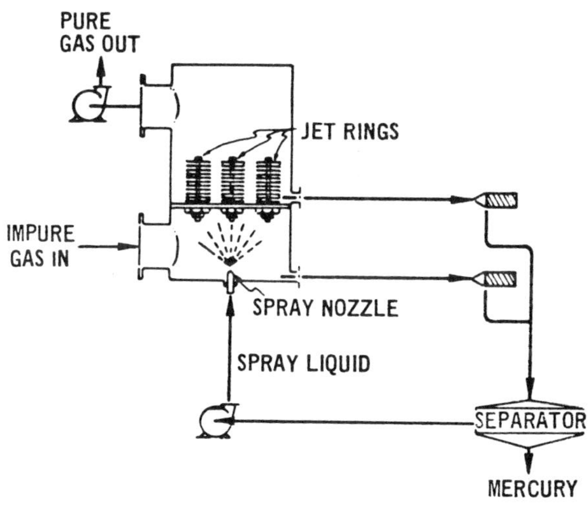

Source: Reference (76)

Depleted brine scrubbing techniques have been applied for mercury removal from hydrogen and end-box ventilation gas streams in only a few instances. The depleted brine scrubbing technique uses the brine discharged from a chlorine cell as a scrubbing liquor. This depleted brine contains about 250 g/l of sodium chloride and 0.6 to 0.9 g/l of available chorine; it has a pH of 2 to 4. It is used as the scrubbing medium in a sieve plate tower or in a packed-bed scrubber. Upon contact with the brine, mercury vapor and mist form soluble mercury complexes. The mercury is later recovered by electrolysis when the scrubbing medium is returned to the mercury cell. Figure 27 presents a simplified flow sheet for a depleted brine scrubbing system. One application of a depleted brine scrubbing system (System A) has been used since the early 1960s for the removal of mercury from the hydrogen stream. The mercury concentration of the treated hydrogen has been reported to be approximately 85 mg/m^3. Mercury losses from the treated hydrogen stream would, therefore, be less than 0.01 ppd on a 100 tpd chlorine basis. In this system, the hydrogen gas is discharged from the decomposer at 180°F. It is then pumped into a direct-contact water scrubber where it is cooled to 100°F.

After partial separation of the entrained mercury mist with a knockout drum, the hydrogen is cooled to 55°F by a direct-contact cooler using chilled brine in a closed-loop system. The cooled hydrogen then flows through a mist eliminator and a depleted brine scrubbing tower. A final alkaline scrubber is used to remove entrained chlorine and acid.

FIGURE 26: COOLING, CONDENSING, AND MIST ELIMINATION

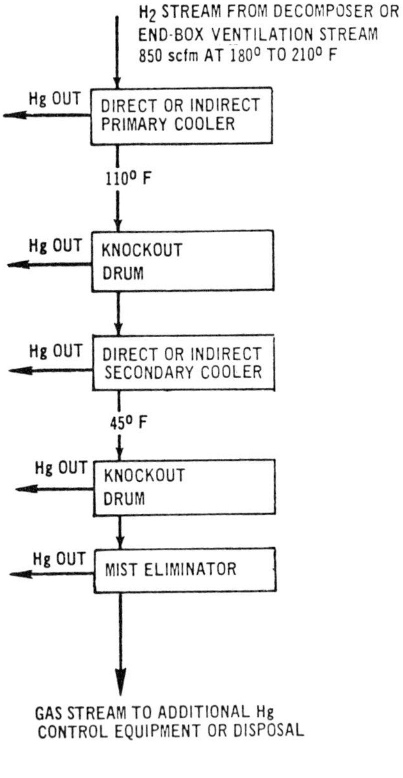

Source: Reference (76)

A second application of depleted brine scrubbing utilizes a system similar to the one discussed above, but deletes the alkaline scrubber (System B). With this system, mercury emissions from the combined process streams are reported to be about 0.6 to 0.9 pound of mercury per day on a 100 tpd chlorine basis.

FIGURE 27: DEPLETED BRINE SCRUBBING SYSTEM

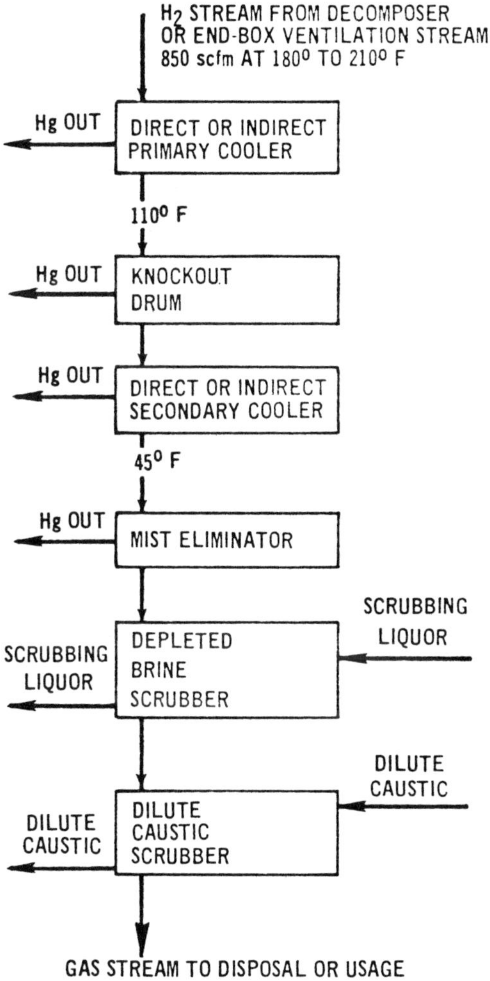

Source: Reference (76)

A second type of chemical scrubbing technique, which has been recently developed, uses a sodium hypochlorite solution as a scrubbing liquor. In one application of this technique, the scrubbing liquor consisted of equal molar amounts of sodium chloride and sodium hypochlorite. The scrubbing medium reportedly required a narrow pH control for optimum mercury removal efficiency. This narrow range of pH was difficult to maintain, and as a result, the sodium hypochlorite scrubber was converted to a depleted brine scrubber.

A second application of a sodium hypochlorite scrubbing technique, which is reported to have solved the problem of pH control, is available for licensing (76). This system employs a dilute solution of sodium hypochlorite with a large excess of sodium chloride over the stoichiometric quantity. The mercury removal efficiency of this system is maintained over a wide enough pH range to make control possible. This system has been employed successfully at two sites to date. A mercury collection efficiency of 95 to 99% has been reported.

Figure 28 presents a general schematic flow sheet of the hydrogen stream for a control system using a hypochlorite scrubber. It is estimated that mercury emissions for the combined hydrogen and end-box streams would range from 0.2 to 0.8 ppd for a 100 tpd plant.

FIGURE 28: HYPOCHLORITE SCRUBBING SYSTEM

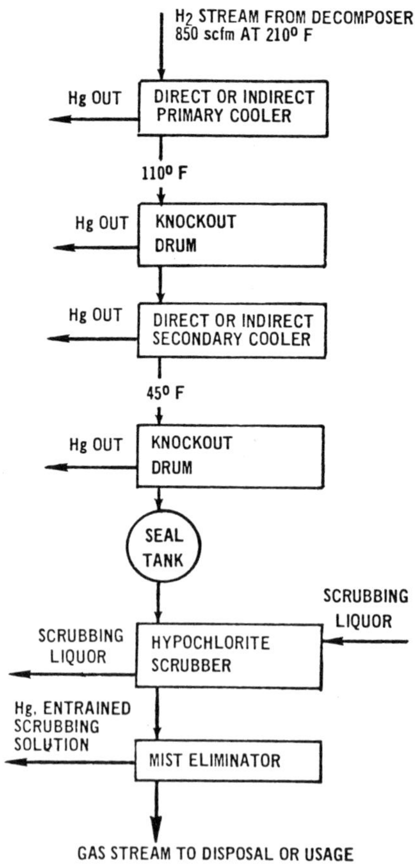

Source: Reference (76)

Control systems containing either sulfur- or iodine-impregnated activated carbon are being utilized by several mercury-cell chlor-alkali plants for reduction of the mercury concentration in the hydrogen stream. In these systems, the mercury vapor is adsorbed by the carbon and chemically reacts with the sulfur or iodine to form mercury compounds. If properly designed, this technique can reduce the mercury concentration in the hydrogen stream to 5 to 10 $\mu g/m^3$. The hydrogen at the inlet to the treated activated carbon bed usually has had 90% of its mercury content removed by primary and secondary cooling followed by efficient mist elimination. Figure 29 is a schematic flow diagram for a typical hydrogen stream employing treated activated carbon. The treated activated carbon can adsorb from 10 to 20% of its weight in mercury before it requires replacement.

FIGURE 29: ACTIVATED CARBON BED SYSTEM

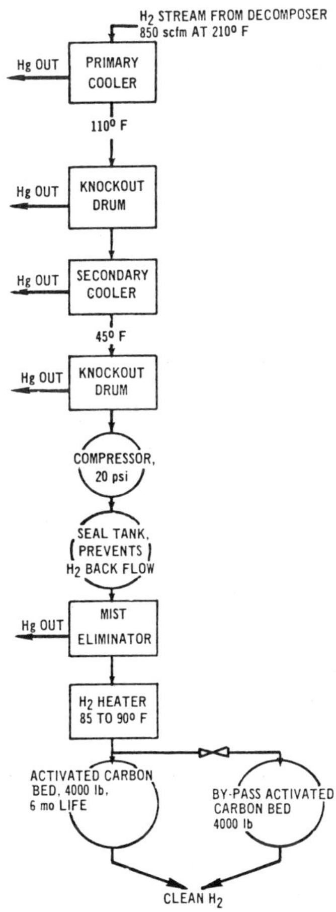

Source: Reference (76)

Destructive distillation of saturated activated carbon in retorts appears practical for recovering the adsorbed mercury. This technique should also be applicable to mercury removal from the end-box ventilation system (76).

The molecular sieve control technique utilizes a sieve-adsorbent blend to adsorb the mercury contained in gas streams. A molecular sieve system currently available for treatment of the hydrogen gas stream of chlor-alkali plants in the PuraSiv-Hg System. The designer of this system guarantees a reduction of the mercury concentration level to 0.50 milligram per cubic meter. This concentration corresponds to a hydrogen stream emission of 0.04 pound of mercury per 100 tons of daily chlorine capacity.

Figure 30 illustrates a simplified PuraSiv-Hg System for the hydrogen stream of a mercury-cell chlor-alkali plant. Hydrogen, laden with mercury vapor and mist, is passed through a secondary cooler and a mist eliminator. The gas stream then passes through one of two adsorption beds, both of which contain a proprietary sieve-adsorbent blend. Eighty to 90% of the treated hydrogen gas, containing 0.50 milligram per cubic meter mercury or less, is vented to the atmosphere, combusted in a burner system, or used in a subsequent production operation.

The remainder of the controlled hydrogen stream is heated to 600°F and used as a recycle-regeneration stream for removing entrapped mercury for the second adsorber bed. After passing through the second adsorption bed, this gas stream, with its high mercury concentration, passes through a cooler and is combined with the incoming mercury-laden hydrogen stream from the primary cooling section. The two adsorption beds alternate in function; while one bed is removing mercury from the gas stream, the other bed is being regenerated.

Since the mercury adsorption characteristics of the end-box ventilation gas stream are similar to those of the hydrogen stream, the molecular sieve system has potential application for the control of mercury emission from the end-box ventilation system. Special materials of construction would be required in most situations because of the more corrosive nature, due to chlorine and mercury salt contaminants, of this gas stream.

Now to swing from a discussion of control measures applicable to specific industrial processes to a discussion of more general control techniques for airborne mercury. A process developed by M. Manes et al (78) achieves mercury removal from gases by impregnating materials which dissolve mercury on activated carbon. It has been found that these products will rapidly and in many cases, quantitatively remove mercury vapor from gases including hydrogen, air, carbon dioxide, nitrogen and oxygen. The high surface area of the activated carbon which is impregnated with the mercury reactant appears to be in part responsible for the greatly improved adsorption of the mercury vapor and the carbon appears to activate the metal or other material which dissolves the mercury.

The action of activated carbon appears to be specific. Thus when silica gel is employed in place of activated carbon as the support for the mercury solvent it is not possible to get a satisfactory coating. Additionally, the silica gel readily picks up water vapor which is undesirable.

As the material employed to impregnate the activated carbon there can be used metals which will amalgamate with mercury. Examples of such metals are gold, silver, cadmium, indium, thallium, aluminum, lead, gallium and copper. Metals which are not easily oxidized such as the noble metals are preferred and silver is an especially preferred metal and copper is one of the poorer materials.

FIGURE 30: PROCESS FLOW SHEET FOR A TWO-BED MOLECULAR SIEVE SYSTEM

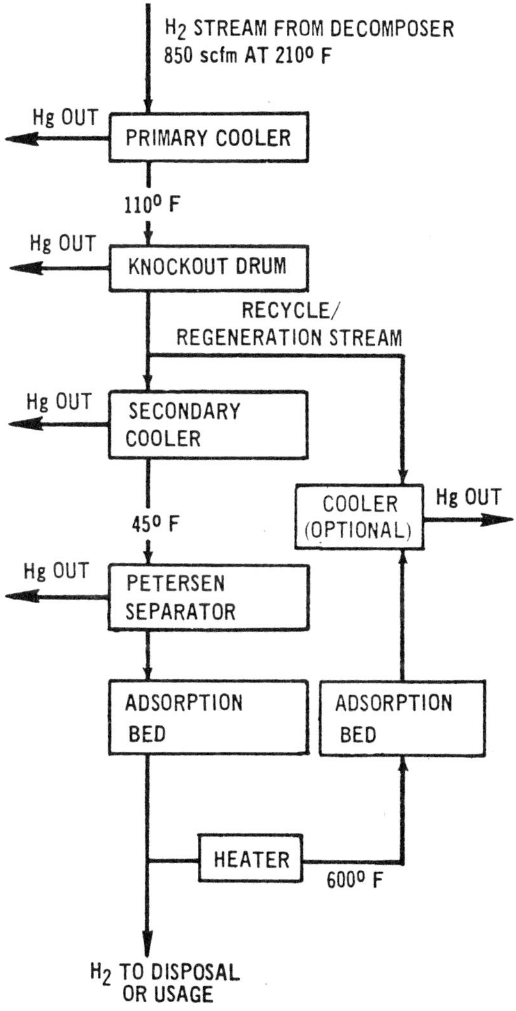

Source: Reference (76)

The metal can be formed on the activated carbon by impregnating the carbon with a solution of a reducible salt of the metal, e.g., aqueous silver nitrate, auric chloride or cupric chloride, and then reducing the salt to the free metal in any conventional manner. After the mercury has been adsorbed on the impregnated carbon it can be regenerated by heating. Thus 100% regeneration of activated carbon impregnated with silver, for example, can be obtained by heating to around 300°C. The mercury is recovered in pure form by this regeneration, e.g., by distillation.

Silver impregnated activated carbon picked up mercury from contaminated air in an amount of over 3% by weight of the carbon. This was 100 times as much mercury as was picked up by the unimpregnated activated carbon. Usually 5 to 50% of the impregnant is used with the activated carbon.

The granular activated carbon employed generally has a particle size between 4 and 60 mesh although this can be varied. In the specific examples there was employed Pittsburgh type carbon BPL which can be made as described in U.S. Patent 2,763,580. There can also be used other activated carbons such as Pittsburgh type CAL, Columbia activated carbon Grade SXAC, Darco activated carbons, etc.

A process developed by S.H. Williston et al (79) utilizes a mercury removal device which comprises a chamber containing a porous absorbent which presents an extended surface of a mercury-wettable metal. The metals desirably used are silver, gold or combinations of these metals, preferably carried in the form of very thin films on an extended support which may take the form of glass or other wool, granular material, or the like.

A process developed by J.C. Park et al (80) is an improved process for reducing the concentration of mercury vapor in hydrogen. Hydrogen produced by electrolysis of brine at a mercury cathode contains substantial quantities of mercury vapor, e.g., more than 20,000 micrograms of mercury per cubic meter, and is usually saturated with water vapor. It is desirable to remove substantially such mercury concentrations from hydrogen since the presence of mercury in hydrogen in amounts greater than about 25 micrograms, and in many cases about 10 micrograms per cubic meter, interferes with the use of the hydrogen in catalytic reductions of organic compounds.

In this process hydrogen gas contaminated with mercury vapor is passed at a temperature below the boiling point of mercury through a bed of particulate adsorbent comprising silver metal on a fused aluminum oxide support at a rate sufficient to substantially reduce the mercury content of the gas.

A process developed by M. Manes (81) involves removing mercury from a gas stream contaminated with mercury by passing the gas stream containing mercury over an activated carbon impregnated with a metallic silver having a crystallite size of not over 250 A. The metallic silver is formed by reducing a silver thiocyanate complex impregnated on the activated carbon.

A process developed by D. Bell (82) involves the removal of mercury from a gas stream by washing the gas with alkaline hypochlorite containing added alkali metal or calcium chloride. The alkaline hypochlorite solution is suitably sodium hypochlorite. Sodium hypochlorite solutions are well-known in commerce and normally contain sodium hypochlorite and sodium chloride in approximately equimolar proportions. When mercury vapor is reacted with such solutions, or with solutions prepared by diluting the commercial solutions with water, a precipitate of insoluble mercury compounds is formed. This is inconvenient because it tends to settle out in vessels and pipelines and the like and makes the recovery of the mercury more difficult.

It has been discovered that if additional alkali metal chloride or calcium chloride is added to the alkaline hypochlorite solution, the mercury remains in solution, possibly in the form of a complex anion. The amount of additional sodium or calcium chloride to prevent mercury compounds precipitating depends upon the composition of the solution particularly with respect to pH.

The mercury vapor can be absorbed effectively from gas streams by dilute solutions of such hypochlorite containing additional alkali metal chloride. The solution should have an available chlorine content of at least about 0.02 g/l. About 25 g/l available chlorine is a usually convenient upper limit, although it is possible to use higher available chlorine contents. The pH should be controlled between approximately 8 and 12, and preferably between pH 9 and 10.5. The minimum sodium chloride content of a solution containing 5 g/l of available chlorine in the preferred pH range is 40 g/l at pH 9 and 140 g/l at pH 10.5 for operation at ambient temperature.

Washing the mercury-containing gas streams may be carried out in any gas-liquid contacting device, for example, a column packed with Raschig rings or on diffuser plates. It may be carried out at ambient temperature or at any other convenient temperature. Mercury may be recovered from the solution either chemically or electrolytically. A suitable electrolytic cell for recovering the mercury contains a graphite or platinized titanium anode and a mercury cathode. The mercury in solution is reduced to mercury at the cathode.

A process developed by I. Ferrara et al (83) is one in which gases contaminated with mercury are purified by bringing the same into intimate contact with a catalytic composition comprised of mercuric sulfide and sulfur. It has been established that such composition has a very high mercury absorption power. It has also been ascertained that this power is due to the catalytic action developed by the mercuric sulfide on the reaction of absorption of the mercury and carried out by the sulfur.

The catalytic composition is fixed on a suitable carrier such as, for example, active carbon or alumina. The quantity of sulfur and mercuric sulfide fixed per unit weight of carrier may vary over a wide range; as a matter of fact, very good results are obtained with catalysts containing only about 10 mg of HgS and about 1 mg of S per gram of carrier, as well as with catalysts containing far greater quantities thereof, for example, on the order of several hectograms HgS and fewer hectograms S per gram of carrier, as well as with catalysts that contain intermediate quantities.

For example, very good results are obtained with different qualities of active carbon or alumina by fixing from 60 to 80 mg of mercuric sulfide and from 5 to 10 mg of sulfur per gram of carrier. The sulfur itself may serve as a carrier. The catalyst in this instance will consist of crystalline sulfur, whose granules have been partially coated with mercuric sulfide.

The catalyst can be used over wide ranges of temperatures and pressures. Good results have been achieved at atmospheric pressure, as well as at pressures of up to about 30 atm. In regard to temperature, one can operate at room temperature, as well as at lower temperatures (from 0° to 10°C, for example) or at higher temperatures (for example at 75°C).

When operating at temperatures higher than room temperature, with certain catalyst formulations, one may encounter a problem where the gas to be purified is contaminated with small quantities of sulfur. Such a drawback can be eliminated by allowing the sulfur to be absorbed by active carbon in a successive absorption tower kept at room temperature.

The concentration of the mercury in the gas to be purified does not, in practice, have any influence on the quality of the results obtained. Good results are obtained regardless of whether it has been operated with a gas having a low content of mercury (i.e., 250 γ/m^3, standard conditions) or with gases saturated or substantially saturated with mercury. The symbol gamma is utilized to indicate microgram.

A process developed by W.H. Revoir et al (84) provides adsorbents for removal of mercury vapor from air or gas including activated carbon granules impregnated with interhalogen compounds, particularly iodine monochloride and iodine trichloride.

A process developed by P. Bryk et al (85) is one in which circulating sulfuric acid adjusted to 80 to 98% by weight is used to wash hot gases containing mercury. The temperature of the acid is maintained between 70° and 250°C and the solid material separating from the circulating wash solution is recovered.

A process developed by B. Kawase et al (86) is one in which mercury vapor contained in various gases, e.g., by-product hydrogen gas generated by a mercury electrolytic cell producing caustic soda, is removed by a simple and effective method which comprises washing the mercury vapor-contaminated gas with a dilute acid solution containing persulfate ions.

A process developed by K.-A. Melkersson et al (87) is one in which mercury-containing gas is passed through a mass of purifying material containing as an active constituent a material consisting of selenium, selenium sulfide or other selenium compounds, or mixtures thereof.

A process developed by L. Raduly et al (88) involves demercuration (removal of mercury) of various gases (air in industrial rooms, hydrogen from an electrolysis plant using a mercury cathode, etc.). According to the process, the polluted gas is passed, at ambient temperature and at atmospheric pressure or higher, through a fixed bed of contact mass. This consists of a porous material (such as active carbon, silica gel or alumina) impregnated with an aqueous solution of a trivalent iron salt. The excess water is run off, and the impregnated mass may be dried superficially before utilization.

A process developed by E. Wygasch (89) is one in which impurities consisting of mercury and/or oxygen are removed from gases by means of an absorbent supported on a carrier. The absorbent used is copper, and the carrier is aluminum oxide which has an internal surface area of 100 to 300 m^2/g, a pore volume of from 0.5 to 0.8 cm^3/g and an average pore diameter of from 40 to 80A.

A process developed by T. Allgulin (90) involves purifying gases containing gaseous elemental mercury. A washing solution is prepared in which there is maintained a content of mercury (II) ions ranging from 0.02 g/l to saturation point and a content of at least one anion selected from the group Cl^-, Br^-, I^- and SO_4^{-2}, these compounds forming sparingly soluble Hg(I) salts. The quantity of the compounds introduced into the washing solution is sufficient to enable formed Hg(I) to be precipitated out but insufficient to cause corresponding Hg(I) salts to be disproportioned to Hg(II) and Hg^o. The contaminated gas is then washed with the thus prepared solution and in those instances when the gas contains reducing agents, Hg(II) is complex bound with Cl^-, Br^- and/or I^- ions to prevent the reduction of Hg(II).

A process developed by C.A. Brockmiller et al (91) is one in which mercury is removed from sulfur dioxide-containing industrial gases preparatory to catalytic oxidation of the sulfur dioxide by admixing hydrogen sulfide with the gas and thereafter subjecting the gas to a suspended solids removal operation.

A process developed by J. Kasai et al (92) is one in which mercury is removed from gases contaminated therewith by passing the gases in contact with activated carbon which has been impregnated with sulfuric acid.

A process developed by Ionescu (93) for the removal of mercury vapor from air containing same comprises introducing the air into the top of a first chamber which is larger at its top than at its bottom and subjecting the air therein to a spray of aqueous soluble sulfide solution. The liquid is thereafter collected along the small floor of this chamber in a tank from which the sodium sulfide solution is recirculated to the spray. The air is induced to flow in an undulating path around vertical baffles and through droplet or mist filters which trap (hold back) any residual droplets before being released to the atmosphere.

REMOVAL FROM WATER

The occurrence of aquatic mercury may be classified into three general and overlapping categories: background, discharged and public. Background concentrations in the aquatic environment do not seem to warrant close attention. The natural level of mercury in water in the northeastern United States is said to be 0.06 ppb (9). Mercury ranged between 0.05 to 0.11 ppb in water in western Canada (10). The U.S. Geological Society analyzed the concentration of mercury in a number of rivers in the United States in 1970. Mercury was reported as 0.1 ppb in most rivers and streams (5).

Discharged sediment levels in many locations may be significantly higher than background levels. Since 1970, substantial decreases in discharges of mercury have taken place and this may be reflected by a lower level of discharged effluents. Sediment immediately downstream of a chlor-alkali plant in Saskatoon (Canada) was reported in 1971 to contain 1,800 ppm of mercury (10). Sediment in Lake Ontario near Hamilton Harbor contained 692 ppm of mercury (10). The concentration of mercury in some water samples ranged from 0.99 to 10.4 ppb (52).

The occurrence of mercury in man's food chain is of primary concern. This introduction of toxic organomercury compounds into man's food chain is documented for fish (2)(17) and pigs (94).

The term "mercury" is conventionally applied to all forms of the element mercury. In an aquatic environment, mercury may exist in the elemental, +1 or +2 oxidation states. For example, it may exist as $HgCl_2$, $Hg(OH)_2$, HgClOH, Hg, or methylmercury (95)(96). Mercury compounds may also adsorb onto suspended matter and sediments. Regardless of form, results of mercury analyses are reported on the basis of micrograms per liter (ppb).

The form of the measured mercury may or may not be stated, although such information is important in terms of toxicological effects. Mercury in fish exists primarily as methylmercury (13)(52); the form is toxic to man if ingested in amounts exceeding about 1 milligram per day (13).

Sources of mercury in the aquatic environment are reviewed by Caton et al (17). The sources include discharged effluents from chlor-alkali plants, wood pulp mills (13) and possible municipal wastes (10). The multiplicity of uses of mercury (over 3,000) creates many sources of contamination and thus a diverse number of routes mercury compounds may take in reaching the environment.

Prior to July 1970, 50 plants were dumping 130 kg (287 lb) of mercury per day into United States waterways; by September 1970, this was reduced to 18 kg (40 lb) (9). Methods used to control mercury discharges into the aquatic environment have been reviewed (1)(17)(94)(97)(98). The methods include precipitation (or cementation) of mercury by a more active metal, e.g., aluminum, copper, iron, zinc. This is effective but it substitutes another metal in solution usually in considerably more than an equivalent amount.

Methods based on the precipitation of mercuric sulfide suffer from the high toxicity and offensive odor of sulfide reagents. Sorption of mercury on preformed metal sulfides, for example, iron sulfide, cadmium sulfide, zinc sulfide, is an effective removal method. However, the mercury in solution is replaced by another ion (97).

A process developed by G.L. Bergeron et al (99) involves recovering mercury from brine effluent from mercury cathode electrolytic cells. The mercury cathode electrolytic cells are constructed with a relatively small gap between a fixed anode and a steel plate or other current conducting material. In the operation of these cells, saturated sodium chloride or potassium chloride brine and mercury are passed through this gap during the electrolysis. The mercury upon entering the cell spreads over the steel plate or other conducting material and acts as a cathode for the cell.

In the process saturated brine solutions are used. After passing the brine once through the cell, the brine discharged from the cell is dechlorinated by air stripping or other means, resaturated, and recycled again through the cell. When a sodium chloride brine is used, the brine generally will contain around 300 grams of sodium chloride per liter, while for potassium chloride solution, the chloride concentration may be up to around 350 grams of potassium chloride per liter. In passing through the cell the chloride concentration of the brine is seldom reduced over 20%. Thus, the brine discharge from the cell is still relatively saturated.

While the mercury cathode cells have many advantages over other conventional cells, a small but significant amount of mercury is lost in the process. A major portion of the mercury loss results from the chlorination of the mercury to a soluble salt which dissolves in the brine as it passes through the cell. This mercury which reacts with the chlorinated brine is often lost in the resaturation step of the process. The brine leaving the cell may contain as much as 50 parts of mercury per million parts of brine and in some cases a considerably greater amount. This loss of mercury while small per pass becomes considerable when a battery of cells is operated over an extended period of time.

It is therefore, a principal object of this process to provide for the recovery of mercury from the brine discharged from mercury cathode cells. A further object is to recover the mercury by an economical method which will not contaminate the brine and make it unsuitable for further electrolysis.

The above objects may be accomplished by adding a soluble sulfide such as sodium sulfide, potassium sulfide or lithium sulfide to the brine to convert the reacted mercury in the brine to a mercury sulfide dispersion and then separating the dispersion from the brine through flocculation. A ferric compound and starch or gum arabic are added to the brine and the mercury sulfide flocculated by adjusting the pH of the mixture. The combined precipitate is then recovered from the brine by mechanical means and processed by usual methods to recover the mercury.

A process developed by M.P. Neipert et al (100) involves intermixing the brine effluent containing the mercury in a reacted form as a mercury compound with at least a stoichiometric amount of an aldehyde elected from the group consisting of formaldehyde and acetaldehyde at a pH of at least 7. The soluble mercury salts in the brine are reduced by the aldehyde to metallic mercury which is then recovered by settling.

A process developed by R.S. Karpiuk et al (101) involves recovering mercury from aqueous solutions containing compounds thereof by passing the solution and a liquid alkali metal amalgam concurrently through a bed, preferably a vertical column, of steel turnings or pieces, preferably first amalgamated, whereby the mercury component of the mercury compounds therein is caused to form additional metallic mercury and/or diluted amalgam, some of which adheres to the surface of the amalgamated steel pieces or turnings and some of which drips from the surface of the turnings to the lower part of the bed or column and collects there from which it can be subsequently removed.

Another process developed by R.S. Karpiuk et al (102) for recovering mercury from aqueous solution comprises passing the solution and concurrently therewith an alkali metal amalgam through a bed, preferably longitudinally, through a vertical column of glass, ceramic, or plastic beads or fragments or Berl saddles or other nonmetallic material which is substantially inactive to the brine, to the alkali metal amalgam, and to compounds normally found in the brine or amalgam, whereby a substantial proportion of the mercury in the brine forms additional dilute amalgam with the alkali metal amalgam and to some extent subsequently forms metallic mercury, mercury thereby being separated from the brine, either as diluted amalgam or as mercury metal.

Both the diluted amalgam and the metallic mercury may be returned to the denuder or decomposition chamber of a mercury chlorine cell (where the amalgam is converted by hydrolysis to mercury and NaOH) or the mercury may be conveniently trapped below the bed of beads and thereafter tapped off as desired.

A process developed by J.F. Gilbert et al (103) is based on the discovery that by bringing an aqueous solution having a pH between 2 and 11 and containing from 1 to 500 ppm of dissolved mercury into intimate contact in a reaction zone with a substantially water-stable solid metallic reducing agent having a greater solution potential than mercury, elemental metallic mercury is liberated. The liberated mercury amalgamates the surfaces of the reducing agent, and also coalesces into droplets on the surfaces.

Depending on the manner of carrying out the process, particles of amalgam and mercury droplets are either allowed to fall from the reducing agent and collected from time to time, or the amalgam and mercury droplets are flushed from the surfaces of the reducing agent along with inert solid settlings formed thereon and recovered from the flushing liquid as by settling or filtration. Impure mercury so recovered is purified by standard methods, such as acid washing or retorting or by a combination of methods. If desired, mercury may also be recovered by removing the reducing agent from the reaction zone periodically along with accumulated reaction products and retorting the entire mass.

As reductor metals, those readily reducing or liberating mercury from solution as metallic mercury include iron, zinc, bismuth, tin, nickel, magnesium, manganese and copper. Of these iron and zinc are to be preferred because of the lower cost of the metals as well as generally lower solution losses and higher reaction rates when used in this process. Iron may be used for those solutions advantageously treated at a moderate pH, for example, to avoid or minimize precipitation of solids, such as oxides and hydroxides at higher pH values.

Optimum mercury removal per pound of iron consumed is obtained upon adjusting the pH of the solution to a value between 6 and 9. At pH values below 6 solution losses of iron become increasingly larger and below a pH of 5 hydrogen evolution reduces the effective surface area of the metal and the competition of hydrogen ions with mercury ions becomes significant.

At a pH of 7 to 8 the consumption of iron in the form of steel turnings may be expected to be of the order of 0.1 pound per thousand imperial gallons (1,200 U.S. gallons) of solution treated by this method. Zinc metal may be used for those solutions that are advantageously treated at a higher pH. Zinc is best used with solutions brought to a pH between 9 and 11. Although zinc readily liberates mercury in less alkaline or in acidic solutions zinc solution losses become increasingly larger at lower pH values.

Zinc amalgams such as are produced by the liberation of mercury at the surface of zinc metal are physically more stable than iron amalgams similarly produced and do not suffer the disadvantage of being rather readily carried away in a fine state of subdivision during a flushing cycle as are iron amalgams. On the other hand, iron has the advantages of having a slightly higher reaction rate and of being a less expensive metal than zinc.

A process developed by R.C. Calkins et al (104) is based on the discovery that mercuric ions can be selectively removed from brines containing halide ions and metal ions, particularly the alkali and alkaline earth metal ions, for example, sodium, calcium, and the like, by contacting the brine at a pH of 9.5 or above, with an insoluble homopolymer or copolymer prepared from at least one monomer of the group consisting of vinylphenyl aliphatic primary and secondary aminomonocarboxylic acids and mixtures containing a predominant amount by weight of at least one such aminomonocarboxylic acid and a minor proportion, for example, up to 20% of divinylbenzene.

It has been further found that the mercuric ions may be readily recovered from the resin chelate by elution with chloride solutions having a pH value below 9.5. It is necessary that the pH of the brine be maintained at 9.5 or above for efficient mercury removal, and is desirably maintained at 10 to 11, although higher pH values may be used. It is desirable to have resin properties which will permit a minimum contact time.

Mercury is recovered from the resin by elution with solutions having pH values below 9.5. Elution operation is advantageously performed at pH values less than 2 in concentrated brine, for example, 5 N sodium chloride. Removal is facilitated by the presence of a high concentration of halide ions, for example, chloride ions, to tie up mercury as the complex chloro anion, $HgCl_4^{=}$, as fast as it is released. Thus, a solution of 1 N HCl in 5 N NaCl is more efficient as an eluting agent than a solution of 1 N HCl alone.

A process developed by H.G. Scholten et al (105) involves passing the brine effluent, containing dissolved mercury salts, through a strongly basic anion exchange resin; washing the basic anion exchange resin with an excess of an aqueous sulfide solution; and separating the mercury from the aqueous sulfide solution.

A process developed by G.E. Edwards et al (106) permits recovery of chlorine by at least partial dechlorination of a brine which contains free and available chlorine and which also permits in addition simultaneous recovery of mercury by at least partial removal of mercury from a brine which contains both free and available chlorine and mercuric chloride.

The chlorine and mercury can be removed simultaneously or concurrently from the effluent brine of mercury cells by flowing the effluent brine through the cathode compartment of for instance a diaphragm cell having a graphite anode and a steel wool cathode where an emf is applied between the anode and the cathode at least equal to the sum of the overvoltage of the chlorine at the anode and of the electromotive force required to overcome the ohmic resistance of the brine.

A process developed by M.G. Deriaz (107) is one in which waste brine can be rendered substantially free from mercury by the precipitation of mercury as mercuric sulfide by treatment in two stages with a material which will introduce sulfide ions into the brine such as a sodium sulfide solution.

The waste brine coming from the mercury cell is acidic and hot. If sodium sulfide solution is added to the brine any chlorine remaining is converted to hydrochloric acid. However, if the addition of sulfide is continued so that a quantity sufficient to precipitate the mercury present in brine is added, no clear end point is obtained with the first small excess of sulfide and then a large excess of sulfide tends to be added to the brine. This is undesirable as the mercuric sulfide separates in a form difficult to settle and separate. On the other hand, if the brine is made alkaline on leaving the cell and the addition of sulfide carried out, a clear end point is obtained, but sulfur separates due to the reaction between hypochlorite and sulfide. This reaction does not occur in two-stage addition as little or no chlorine is present having already been converted to hydrochloric acid before the brine is made alkaline.

According to the process, the substantial removal of mercury from waste brine comprises the following steps: the addition to brine containing chlorine of sulfide ions until the redox potential relative to the saturated KCl-calomel electrode is in the range +0.85 volt to +0.93 volt, raising the pH by the addition of alkali to a pH in the range 7 to 9, preferably 8, then continuing the introduction of sulfide ions into the brine until the redox potential is in the range 0 to 0.2 volt, flocculating the precipitate obtained, settling the precipitate and separating the substantially mercury-free brine.

A process developed by G.E. Crain et al (108) involves the recovery of mercury from effluent brine by adsorptive contact with a strong anion exchange organic resin of the quaternary ammonium, crosslinked type.

A process developed by J.W. Town (109) involves the selective precipitation of mercury from a solution containing mercury and antimony. Sulfide ores or concentrates containing both mercury sulfide (HgS) and antimony sulfide (Sb_2S_3) are conventionally leached with sodium sulfide solution or with a sodium sulfide-sodium hydroxide solution to extract both the mercury and antimony. Such solutions contain mercury and antimony in concentrations of from 5 g/l to saturation depending on the sodium sulfide concentration of the leach solution.

Typically, the sodium sulfide concentration in the leach solution ranges from 25 to 200 g/l, with sodium hydroxide used to replace up to about 50% of the sodium sulfide. A solution concentration of 150 g/l sodium sulfide and 50 g/l sodium hydroxide is generally optimum for extracting the mercury sulfide and antimony sulfide from ore or concentrate.

Aluminum is then conventionally added to the mercury and antimony-containing extracts to precipitate the two metals. The resulting precipitate, containing mercury, antimony, sodium aluminate and aluminum, requires considerable washing before the mercury will coalesce. Some mercury is thus lost along with the antimony and unreacted aluminum.

It has been found that this disadvantage may be overcome by treating the mercury and antimony-containing solution with antimony metal to precipitate the mercury in a clean metallic form, without precipitation of antimony. The reaction is illustrated by the following equation:

$$3Na_2HgS_2 + 2Sb \longrightarrow 3Hg + 2Na_3SbS_3$$

The mercury coalesces readily for recovery and is tapped from the bottom of the reaction vessel. When antimony is used as the precipitant no additional elements (such as aluminum) are added to the solution and the antimony can be electrodeposited from solution for reuse, thus reducing reagent consumption. The amount of antimony required to precipitate the mercury depends on the concentration of mercury in solution, temperature of solution, time allowed for reaction, particle size of antimony precipitant and concentration of antimony in solution and is best determined experimentally. An excess of antimony is usually required to obtain high mercury precipitation and the unused antimony is recovered for reuse.

A process developed by D.W. Rhodes et al (110) involves recovering mercury metal, present in acidic aqueous solutions as the mercuric ion, by adding hydrazine hydrate to the solution and refluxing the solution until the mercuric ion is reduced to mercury metal.

A process developed by J.B. MacMillan (111) is a process for removing mercury from a mercury cell caustic soda liquor whereby the crude liquor is allowed to pass in a generally upward direction through a composite bed of particulate material selected from polyethylene shreds, polytetrafluoroethylene shreds, graphite, charcoal and activated carbon and pieces of a metal selected from nickel, stainless steel and tantalum. Caustic soda solutions are obtained containing as low a concentration of mercury as 0.1 ppm and 75% of the mercury suspended in the crude caustic liquor can be recovered.

A process developed by E.J. Botwick et al (112) involves the recovery of mercury chloride from weak brines effluent from mercury cathode electrolytic cells by contacting the effluent with activated carbon, washing the carbon with water, then heating in a stream of inert gas and condensing mercury chloride by cooling the gas.

A process developed by K.O.H. Fuxelius (113) involves the purification of polluted water from heavy metal ions such as lead and mercury, present therein. The process is characterized by treating the water with an ion exchange composition on the basis of epoxidized black liquor obtained from the sulfate pulp process. The product as used for this purification may be a reaction product of (a) black liquor, (b) epichlorohydrin or 1,2-dichlorohydrin and if desired (c) a thiol compound.

A process developed by W.E. Dean et al (114) for removing mercury from an alkaline solution having a pH of at least 7 is provided comprising intermixing with the solution a sulfur compound in an amount sufficient to provide sulfide ion in an amount greater than the stoichiometric equivalent of mercury, oxidizing the soluble mercury polysulfide and recovering the insoluble mercury sulfide.

A process developed by W. Knepper et al (115) is one in which mercury recovery from wastewaters is accomplished by converting the mercury to low water-solubility compounds and/or to metallic mercury by the use of precipitants and subsequent separation thereof from the wastewater.

The mercury is preferably converted to mercuric sulfide by precipitation with the preferred compound hydrogen sulfide or other compounds which yield sulfide ions. The mercury can also be converted to metallic mercury by reduction with reducing agents such as nonnoble metals and low molecular aldehydes. The precipitant is used in chemical equivalent amounts directly proportional to the dissolved mercury content of the wastewater, and it is advisable to add an excess amount of precipitant up to 10^4 molar and preferably up to 100 molar. After addition of the precipitant, a holding period of at least 5 seconds, preferably 15 to 500 seconds, and especially 30 to 120 seconds is required to permit the precipitation to take place before separation of the formed low water-solubility mercury compounds and/or metallic mercury. Conventional means such as settling tanks, filters, and the like can be used for separating the precipitated mercury compounds and/or metallic mercury from the wastewater.

A process developed by P.F. Waltrich (116) involves treatment of effluent streams containing very low concentrations and even trace amounts of mercury in both elemental form and as mercuric ion. The stream is passed through a closed vessel having a bed of reactive metal particles, such as zinc, magnesium and aluminum, and iron, so that the mercury forms an amalgam on the metal surface. The bed is regenerated and the mercury recovered by heating the bed while the vessel is under vacuum and condensing the vaporized mercury. The lower temperature permitted by the reduced pressure and the absence of air precludes oxidation and clumping of the metal granules.

A process developed by M. Friedman (117) is one in which mercury is effectively removed from water or other liquids by contact with a keratin derivative, for example, reaction products of reduced wool with N-vinylimidazole, N-vinylpyrrolidone, or 2-vinylpyridine.

A process developed by P. De Angelis et al (118) involves contacting water containing metallic mercury with finely divided anthracite coal. The mercury laden water can be first treated with sodium borohydride to reduce dissolved mercury to the metallic form. The water treated with reducing agent, is filtered through a pressure leaf filter coated with a filter aid, contacted with anthracite coal and contacted with chelating resin.

A process developed by W.E. Dean et al (119) involves removing mercury from solution and particularly from mercury electrolytic cell effluent. The process comprises intermixing with the solution a sulfur compound in an amount sufficient to provide sulfide ion to react with the mercury and other ions present therein which react with sulfide ion and treating the resultant solution with an adsorbent, preferably activated carbon, to remove the soluble mercury.

A process developed by J.P. Guptill et al (120) involves removing mercury from a mercury cell caustic soda or caustic potash solution. A vapor or gas is passed through the solution in the presence of a small amount of a reducing agent. Caustic solutions are obtained containing as low a concentration of mercury as 0.01 ppm. Special means are provided for recovery of the mercury.

A process developed by B. Hultman et al (121) involves recovering mercury from mercury-containing liquids by stripping the mercury as vapor from the liquid with an inert gas, and washing the mercury vapor-containing inert gas with a solution containing an oxidizing agent, such as active chlorine, so as to absorb the mercury vapor and oxidize elemental mercury to mercury ion which can then be dissolved in the solution in the form of complex mercury ions such as mercury chloride complex ions.

A process developed by E.L. Cadmus (122) contemplates the removal of soluble inorganic mercury compounds from waste streams by mixing with the stream an amount of sodium borohydride about 100 to 150% in excess of the stoichiometric amount required to reduce its inorganic mercury content to metal as a precipitate. If organic mercurials are present in the stream, they are converted to soluble inorganic mercury salts by chlorination of the stream before mixing the stream with borohydride.

A process developed by R.P. Williams (123) is one in which mercury compounds can be removed from aqueous solutions by extracting the solution with water-insoluble organic sulfides or disulfides, including mixtures thereof. The insoluble organic phase can be readily separated from the aqueous phase.

A process developed by G. Nolte et al (124) is one in which mercury is recovered in metallic form from an aqueous solution, particularly from a salt brine, having mercury compounds therein. The solution is treated with an iron-containing alloy, the potential of which, determined in the solution or brine, is by at least 100 mv more positive than that of iron.

A process developed by K. Motani et al (125) involves regenerating a chelate resin having mercury adsorbed thereto, which comprises contacting the resin with a solution of a sulfide of an alkali metal, barium or ammonium having a pH of preferably at least 13.0, to liberate the mercury from the chelate resin.

A technique developed by M. Krieg (126) is one whereby losses of mercury arising during filtration in the electrolytic production of chlorine according to the amalgamation process with electrolyte cycle and alkaline filtration of the electrolyte are reduced by adding hypochlorite to the electrolyte before filtration.

A mercury removal apparatus developed by R.F. Gerow et al (127) comprises a reductor metal coated on an inert substrate and capable of forming an amalgam with mercury contained in a mercury-bearing liquid in contact with the composition.

A process developed by M.O. Coulter and D. Bell (128) is one in which mercury is recovered from aqueous solutions containing elemental mercury in finely divided or colloidal form or in the form of a soluble inorganic compound by passing a stream of inert gas (air or nitrogen) through the aqueous solution in the presence of a reducing agent, preferred agents being ferrous sulfate at a pH greater than 5, hydrazine, hydroxylamine, D-glucose and sodium borohydride in alkaline solution.

A process developed by M.O. Coulter (129) involves the recovery of mercury from aqueous solutions containing elemental mercury in finely divided or colloidal form or in the form of a soluble inorganic compound by treating the solution with a chemical reducing agent such as ferrous sulfate at a pH greater than 5, iron or steel turnings or chips, hydrazine, hydroxylamine, D-glucose and sodium borohydride with simultaneous removal of the resulting finely divided or colloidal mercury from the solution as a vapor and recovering the mercury.

Methods of removing mercury as a vapor are by passing a stream of inert gas through the solution and steam distillation, which latter term includes injection of live steam and using the technique of single or multiple stage flash distillation under vacuum. A combination of the two methods is found most effective.

A process developed by A.E. Hatch (130) is one in which heavy metals such as mercury dissolved in aqueous streams are removed by contacting the aqueous stream with a polyelectrolyte in the presence of particulate matter, allowing the heavy metal to be bound to the particulate matter and separating the particulate matter containing the heavy metal from the aqueous stream, thereby purifying it.

A process developed by W.N. Marchant (131) is one in which mercury is removed from aqueous solution by contacting the solution with an absorbent prepared by incorporating vicinal thiol groups into a cellulose matrix.

A device developed by H.J. Davis et al (132) utilizes a porous film which is capable of removing heavy metal pollutants, such as mercury, from aqueous media. The film is basically prepared from an acid-containing polymer which has been rendered porous or microporous by standard methods. Reacted with the acid groups in the polymeric film backbone is a chelate-divalent metal complex which chelate is capable of forming a complex with the heavy metal pollutants to be removed from the aqueous media.

A process developed by J.W. Fitzpatrick et al (133) is one in which the mercury content of an aqueous solution contaminated by mercury can be reduced by treating the solution with a mercury-reactive factor, such as a metal or a thio-compound, and an adsorbent such as activated charcoal.

A process developed by M.L. Iverson (134) involves removing mercury from an aqueous solution containing the same, such as a wastewater, including the steps of passing the solution through a bed of zinc or zinc-coated particles, amalgamating the mercury in the solution and the zinc metal of the particles by plating the mercury on the zinc in an electrochemical cell, and preferably then separating the mercury from the zinc or zinc-coated particles.

A technique developed by H. Isfort et al (135) involves eliminating emission of mercury in an alkaline chloride electrolysis plant by converting prepurified wastewater to flushing water through vacuum distillation. Hot brine from the electrolysis plant serves as a heat carrier for the vacuum distillation step.

A process developed by A.M. Chakrabarty et al (136) involves treatment by the use of microorganisms of a liquid stream containing mercury, mercury salts or organomercury derivatives as a pollutant or impurity. The stream is directed through a bed of genetically engineered Pseudomonas bacteria which bind mercury through elaboration of mercury-binding protein, the coding for which is specified by a transmissible plasmid. Not only are the Pseudomonas organisms resistant to unusually high concentrations of mercury, but the uptake of mercury, which becomes protein-bound is rapid and very specific even in the presence of high concentrations of a number of divalent metal ions.

A process developed by W.H. Schuller (137) is one whereby dilute solutions of mercury in aqueous solutions can be lowered drastically in mercury concentration by adding certain naval stores products such as tall oil soap, acidified tall oil soap, tall oil rosin, and others named herein, shaking the mixture, and filtering the solution.

Articles by M. Rosenzweig (138) and N.R. Iammartino (139) survey various routes for mercury cleanup from aqueous solutions.

SOLID WASTE DISPOSAL

A process developed by K. Yamori et al (140) involves recovering mercury from sludge from a purification tank for the purification of saturated alkali chloride solution obtained in the production of caustic alkali and chlorine by the electrolysis of alkali chloride solution in the so-called mercury process. The process comprises dissolving the sludge or a cake filtered from the sludge into an acidic liquid, for the sake of converting the mercury in the sludge into soluble salt form or the like, after which the mercury solution obtained is contacted with anion exchanger, with or without prior neutralization.

The mercury which is adsorbed on the anion exchanger is then eluted, preferably with hydrochloric acid. When the eluate is used for adjusting the pH value of the purified, saturated alkali chloride solution, the recovered mercury is collected as metallic mercury by means of electrolytic reduction in the electrolytic cell.

A process developed by V.A. Stenger (141) involves recovering mercury from sediment at the bottom of a body of water by overlaying the sediment with a layer of metal containing at least one of aluminum, magnesium and zinc. Preferably, the metal is in sheet form for more efficient interception of mercury vapor rising from the sediment. After removal of the metal from the water body, the mercury, which forms an amalgam on the surface of the metal, may be recovered by distillation from the metal surface.

A study of mercury pollution control in stream and lake sediments has been made by the Advanced Technology Center (142). The study was directed at a means for removing mercury from sediments without destroying the environment or otherwise upsetting the ecological balance. The problem was addressed by evaluating a number of recoverable sulfur-based getter systems. Examples of the systems tested were elemental sulfur employed as a coating on a recoverable meshwork and thiourea-type organic compounds dispersed in a recoverable matrix.

The control of mercury pollution in sediments has been reviewed by the Environmental Protection Agency (143).

ECONOMIC IMPACT OF CONTROLS

Control costs for mercury extraction and for chlor-alkali plants have been published (76). Data on the economic impact of the application of new regulations on the chlor-alkali industry, the paint industry, and the battery industry have also been published (26).

REFERENCES

(1) Jones, H.R., *Mercury Pollution Control,* Park Ridge, New Jersey, Noyes Data Corp. (1971).

(2) Wallace, R.A., Fulkerson, W., Shults, W.D. and Lyon, W.S., "Mercury in the Environment–The Human Element," Oak Ridge National Laboratory Report ORNL-NSF-EP-1, National Technical Information Service, Springfield, Virginia (March 1971).

(3) Kothny, E.L., "The Three-Phase Equilibrium of Mercury in Nature," presented at the 162nd National American Chemical Society Meeting, Washington, D.C. E.L. Kothny, Air and Industrial Hygiene Laboratory, State of California Department of Public Health, 2151 Berkeley Way, Berkeley, California 94704.

(4) Friberg, L. and Vostal, J., *Mercury in the Environment–A Toxicological and Epidemiological Appraisal,* Chemical Rubber Company Press, Cleveland, Ohio 44128 (1972).

(5) Goldwater, L.J., "Mercury in the Environment," *Sci. Am.* 224, 5, 15 (1971).

(6) *Mercury in the Environment,* Geological Survey Professional Paper 713, U.S. Government Printing Office, Washington D.C. 20402 (1970); compilation of papers on the abundance, distribution and testing of mercury in rocks, soils, waters, plants and the atmosphere.
(7) Hammond, A.L., "Mercury in the Environment: Natural and Human Factors," *Science* 71, 788 (1971).
(8) Wood, J.M., "A Progress Report on Mercury," *Environment* 14 (1), 33 (1972).
(9) "Mercury in the Environment," *Env. Sci. Tech.* 4 (1), 890 (1970).
(10) Aronson, T., "Mercury in the Environment," *Environment* 13 (4), 16 (1971).
(11) "Mercury: Anatomy of a Pollution Problem," *Chem. Eng. News* 49 (27), 22 (1971).
(12) Grant, N., "Mercury in Man," *Environment* 13 (4), 2 (1971).
(13) Interagency Committee on Environmental Mercury, "Mercury in the California Environment," California State Department of Public Health, Environmental and Consumer Protection Agency, 2151 Berkeley Way, Berkeley, California 94704 (July 1971).
(14) Jonasson, I.R., *Mercury in the Natural Environment: A Review of Recent Work,* Geological Survey of Canada Paper 70-57, Geological Survey of Canada, 601 Booth St. Ottawa, Canada (1972).
(15) D'Itri, F.M., *The Environmental Mercury Problem,* Chemical Rubber Company Press, Cleveland, Ohio 44128 (1972).
(16) Jensen, S. and Jernelov, A., "Biological Methylation of Mercury in Aquatic Organisms," *Nature* 223, 753 to 754 (1969).
(17) Caton, G.M., Oliveira, D.P., Oen, C.J. and Ulrikson, G.U., *Mercury in the Environment–An Annotated Bibliography,* Oak Ridge National Laboratory Report ORNL-EIS-71-8; Oak Ridge National Laboratory, Oak Ridge, Tennessee 37830 (May 1972).
(18) Stahl, Q.R., *Preliminary Air Pollution Survey of Mercury and Its Compounds,* EPA Publication APTD 69-40; APTIC, EPA, Research Triangle Park, North Carolina 27711 (October 1969).
(19) *Background Information–Proposed National Emission Standards for Hazardous Air Pollutants: Asbestos, Berylium, Mercury,* EPA Publication APTD-0753; APTIC, EPA, Research Triangle Park, North Carolina 277-1 (December 1971).
(20) "Mercury in the Air," *Environment* 13 (4), 28 (1971).
(21) Lawrence Berkeley Laboratory, *Instrumentation for Environmental Monitoring–Air,* Publ. LBL-1, Volume 1, Berkeley, University of California (February 1, 1973).
(22) Dow Chemical Co., *Environmental Aspects of Mercury Usage,* 3 vols. 940 pp, Midland, Michigan (1970).
(23) Imperial Chemical Industries, *Mercury as an Environmental Pollutant, A Bibliography,* Brixham Laboratory, England (August 1973).
(24) Hartung, R. and Dinman, B.D., *Environmental Mercury Contamination,* Ann Arbor, Michigan, Ann Arbor Science Publishers (1972).
(25) Organization for Economic Cooperation and Development, *Mercury and the Environment,* Paris (1974).
(26) Office of Toxic Substances, *Materials Balance and Technology Assessment of Mercury and Its Compounds on National and Regional Bases,* Washington, D.C., Environmental Protection Agency (October 1975).
(27) Luckey, T.D., Venugopal, B. and Hutcheson, D., *Heavy Metal Toxicity, Safety and Hormonology,* New York, Academic Press (1975).
(28) Schroeder, H.A., *The Poisons Around Us,* Bloomington, Indiana, University Press (1974).
(29) Bidstrup, P.L., *Toxicity of Mercury and Its Compounds,* Amsterdam, Elsevier (1964).
(30) Stecher, P., Ed., *The Merck Index,* 8th Edition, Merck & Co., Inc., Rahway, New Jersey (1968).
(31) McKee, J.E. and Woff, H.W., *Water Quality Criteria,* 2nd Edition, California State Water Works Resources Control Board, Sacramento, California (1963).
(32) Moffitt, A.E., Jr. and Kupel, R.E., "A Rapid Method Employing Impregnated Charcoal and Atomic Absorption Spectrophotometry from the Determination of Hg in Atmospheric, Biological and Aquatic Samples," *At. Abs. Newsletter,* 9 (6), 113 (1970).
(33) Goldberg, A.J., *A Survey of Emissions and Controls for Hazardous and Other Pollutants,* Washington, D.C., Environmental Protection Agency (February 1973).

(34) National Institute for Occupational Safety and Health, *Criteria for a Recommended Standard...Occupational Exposure to Inorganic Mercury,* Washington, D.C. (1973).

(35) Benning, D., "Outbreak of Mercury Poisoning in Ohio," *Ind. Med. Surg.* 27, 354-63 (1958).

(36) Copplestone, J.F. and McArthur, D.A., "An Inorganic Mercury Hazard in the Manufacture of Artificial Jewelry," *Br. J. Ind. Med.* 24, 77-80 (1967).

(37) Bidstrup, P.L., Bonnell, J.A., Harvey, D.G. and Locket, S., "Chronic Mercury Poisoning in Men Repairing Direct Current Meters," *Lancet* 2, 856-61 (1951).

(38) Smith, R.G., Vorwald, A.J., Patil, L.S. and Mooney, T.F., "Effects of Exposure to Mercury in the Manufacture of Chlorine," *Am. Ind. Hyg. Assoc. J.* 31, 687-700 (1970).

(39) Irukayama, K., "The Pollution of Minamata Bay and Minamata Disease," *Proc. 3rd Int'l. Conf. Wat. Poll. Res.,* Munich, 3, 153-180 (1966).

(40) Krenkle, P.A., "Report International Conference on Environmental Mercury Contamination–Summary," *Wat. Res.,* 5, 1121-1122 (1971).

(41) Turney, W.G., "Mercury Pollution: Michigan's Action Program," *Jour. Wat. Poll. Control Fed.,* 43, 1427-1438 (1971).

(42) Cranston, R.E. and Buckley, D.E., "Mercury Pathways in a River and Estuary," *Env. Sci. Techn.,* 6, 274-278 (1972).

(43) D'Itri, F.M., *Mercury in the Aquatic Ecosystem,* Technical Report No. 23, Institute of Water Research, East Lansing, Michigan, Michigan State University (1972).

(44) Ottinger, R.S., Blumenthal, J.C., Dal Porto, D.F., Gruber, G.I., Santy, M.J. and Shih, C.C., *Recommended Methods of Reduction, Neutralization, Recovery or Disposal of Hazardous Waste,* Report PB 224 580, Springfield, Virginia, National Technical Information Service (August 1973).

(45) Campbell, E.E., Eutsler, B.C. and Trujillo, P.E.; U.S. Patent 3,888,124; June 10, 1975; assigned to U.S. Energy Research and Development Administration.

(46) Trujillo, P.E., Campbell, E.E. and Eutsler, B.C.; U.S. Patent 3,933,431; January 20, 1976; assigned to U.S. Energy Research and Development Administration.

(47) Hadieshi, T., Church, D.A., McLaughlin, R.D., Zak, B.D., Nakamura, M. and Chang, B., "Mercury Monitor for Ambient Air," *Science* 187, 348-49 (January 31, 1975).

(48) *Methods for Chemical Analysis of Water and Wastes,* Environmental Protection Agency, Cincinnati, Ohio 45268 (1971).

(49) *Apparatus for Flameless Mercury Determinations,* Brochure from Perkin-Elmer Corp., Norwalk, Connecticut 06852 (1971).

(50) Braman, R.S., "Membrane Probe-Spectral Emission Type Detection System for Mercury in Water," *Anal. Chem.* 43 (11), 1462 (1971).

(51) Chau, Y.-K. and Saitoh, H., "Determination of Submicrogram Quantities of Hg in Lake Waters," *Env. Sci. Tech.,* 4 (10), 839 (1970).

(52) Kalb, G.W., "The Determination of Hg in H_2O and Sediment Samples by Flameless Absorption," *At. Absorption Newsletter,* 9 (4), 84 (1970).

(53) Brandenberger, H. and Bader, H., "The Determination of Mercury by Flameless Atomic Absorption," *At. Abs. Newsletter,* 7 (3), 53 (1968).

(54) Fishman, M.J., "Determination of Mercury in Water," *Anal. Chem.,* 42 (12), 1462 (1970).

(55) Doherty, P.E. and Dorsett, R.S., "Determination of Trace Concentrations of Mercury in Environmental Water Samples," *Anal. Chem.* 43 (13), 1887 (1971).

(56) Muscat, V.I., Vichers, T.J. and Andren, A., "Sample and Versatile Atomic Fluorescence System for Determination of Nanogram Quantities of Mercury," *Anal. Chem.,* 44 (2), 218 (1972).

(57) Mesman, B.B. and Smith, B.S., "Determination of Hg in Urine by Atomic Absorption Utilizing the APDC/MIBK Extraction System and Boat Technique," *At. Abs. Newsletter,* 9 (4), 81 (1970).

(58) Bailey, B.W. and Lo, F.C., "Automated Method for Determination of Mercury," *Anal. Chem.,* 43 (11), 1525 (1971).

(59) Fishman, M.J. and Erdmann, D.E., "Automation of Atomic Absorption Analysis," *At. Absorption Newsletter,* 9 (4), 88 (1970).

(60) Stack, V.T., Jr., "Water Quality Surveillance," *Anal. Chem.,* 44 (8), (1972).

(61) Capuano, I.A.; U.S. Patent 3,826,614; July 30, 1974; assigned to Ionics, Inc.

(62) Hammitt, D.L.; U.S. Patent 3,844,719; October 29, 1974; assigned to PPG Industries, Inc.

(63) Grengg, W.M.; U.S. Patent 3,852,604; December 3, 1974; assigned to Pollution Control Technology, Inc.

(64) April, R.W. and Hume, D.N., "Environmental Mercury: Rapid Determination in Water at Nanogram Levels," *Science,* 170, 849 (1970).

(65) Hatch, W.R. and Ott, W.L., "Determination of Sub-Microgram Quantities of Mercury by Atomic Absorption Spectrophotometry," *Anal. Chem.,* 40 (14), 2085 (1968).

(66) Braman, R.S., Justen, L.L. and Foreback, C.C., "Direct Volatilization–Spectral Emission Type Detection System for Nanogram Amounts of Arsenic and Antimony," *Anal. Chem.,* 44 (13), 2195 (1972).

(67) Manning, D.C. and Fernandez, F., "Cobalt Spectral Interference in the Determination of Mercury," *At. Absorption Newsletter,* 7 (1), 24 (1968).

(68) Rottschafer, J.M., Jones, J.D. and Mark, H.B., Jr., "A Simple Rapid Method for Determining Trace Mercury in Fish via Neutron Activation Analysis," *Env. Sci. Tech.,* 5 (4), 336 (1971).

(69) Sivasankara Pillay, K.K., Thomas, C.C., Jr., Sondel, J.A. and Hyche, C.M., "Determination of Mercury in Biological and Environmental Samples by Neutron Activation Analysis," *Anal. Chem.,* 43 (11), 1419 (1971).

(70) Tanner, J.T., Friedman, M.H., Lincoln, D.N., Ford, L.A. and Jaffee, M., "Mercury Content of Common Foods Determined by Neutron Activation Analysis," *Science,* 177, 1102 (1972).

(71) Becknell, D.E., Marsh, R.H. and Allie, W., Jr., "Use of Anion Exchange Resin-Loaded Paper in the Determination of Trace Mercury in Water by Neutron Activation Analysis," *Anal. Chem.,* 42 (10), 1230 (1971).

(72) Sittig, M., *Pollution Detection and Monitoring Handbook,* Park Ridge, New Jersey, Noyes Data Corp. (1974).

(73) *Threshold Limit Values of Airborne Contaminants and Physical Agents with Intended Changes Adopted by ACGIH for 1974,* American Conference of Government Industrial Hygienists, P.O. Box 1937, Cincinnati, Ohio 45201 (1974).

(74) "Part 1910–Occupational Safety and Health Standards, National Consensus Standards and Established Federal Standards," *Federal Register* 36 (105), 10466 (May 29, 1971); notice from the Occupational Safety and Health Administration, Department of Labor.

(75) Berman, E., "Determination of Cd, Tl and Hg in Biological Materials by Atomic Absorption," *At. Abs. Newsletter,* 6 (3), 57 (1967).

(76) Office of Air and Water Programs, *Control Techniques for Mercury Emissions from Extraction and Chlor-Alkali Plants,* Publ. No. AP-118, Research Triangle Park, N.C., Environmental Protection Agency (February 1973).

(77) Sittig, M., *Pollutant Removal Handbook,* Park Ridge, New Jersey, Noyes Data Corp. (1973).

(78) Manes, M. and Grant, R.J.; U.S. Patent 3,193,987; July 13, 1965; assigned to Pittsburgh Activated Carbon Co.

(79) Williston, S.H. and Morris, M.H.; U.S. Patent 3,232,033; February 1, 1966; assigned to Cordero Mining Co.

(80) Park, J.C. and Winstrom, L.O.; U.S. Patent 3,257,776; June 28, 1966; assigned to Allied Chemical Corp.

(81) Manes, M.; U.S. Patent 3,374,608; March 26, 1968; assigned to Pittsburgh Activated Carbon Co.

(82) Bell, D.; U.S. Patent 3,647,359; March 7, 1972; assigned to BP Chemicals (UK) Limited, England.

(83) Ferrara, I. and Castellani, I.; U.S. Patent 3,661,509; May 9, 1972; assigned to Simcat-Societa Industriale Catanese SpA, Italy.

(84) Revoir, W.H. and Jones, J.A.; U.S. Patent 3,662,523; May 16, 1972; assigned to American Optical Corp.

(85) Bryk, P., Haani, M., Honkasalo, J.B., Kinnunen, J., Lindsjo, O., Nyholm, E., Poijarvi, J., Rastas, J. and Kangas, J.; U.S. Patent 3,677,696; July 18, 1972; assigned to Outokumpu Oy, Finland.

(86) Kawase, B., Kojima, I. and Otani, K.; U.S. Patent 3,725,530; April 3, 1973; assigned to Showa Denko KK, Japan.
(87) Melkersson, K.-A. and Hedenas, B.G.V.; U.S. Patent 3,786,619; January 22, 1974; assigned to Boliden Aktiebolag, Sweden.
(88) Raduly, L., Urmosi, I., Todereanu, N. and Chiroiu, N.; U.S. Patent 3,803,803; April 16, 1974; assigned to Grupul Industrial de Chimie, Romania.
(89) Wygasch, E.; U.S. Patent 3,814,799; June 4, 1974; assigned to Badische Anilin- & Soda-Fabrik AG, Germany.
(90) Allgulin, T.; U.S. Patent 3,849,537; November 19, 1974; assigned to Boliden Aktiebolag, Sweden.
(91) Brockmiller, C.A., Lund, R.E. and Fitzsimmons, J.E.; U.S. Patent 3,855,387; Dec. 17, 1974; assigned to St. Joe Minerals Corporation.
(92) Kasai, J., Miyazaki, K., Kawasaki, K. and Kawase, B.; U.S. Patent 3,876,393; April 8, 1975; assigned to Showa Denko KK, Japan.
(93) Ionescu, V.; U.S. Patent 3,919,389; November 11, 1975; assigned to i Proiectari Automatizari si Proiectari Uzine si Instalatil Pentru Industria Electrotehnica Institutul de Cercetari, Romania.
(94) Putnam, J.J., "Quicksilver and Slow Death," *National Geographic,* Volume 124, No. 4 (October 1972).
(95) Sumner, A.K. and Saha, J.G., "Methylmercury Concentrations in Muscle Tissues of Fish from the Saskatchewan River," *Environ. Lettrs.,* 2 (3), 167 (1971).
(96) Gilmour, J.T., "Inorganic Complexes of Divalent Mercury in Natural Water Systems," *Environ. Letters,* 2 (3), 143 (1971).
(97) Moore, F.L., "Solvent Extraction of Mercury from Brine Solutions with High-Molecular-Weight Amines," *Env. Sci. Tech.,* 6 (6), 525 (1972).
(98) Maass, W.B., "Water Pollution and Industrial Organic Finishing," *Metal Finishing,* page 61 (January 1972).
(99) Bergeron, G.L. and Bon, C.K.; U.S. Patent 2,860,952; November 18, 1958; assigned to The Dow Chemical Co.
(100) Neipert, M.P. and Bon, C.K.; U.S. Patent 2,885,282; May 5, 1959; assigned to The Dow Chemical Co.
(101) Karpiuk, R.S. and Hoekstra, J.J.; U.S. Patent 3,029,143; April 10, 1962; assigned to The Dow Chemical Co.
(102) Karpiuk, R.S. and Hoekstra, J.J.; U.S. Patent 3,029,144; April 10, 1962; assigned to The Dow Chemical Co.
(103) Gilbert, J.F. and Rallis, C.N.; U.S. Patent 3,039,865; June 19, 1962; assigned to The Dow Chemical Company.
(104) Calkins, R.C., Mock, R.A. and Morris, L.R.; U.S. Patent 3,083,079; March 26, 1963; assigned to The Dow Chemical Co.
(105) Scholten, H.G. and Prielipp, G.E.; U.S. Patent 3,085,859; April 16, 1963; assigned to The Dow Chemical Co.
(106) Edwards, G.E. and LePage, N.J.; U.S. Patent 3,102,085; August 27, 1963; assigned to Imperial Chemical Industries Ltd., England.
(107) Deriaz, M.G.; U.S. Patent 3,115,389; December 24, 1963; assigned to Imperial Chemical Industries Limited, England.
(108) Crain, G.E. and Judice, R.H.; U.S. Patent 3,213,006; October 19, 1965; assigned to Diamond Alkali Co.
(109) Town, J.W.; U.S. Patent 3,361,559; January 2, 1968; assigned to U.S. Secretary of Interior.
(110) Rhodes, D.W. and Wilding, M.W.; U.S. Patent 3,463,635; August 26, 1969; assigned to U.S. Atomic Energy Commission.
(111) MacMillan, J.B.; U.S. Patent 3,502,434; March 24, 1970; assigned to Canadian Industries Limited, Canada.
(112) Botwick, E.J. and Smith, D.B.; U.S. Patent 3,600,285; August 17, 1971; assigned to Olin Corp.
(113) Fuxelius, K.O.H.; U.S. Patent 3,617,563; November 2, 1971; assigned to Research Ab Sundbyberg, Sweden.

(114) Dean, W.E. and Dorsett, C.M.; U.S. Patent 3,674,428; July 4, 1972; assigned to PPG Industries, Inc.
(115) Knepper, W. and Austin, S.; U.S. Patent 3,695,838; October 3, 1972; assigned to Chemische Werke Huels, AG, Germany.
(116) Waltrich, P.F.; U.S. Patent 3,704,875; December 5, 1972; assigned to Pennwalt Corp.
(117) Friedman, M.; U.S. Patent 3,725,261; April 3, 1973; assigned to the U.S. Secretary of Agriculture.
(118) De Angelis, P., Morris, A.R. and MacMillan, A.L.; U.S. Patent 3,736,253; May 29, 1973; assigned to Sobin Chlor-Alkali, Inc.
(119) Dean, W.E. and Dorsett, C.M.; U.S. Patent 3,749,761; July 31, 1973; assigned to PPG Industries, Inc.
(120) Guptill, J.P. and Foley, G.W.; U.S. Patent 3,764,495; October 9, 1973; assigned to Allied Chemical Corp.
(121) Hultman, B. and Norgren, E.B.W.; U.S. Patent 3,764,496; October 9, 1973; assigned to Mo och Domsjo Aktiebolaget, Sweden.
(122) Cadmus, E.L.; U.S. Patent 3,764,528; October 9, 1973; assigned to Ventron Corp.
(123) Williams, R.P.; U.S. Patent 3,769,205; October 30, 1973; assigned to Phillips Petroleum Co.
(124) Nolte, G., Rabowsky, H., Krause, J. and Narres, H.-D.; U.S. Patent 3,788,842; Jan. 29, 1974; assigned to Knapsack AG, Germany.
(125) Motani, K., Kiyotoshi, T. and Nakahara, A.; U.S. Patent 3,790,535; February 5, 1974; assigned to Tokuyama Soda KK, Japan.
(126) Krieg, M.; U.S. Patent 3,801,480; April 2, 1974; assigned to Farbwerke Hoechst Aktiengesellschaft, Germany.
(127) Gerow, R.F. and Soule, S.B.; U.S. Patent 3,802,910; April 9, 1974; assigned to GAF Corp.
(128) Coulter, M.O. and Bell, D.; U.S. Patent 3,847,598; November 12, 1974; assigned to BP Chemicals Limited, England.
(129) Coulter, M.O.; U.S. Patent 3,857,704; December 31, 1974; assigned to BP Chemicals Limited, England.
(130) Hatch, A.E.; U.S. Patent 3,859,210; January 7, 1975.
(131) Marchant, W.N.; U.S. Patent 3,864,327; February 4, 1975; assigned to the U.S. Secretary of the Interior.
(132) Davis, H.J. and Lee, L.A.; U.S. Patent 3,872,001; March 18, 1975; assigned to Celanese Corp.
(133) Fitzpatrick, J.W., Berninger, C.J. and Lewis, D.O.; U.S. Patent 3,873,581; March 25, 1975; assigned to Toms River Chemical Corp.
(134) Iverson, M.L.; U.S. Patent 3,899,404; August 12, 1975; assigned to Rockwell International Corp.
(135) Isfort, H. and Ilgner, H.-P.; U.S. Patent 3,922,210; November 25, 1975; assigned to Friedrich Uhde GmbH, Germany.
(136) Chakrabarty, A.M., Friello, D.A. and Mylroie, J.R.; U.S. Patent 3,923,597; Dec. 2, 1975; assigned to General Electric Company.
(137) Schuller, W.H.; U.S. Patent 3,929,634; December 30, 1975; assigned to the U.S. Secretary of Agriculture.
(138) Rosenzweig, M., *Chem. Eng.,* pp. 60-61 (January 20, 1975).
(139) Iammartino, N.R., *Chem. Eng.,* pp. 36-37 (February 3, 1975)
(140) Yamori, K., Takatoku, M., Miyahara, A., Omagari, T. and Kitamura, M.; U.S. Patent 3,536,597; October 27, 1970; assigned to Toyo Soda Manufacturing Co., Ltd. and Japan Organo Co. Ltd., Japan.
(141) Stenger, V.A.; U.S. Patent 3,679,396; July 25, 1972; assigned to The Dow Chemical Co.
(142) Advanced Technology Center, Inc., *Mercury Pollution Control in Stream and Lake Sediments,* Washington, D.C., U.S. Government Printing Office (1972).
(143) Environmental Protection Agency, *Control of Mercury Pollution in Sediments,* Report EPA-R2-72-043, Washington, D.C. (September 1972).

MOLYBDENUM

Molybdenum is a transition element in group VI-b of the periodic table. It can occur in 6 valence states: 0, +2, +3, +4, +5 and +6, making its chemistry very complicated. Molybdenum does not occur in the native state, but is obtained from minerals, such as molybdenite (MoS_2), wulfenite ($PbMoO_4$), ferrimolybdate ($FeMoO_3 \cdot xH_2O$) and jordisite (amorphous MoS_2). Many other minerals contain molybdenum, but these four are the ones used for commercial exploitation (1).

Molybdenite is the greatest source of molybdenum, the largest deposit being the one in Climax, Colorado. 75% of the world's known reserves of molybdenum are in the western parts of North and South America. Large deposits are also known to exist in the U.S.S.R., Canada and Chile. In the U.S. the main part of the produced molybdenum comes as a primary product from the processing of molybdenite, but a substantial fraction is obtained in connection with processing of copper, tungsten or uranium ores. Molybdenum is also obtained by recycling of scrap materials.

Molybdenum is considered to be an essential element. It is a constituent of some enyzmes including xanthine oxidase which oxidizes xanthine or hypoxanthine to uric acid. Although there are no minimum requirement data for human beings, information from other mammals tends to show that the necessary amount of molybdenum is very small. Therefore, the concentration of molybdenum in a normal diet with all probability will be sufficient for maintaining physiological levels of the metal in humans.

TOXICITY

A limited amount of data suggests that molybdenum has a relatively low order of toxicity. Despite the fact that human beings are exposed to relatively high amounts of molybdenum (100-500 μg/day), primarily via food, there is no substantial age-related accumulation of molybdenum in the body. Concentrations in those organs in which molybdenum is mainly found, i.e., liver, kidney and bones, are relatively low, in the order of below 1 μg/g wet weight. Molybdenum occurring naturally in the diet is absorbed to a high degree from the gastrointestinal tract (probably 25-75%) but is eliminated rapidly, primarily via the urine. The biological half-time is not well established in humans but the major part of the absorbed molybdenum is eliminated within days or, at a maximum, a few weeks (1). Animal data show that the metabolism of molybdenum is closely related to the metabolism of copper and sulfur compounds.

Most observations on the toxicity of molybdenum are based on data from animals. It is clearly shown that an excessive exposure to molybdenum via food can give rise to a severe disease in ruminants which involves diarrhea, anemia and emaciation and may progress to death. The disease may be prevented or cured by the administration of copper compounds or removal of animals from the areas high in molybdenum. The disease is termed teart disease and has been possible to reproduce experimentally.

It has also been possible to show toxic effects resembling teart disease in several of the common laboratory animals. In addition, several forms of bone deformities, including exostoses, have been induced. The doses required are excessive and much higher than those to which human beings eating a normal diet may be expected to be exposed, even in molybdenum-rich areas. In several studies 200 to 400 mg of soluble hexavalent molybdenum compounds per kg diet were required over prolonged periods.

Information into possible toxic effects in human beings is scarce. There are data from the U.S.S.R. literature pointing towards the possibility that exposure to some molybdenum dusts may give rise to pulmonary disorders in the form of pneumoconiosis. Similar effects have been reproduced in animals. No conclusions concerning the necessary exposures to give rise to this condition in humans can be drawn at this time.

There have also been reports on effects of molybdenum on purine metabolism. It has been shown in studies from the U.S.S.R. that people exposed to molydenum (concentration not known) in industry have increased levels of uric acid in urine and in blood. Moreover, a report from the U.S.S.R. from 1961 has indicated that people living in an area of that country with high dietary intake of molybdenum (by the authors calculated to 10-15 mg/day as compared to 1-2 mg in the control area) show an increased prevalence of joint-disorders claimed to be gout-like as well as an increase in blood levels of uric aicd and an increase in the urinary excretion of uric acid.

The reason for the high daily intake value in the control area is not known. The results are very difficult to evaluate since no details are given concerning the study population and the selection of subjects; furthermore, the control group consisted of only five subjects. Nevertheless, the data fully motivate further studies, not only in the U.S.S.R. but also in regions of other countries where the exposure to molybdenum via diet is high, one reason being that from the theoretical standpoint, effects like those reported might well appear. Thus, molybdenum exposure could give rise to an increase in xanthine oxidase activity which in turn should give rise to an increase in uric acid formation.

The reason that effects have not been reported from other parts of the world may be a lack of studies focusing on these effects. Target groups of special importance are: [1] persons living in areas with a high molybdenum content in food, particularly in such areas in which the inhabitants are dependent upon locally produced food; [2] persons from noncontaminated areas with consumption habits favoring a high molybdenum intake and an imbalance in relation to other dietary constituents of importance, e.g., copper (1). The toxicity of molybdenum has been discussed briefly by Luckey et al (2) and at more length by Underwood (3).

EXTENT OF EXPOSURE

Table 44 summarizes information on sources and emissions of molybdenum (4). Concentrations of molybdenum in ambient air may range from below detectable amounts to maximum values of the order of 0.34 $\mu g/m^3$ of air.

DETECTION

A tentative method for the analysis for molybdenum content of atmospheric particulates has been published (5). The method depends on the formation of a stable molybdenum thiocyanate complex. The complex is extracted with MEK and its absorbance is measured. A concentration of 0.055 $\mu g/m^3$ of molybdenum can be measured by the method if a 20 cubic meter sample can be provided.

The determination of trace amounts of molybdenum in water has been reported by R. Armstrong and C.R. Goldman of the Institute of Ecology of the University of California at Davis (6). They reported that molybdenum may be determined directly in solutions containing above 100 ppb. A spectrophotometric method using the extractable complex with diacetyldithiol was found to be the most convenient and the least subject to interference of a number of methods. Preconcentration of dissolved molybdenum by coprecipitation with hydrated manganese dioxide followed by spectrophotometry of the dithiol complex allows the determination of 1 μg of molybdenum in a 5 liter sample (0.2 ppb) with a relative error of 3%.

Particulate matter may be removed from water samples by HA Millipore filtration or by continuous flow (150 ml/min) centrifugation at 28,000 g. The latter method is more convenient for processing the large (10 to 50 liter) water samples needed for estimating the trace metal content of suspended particles in unpolluted waters. Molybdenum is most conveniently released from organic materials by wet oxidation with perchloric acid, the residue being taken up in 3 N hydrochloric acid. Hydrofluoric acid digests of silicate samples were found to give higher recoveries than perchloric acid digests and were found to be more reproducible than fusion methods. Molybdenum may also be determined in water by atomic absorption at a detection limit of 14 ppb.

Colorimetric methods have been used for decades for determination of molybdenum, but during later years neutron activation, atomic absorption and x-ray fluorescence have also come into the picture. For more detailed discussions on the determination of molybdenum reference is made to recent reviews by Meglen and Glaze (7)(8). More than 100 years ago it was discovered that reduced molybdenum in an acid solution formed a colored complex with thiocyanate, which could be extracted into an organic solvent. This formed the principle for the most common method for determination of molybdenum, the thiocyanate color-

TABLE 44: SOURCES AND ESTIMATES OF MOLYBDENUM-CONTAINING EMISSIONS

Source	Uncontrolled Particulate Emission Factor (lb/ton)	(kg/10^3 kg)	Reliability Code	Production Level (tpy)	% Mo in Emissions	Mo Emissions Before Controls (tpy)	Estimated Level of Emissions Control	Mo Emissions Following Controls (tpy)
Mining:								
Open pit	10	15	D	13,000	*	65	0	65
Underground	0.5	0.25	D	25,986	*	6	0	6
Copper open pit	10	5	D	16,690	*	83	0	83
Beneficiation	52	26	D	34,821	*	906	98.8%	11
Roasting	100	50	C	34,821	*	1,740	98%	35
Metallurgical:								
Ferromolybdenum-electric arc	200	100	D	6,221	*	622	90%	62
Molybdenum metal	2,000	1,000	B	803	*	803	99.9%	1.0
Chemical production	Nil	Nil	-	1,148	*	Nil	Nil	Nil
Steel and alloy production:								
Steel	25	12.5	C	14,712	*	184	78%	41
Cast iron	15	7.5	C	1,958	*	15	99%	0.1
Super alloys	200	100	C	1,254	*	126	78%	28
Alloys	25	12.5	C	395	*	5	78%	1
Mill products	0.5	0.25	C	858	*	0	-	0
Nonmetallurigical uses	2	1	D	1,861	*	2	0	2
Inadvertent sources:								
Coal burning	NA	NA	-	33,800,000 (uncontrolled emissions)	0.01 (B)	3,380	82%	610
Oil combustion	NA	NA	-	287,000 (uncontrolled emissions)	0.01 (C)	29	0	29
Mineral processing	NA	NA	-	6.9×10^6 (controlled emissions)	0.00023 (D)	16	NA	16

*Emission factor multiplier equal to tons of Mo processed annually.
Note: NA = not applicable.
Reliability code: see Table 19, p. 99.

Source: Reference (4)

imetric method. Another colorimetric method is the dithiol method, according to which hexavalent molybdenum is combined with toluene-3,4-dithiol to form a colored compound, which can then be extracted into an organic solvent. This method gives results similar to the thiocyanate method for determination of molybdenum in water. Spectrographic methods have been used to determine molybdenum in plants and water. Molybdenum has been precipitated with, e.g., 8-hydroxyquinoline as a preconcentration step. Such a method is seen to give results similar to other methods for determination of molybdenum in water. By concentrating and extracting from large samples concentrations of a few ng/g can be measured. Molybdenum has been concentrated by complexing with tert-carbate (N-pyrrolidinodithiocarbamic acid-sodium salt) and extraction into chloroform.

Neutron activation analysis has been employed in several investigations. After irradiation molybdenum is separated by a complexing agent or ion-exchange. Both ^{99}Mo and ^{101}Mo may be used. The detection limit is reported to be 0.1 μg for ^{101}Mo and 0.2 ng for ^{99}Mo. Good agreement was found when the ^{101}Mo method was compared to a colorimetric method.

Molybdenum is not well suited for analysis by atomic absorption in flame. A claimed sensitivity of around 0.01 μg/ml has not been supported by everyone. The sensitivity is generally relatively low. Molybdenum is not easily atomized, forming refractory oxides in the flame. After extraction into organic solvents concentrations of 1 to 3 μg/g in fertilizers could be determined. In fresh water molybdenum has been determined at concentrations of from 230 to 3,200 μg/l, in which range the agreement with other methods is good. In this case the analysis was carried out on a dithiol-MIBK extract.

Flameless methods have been recommended, but data are not available as to how they compare to other methods. A sensitivity of 0.035 μg/g has been reported. Molybdenum in seawater has been determined by ion-exchange with a graphite atomizer after separation and from these data it can be calculated that about 0.03 μg/ml can be detected in the concentrated solution.

ENVIRONMENTAL STANDARDS

Threshold limit values have been set by the American Conference of Governmental Industrial Hygienists as follows: 5 mg/m^3 soluble molybdenum compounds, and 10 mg/m^3 insoluble molybdenum compounds. The tolerance for molybdenum in irrigation waters is only 0.005 mg/l for continuous use, however. The tolerance on a short term use basis in fine textured soil is 0.05 mg/l.

HANDLING PROCEDURES

Suitable precautions in the form of respirator use should be taken against the inhalation of the more soluble molybdenum compounds or of the oxide.

REMOVAL FROM AIR

The conversion of molybdenite to molybdic oxide involves roasting which releases sulfur that is exhausted as sulfur oxides into the atmosphere. Pollution from roasting is more widespread because molybdenite is shipped to a number of subsidiary companies and other consumers in many parts of the country for processing. The growing national concern for air pollution will probably alter molybdenite processing practices in the future. There are undoubtedly other pollution considerations that must be resolved in the recovery of molybdenum. The problems are relatively local, however, and undefined, according to the Bureau of Mines (9).

REMOVAL FROM WATER

Information on molybdenum removal from wastewaters is not available in the literature.

SOLID WASTE DISPOSAL

The mining of molybdenum primary ores entails the disturbance of some land areas which in the future may require reclamation. The problem is minimal, however, because of the very few acres involved per unit of production. Molybdenum recovered in conjunction with copper production entails the disturbance of significantly larger land areas (9).

ECONOMIC IMPACT OF CONTROLS

No data were available on the economic impact of controls on molybdenum pollution.

REFERENCES

(1) L. Friberg, P. Boston, G. Nordberg, M. Piscator and K.H. Robert, *Molybdenum—A Toxicological Appraisal,* Report EPA-600/1-75-004, Research Triangle Park, N.C., Environmental Protection Agency (November 1975).

(2) T.D. Luckey, B. Venugopal and D. Hutcheson, *Heavy Metal Toxicity, Safety and Hormonology,* New York, Academic Press (1975).

(3) E.J. Underwood, *Trace Elements in Human and Animal Nutrition,* New York, Academic Press (1971).

(4) GCA Corp., *National Emissions Inventory of Sources and Emissions of Molybdenum,* Report PB 230,035, Springfield, Va., Nat. Tech. Information Service (May 1973).

(5) American Public Health Association, *Methods of Air Sampling and Analysis,* Washington, D.C. (1972).

(6) R. Armstrong and C.R. Goldman, Paper presented before ASTM 71st Annual Meeting, San Francisco, California (June 1968).

(7) R.R. Meglen and M.L. Glaze, *Analytical Methods,* Report to the Molybdenum Project No. Mo P/g-73/MW, Boulder, University of Colorado and Fort Collins, Colorado State University (April 24, 1973).

(8) R.R. Meglen and M.L. Glaze, in *Transport and Biological Effects of Molybdenum in the Environment,* Boulder, University of Colorado and Fort Collins, Colorado State University (January 1, 1974).

(9) E.T. Sheridan, in *Mineral Facts and Problems,* Washington, D.C., U.S. Bureau of Mines (1970).

NICKEL

Nickel is widely used in ferroalloys, stainless steels and metal plating. In steel alloys and as a plating material nickel offers superior corrosion resistance and excellent low-temperature properties. Over 75% of the nickel consumed yearly in the United States is imported; the domestic production (as ferro-nickel) occurs at Riddle, Oregon and amounts to about 5% of the nickel consumption. The remainder, about 20% of the consumption, arises as a smelter by-product (minor) or through recovery from scrap (1). Nickel and its compounds are of concern as air pollutants because harmful effects of exposure to these materials have been observed among industrial workers as described by Sullivan (2).

TOXICITY

Nickel is one of the relatively nontoxic metals found in the tissues of man, ranking in this respect with the essential elements, iron, cobalt, copper and zinc, according to Cothern (3). Its physiological role, if any, has not been established. The toxicity for mammals is low, ranking with the essential metals and with chromium, barium and silver (4).

Nickel carbonyl is a flammable liquid which boils at 43°C. Because it is volatile and easily decomposable into metallic nickel and carbon monoxide, it is very toxic. It is considered to be five times as toxic as carbon monoxide and can also bring a large concentration of active nickel into the body and deposit it in the lungs. The toxicity of nickel and its compounds has been reviewed by Luckey et al (5).

Nickel and its compounds are of concern as air pollutants because harmful effects of exposure to these materials have been observed among industrial workers. Exposure to airborne nickel dust and vapors may have produced cancers of the lung and sinus, other disorders of the respiratory system, and dermatitis.

There is a substantially higher mortality rate among nickel workers due to sinus cancer—up to 200 times the expected number of deaths. However, since other metal dusts have also been present in industrial exposures to nickel, it has not been possible to determine whether nickel is the carcinogen. Yet experiments have shown that nickel carbonyl and nickel dusts can induce cancer in animals. Nickel contact dermatitis was found in 77% of the females and 10% of the males suspected of having allergenic reactions to metals (6).

Chronic Toxicity

458 workers in the nickel electrolytic refining industry in eastern Europe were examined by x-ray (1). Inflammatory processes of subatrophic and atrophic character were associated with the chronic inflammation of accessory sinus cavities in 33.07% of the workers.

These processes were apparent in 17.75% of workers suffering from acute sinusitis and in 22.05% with cysts and cyst-like formations. High incidence of various blood irregularities were noted in eastern European workers producing nickel ferrite powders. 25% developed thrombocytopenia, anemia and leukopenia, 35% exhibited eosinophilia, 14% had monocytosis and 50% had decreased osmotic erythrocyte resistance.

Occupational diseases of workers in the electrolytic refining of nickel are apparently caused by the inhalation of an atmosphere of aerosol solutions of nickel chlorides and sulfates. The microclimate of such plants is also characterized by heat and moisture. Pure nickel production via the carbonyl cycle is related to the danger of contamination of the air by nickel carbonyl.

Nickel carbonyl poisoning by inhalation in 46 workers in a soda and aniline factory in Ludwigshafen, Germany, on the Rhine during the period 1952 to 1967 consisted of 39 light poisonings, 7 severe cases and 2 deaths (1). The clinical symptoms were headache, dizziness, chest pain, nausea, fever, coughing, and bronchial pneumonia; pulmonary edema occurred from 1 to 48 hours after exposure. The urine nickel concentration was a good index of degree of poisoning; greater than 10 micrograms per deciliter indicated poisoning and greater than 50 micrograms per deciliter indicated that hospitalization was required.

Acute Toxicity

The symptoms of acute $Ni(CO)_4$ toxicity in humans include the abnormally high position of the diaphragm, limited or absent respiratory motions, a significant decrease in pneumatization of pulmonary areas, intensification and lubricability of pulmonary patterns, the bilateral focality of various degrees of manifestation and a high position of the cardiovascular bundle (1).

In several young people pulmonary changes following $Ni(CO)_4$ intoxication were observed via x-ray. In two patients, intensification of the overall lung picture, as well as the roots of the lungs, was observed. Bilateral foci of exposure were observed in all. Retrograde development of changes in the x-ray patterns began 7 to 12 days after initial exposure. The principal $Ni(CO)_4$ effect appears to be the disturbance of lung capillaries and arterioles.

Nickel carbonyl intoxication apparently does not leave any aftereffects in healthy people. Treatment of more than 300 workmen with acute nickel carbonyl poisoning with the chelating agent, diethyldithiocarbamate (Dithiocarb), prevented death in 100% of the cases. In one severe case of total nickel carbonyl poisoning, diethyldithiocarbamate sodium did not prevent death, probably due to severe pulmonary and cerebral edema. Dithiocarb has been shown to be effective in the mobilization of copper and nickel in patients with hepatolenticular degeneration (Wilson's disease).

Allergies and Sensitization

Nickel dermatitis can be induced in humans through contact with nickel-plated articles, industrial exposures to nickelous dusts or baths, or internal exposure to nickel (1). The symtoms comprise two levels: [1] a simple dermatitis at the area of contact, the typical "nickel itch" consisting of burning and itching of the exposed skin; and [2] a chronic eczematous reaction, in which erythema occurs, and later nodules, which may eventually form pustules, appear in the web of the fingers and the forearm.

Symptoms may be relieved within a week by removal from exposure, but a rash may persist for some weeks. Specific desensitization to allergic dermatosis induced by nickel, chromium or cobalt, can be achieved by increasing doses of the specific allergen or one of the other two metals. The duration of the rash after avoidance of contact with an allergen is highly variable, but it seems to last longer in cases of nickel sensitivity than in other types.

Nickel sensitivity is usually determined by patch testing with a 5% $NiSO_4$ solution. Out of 200 patients tested in one study, 13.5% exhibited positive tests, 85.2% of whom had a previous history of nickel contact dermatitis. In a study of 5,416 persons tested for nickel, chromium or cobalt allergy, 9.9% showed a positive reaction. 77% of the allergic women were sensitive to nickel; 87% of the allergic men were sensitive to chromium. This difference in sensitivity is presumably due to differences in exposures to sensitizing contacts. In another study on 1,000 patients, 64% of sensitive women were suspected of nickel allergy. Nickel sensitivity and atopy are apparently unrelated. Only 11% of the persons with a personal or family history of proven nickel dermatitis exhibited atopy.

A study of 25 workers engaged in electroplating with nickel, chromium, copper, and zinc revealed seven who suffered dermatosis. The dermatosis disappeared when workers took their annual vacation but recurred when they returned to work. The proportion of nickel dermatosis was much greater when the bath temperature was 60°C than when the temperature was 35° to 40°C (1).

Apparent cross-sensitization between nickel and chromium or cobalt seems to be due to the presence of several allergenic metals in the sensitizing contact rather than to true cross-sensitization. In a study of nickel platers in Finland from 1949 to 1964, only 16% of nickel-sensitive workers also showed cobalt sensitivity; in all but two of these cases, the cross-sensitization occurred during a period when cobalt was added to the nickel plating bath as a brightener (1). 63% of 87 patients with suspender eczema exhibited cross-sensitivity. Suspenders contain both metals. More people with cement eczema were sensitive to cobalt than to nickel, but cement also contains both elements.

The nickel sensitivity threshold in eczematous patients with nickel allergy was investigated by patch testing. The mean threshold (n = 53) was 0.43% when distilled water was the solvent, and 0.51% when the solvent was petrolatum. Patients with combined nickel-cobalt sensitivity had the lowest thresholds (below 0.039%). No correlation was observed between degree of sensitivity and severity of skin disease.

The eliciting safety limit, or the concentration of nickel which does not elicit any contact allergic reaction in highly sensitized persons, is one micromole of nickel per liter. This is also assumed to be the sensitizing safety limit of nickel (1).

Evidence of cellular hypersensitivity in cultured lymphocytes and leucocytes, but not skin explants, from nickel-sensitive patients as compared with normals has been reported. When treated with nickel sulfate or nickel acetate (10^{-4} meq Ni^{++}/ml), cultured lymphocytes from nickel-sensitive patients exhibited an increase in ^{14}C-thymidine uptake, indicating DNA synthesis. Macrophage migration inhibition was also observed. However, no differences were observed in the effects of different nickel concentrations on mitotic capacity of leucocytes from nickel-sensitive and normal patients.

Carcinogenicity

The blastomogenic properties of nickel compounds in humans and in rats have been well established (1). Various environmental exposures to nickel seem to correlate with increased incidence of tumors in man. For example, cancer in the maxillary antrum of snuff users may result from using plant material grown on soil high in nickel.

In Bantus, carcinoma of the maxillary antrum accounts for 45.5% of all respiratory cancers observed. The Swazi snuff used by these people contains high concentrations of nickel, chromium and zinc. The incidence of lung cancer was studied in 845 men employed by a nickel refinery in South Wales for at least 5 years beginning during or before April 1944.

In men first employed before 1925, the incidence of death due to lung cancer was 5 to 10 times the national average, and to nasal cancer, 100 to 900 times the expected mortality. After 1925, death rates due to lung and nasal cancer were comparable to those of other men in the same geographical region. Susceptibility to nasal cancer varied with age at first exposure, but susceptibility to lung cancer varied irregularly.

In a study of the high mortality of nickel-concentrating and smelting workers due to various cancers and sarcomas compared with the general population, the highest incidence of pulmonary carcinoma was observed among 40 year old males. The incidence was higher for workers in roasting, reduction and cobalt departments who were exposed to nickel sulfides and oxides, as well as cobalt and arsenic compounds. Femoral and lung sarcomas were most prevalent among these workers, the males alone being affected.

A recent increase in the incidence of cancer of the nose and accessory sinuses in workers in the industry producing pure nickel by electrolysis warranted a reevaluation of maximum allowable exposures (1).

A recent review by the National Academy of Sciences (7) discusses the biological effects of nickel as an environmental pollutant. It concludes that currently available data on the amounts of nickel in water, air and food do not suggest any public risks; indeed, the metal is probably an essential trace component of a person's biological system. But the NAS report goes on to point out that industrialization and energy consumption are raising the level of environmental nickel, and this fact should be viewed with caution, especially since the metal's role as a hazard may be subtle. For instance, the nickel present in asbestos goods perhaps boosts that substance's own cancer-causing potential.

EXTENT OF EXPOSURE

The most likely sources of nickel in the air appear to be emissions from metallurgical plants using nickel, engines burning fuels containing nickel additives, and plating plants, as well as from the burning of coal and oil, and the incineration of nickel products (6).

In the case of the smelter at Sudbury, Canada, east and south of the smelters, soil and vegetation, plant leaves, dust, and rainfall were sampled and analyzed. Elevated levels of nickel were detected up to 50 kilometers from the smelters and toxic water levels extended to 15 kilometers. The soil contamination had a pattern indicative of an airborne smelter source. Nickel levels were 2,835 ppm at 0.8 kilometer, 1,522 ppm at 5.5 to 8 kilometers, 306 ppm at 20 kilometers and 83 ppm at 50 kilometers from the source. Average soil levels were 40 ppm for nickel (1).

Workers in plants for the electrolytic refining of nickel contracted diseases of the upper respiratory passage as a consequence of the inhalation of air containing nickel chloride and nickel sulfate. Nickel production using the Mond process involves the use as an intermediate of nickel carbonyl, which is an extremely toxic material.

The last extensive survey of the air quality over the United States was conducted in 1968 to 1969 by the Quality Assurance and Environmental Monitoring Laboratory of the National Environmental Research Center, in Research Triangle Park, with the assistance and cooperation of state and local agencies.

Among the data compiled by the National Air Surveillance Networks of the U.S. Environmental Protection Agency were the concentrations of nickel particulates over urban and nonurban regions for various seasons. The maximum value reported was 1.30 $\mu g/m^3$ for Portland, Maine. The nickel concentrations were considerably lower for nonurban vs urban areas. There was a greater concentration of nickel in the air over urban vs nonurban areas. This is a direct result of greater urban industrialization.

Nickel concentration appeared to be greater in the first and fourth quarters of the year (the colder months) than in the second and third quarters. Presumably this could reflect the use of coal and oil for heating (1).

Table 45 and Figure 31 give details of EPA emission estimates for nickel (8). Table 46 summarizes nickel emission factors from various industrial sources (9).

Nickel is present in most crude oils and remains in the residuals after the crude oil is refined. The residual fuel oil burned by utilities and employed for commercial heating contains essentially the same amount of nickel as was present in the original crude.

TABLE 45: NICKEL EMISSIONS

Source	Amount in Tons	% This Pollutant
Nickel mining	2	0.03
Iron and steel:		
Blast furnace	100	1.67
Gray iron foundry:		
Cupola	79	1.32
Ferroalloys:		
Blast furnace	491	8.20
Electric furnace	98	1.64
Nonferrous alloys:		
Furnaces	64	1.07
Material handling	Neg	Neg
Power plant boilers:		
Pulverized coal	87	1.45
Stoker coal	10	0.17
Cyclone coal	3	0.05
All oil	1,441	24.08
Industrial boilers:		
Pulverized coal	7	0.12
Stoker coal	23	0.38
Cyclone coal	3	0.05
All oil	1,139	19.03
Residential/commercial:		
Coal	3	0.05
Oil	2,435	40.69
Total	5,985	100.00

Source: Reference (8)

TABLE 46: EMISSION FACTORS FOR NICKEL FROM INDUSTRIAL SOURCES

Source[a]	Emission Factor	Emission Factor Symbol[b]
Mining and metallurgical	9 kg/10^3 kg (17 lb/ton) of Ni produced	MB
Processing of nickel and its compounds:		
Stainless steel	5 kg/10^3 kg (10 lb/ton) of Ni charged or 0.3 kg/10^3 kg (0.6 lb/ton) of stainless steel produced	Q
Nickel alloy steels[c]	5 kg/10^3 kg (10 lb/ton) of Ni charged	MB
Iron and steel scrap	0.0008 kg/10^3 kg (0.0015 lb/ton) of steel and iron	MB
Nickel alloys (other)[d]	1 kg/10^3 kg (2 lb/ton) of Ni charged	E
Copper base alloys[d]	1 kg/10^3 kg (2 lb/ton) of Ni charged	E
Electrical alloys[d]	1 kg/10^3 kg (2 lb/ton) of Ni charged	E
Cast iron	10 kg/10^3 kg (20 lb/ton) of Ni charged	E
Electroplating	Negligible	-

(continued)

TABLE 46: (continued)

Source[a]	Emission Factor	Emission Factor Symbol[b]
Batteries	4 kg/10^3 kg (8 lb/ton) of Ni processed	Q
Catalysts	Negligible	-
Cement plants:		
Dry process:		
Kiln[d]	0.2 kg/10^6 kg (0.3 lb/10^3 tons) of feed	ES (1)
Feed to raw mill[d]	0.005 kg/10^6 kg (0.01 lb/10^3 tons) of feed	ES (1)
Air separator after raw mill[d]	0.0005 kg/10^6 kg (0.001 lb/10^3 tons) of feed	ES (1)
Raw mill[d]	0.0003 kg/10^6 kg (0.0006 lb/10^3 tons) of feed	ES (1)
Air separator after finish mill[d]	0.002 kg/10^6 kg (0.003 lb/10^3 tons) of feed	ES (1)
Feed to finish mill[d]	0.005 kg/10^6 kg (0.01 lb/10^3 tons) of feed	ES (1)
Wet process:		
Kiln[e] (2 different plants)	0.1 to 1 kg/10^6 kg (0.2 to 2 lb/10^3 tons) of feed	ES (2)
Clinker cooler[d]	0.002 kg/10^6 kg (0.004 lb/10^3 tons) of feed	SSMS, OES (2)
Clinker cooler[e]	0.05 kg/10^6 kg (0.1 lb/10^3 tons) of feed	ES (2)
Clinker cooler[f]	0.1 kg/10^6 kg (0.2 lb/10^3 tons) of feed	ES (2)
Finishing mill after air separator[d]	0.002 kg/10^6 kg (0.004 lb/10^3 tons) of feed	SSMS, OES (1)

[a]Uncontrolled unless otherwise specified.
[b]Defined in Table 34, p 191; numbers in parentheses indicate number of samples analyzed, MB = material balance.
[c]Considered controlled.
[d]Exit from baghouse.
[e]Exit from electrostatic precipitator.
[f]Exit from two baghouses (in parallel).

Source: Reference (9)

The amount varies widely for domestic and imported crude oils. The extremes for 11 sources of domestic crude were 1.4 to 64.0 ppm nickel, for 24 imported crudes from 0.3 to 29.5 ppm nickel. For the most part controls are not used in oil-fired units. Assuming an average nickel content of 10 ppm, the emissions to the atmosphere for 1968 from the combustion of residual fuel oil was estimated at 4,509 metric tons of nickel. It should be noted that this Versar, Inc. (1) estimate differs somewhat from the estimate by MITRE Corp. (8).

The emissions from a coal-fired power plant were collected and analyzed. The nickel concentrations found in the fly ash ranged from 1.3 to 7.0 x 10^{-4} gram per cubic meter. Based on 272,000,000 metric tons of coal consumed by power plants in 1968, 75% particulate control, nickel concentration of 1.3 x 10^{-4} gram per cubic meter, the nickel emissions were 90 metric tons. Again, this Versar Inc. estimate (1) differs slightly from a MITRE Corp. estimate (8).

Nickel carbonyl is formed from a reaction between nickel and hot carbon monoxide probably in incinerators where nickel is found in sewage sludge or solid waste. This could occur in internal combustion engines if an additive containing nickel was used. Such additives are banned in the United States and should not be permitted in the future, according to Versar, Inc. (1).

Wastewaters containing nickel originate primarily from metal industries; particularly plating operations. High levels of nickel have also been reported in wastes from silver refineries (10). In addition, basic steel works and foundries, motor vehicle

FIGURE 31: MATERIAL FLOW THROUGH THE ECONOMY SHOWING PRIMARY EMISSION SOURCES—NICKEL

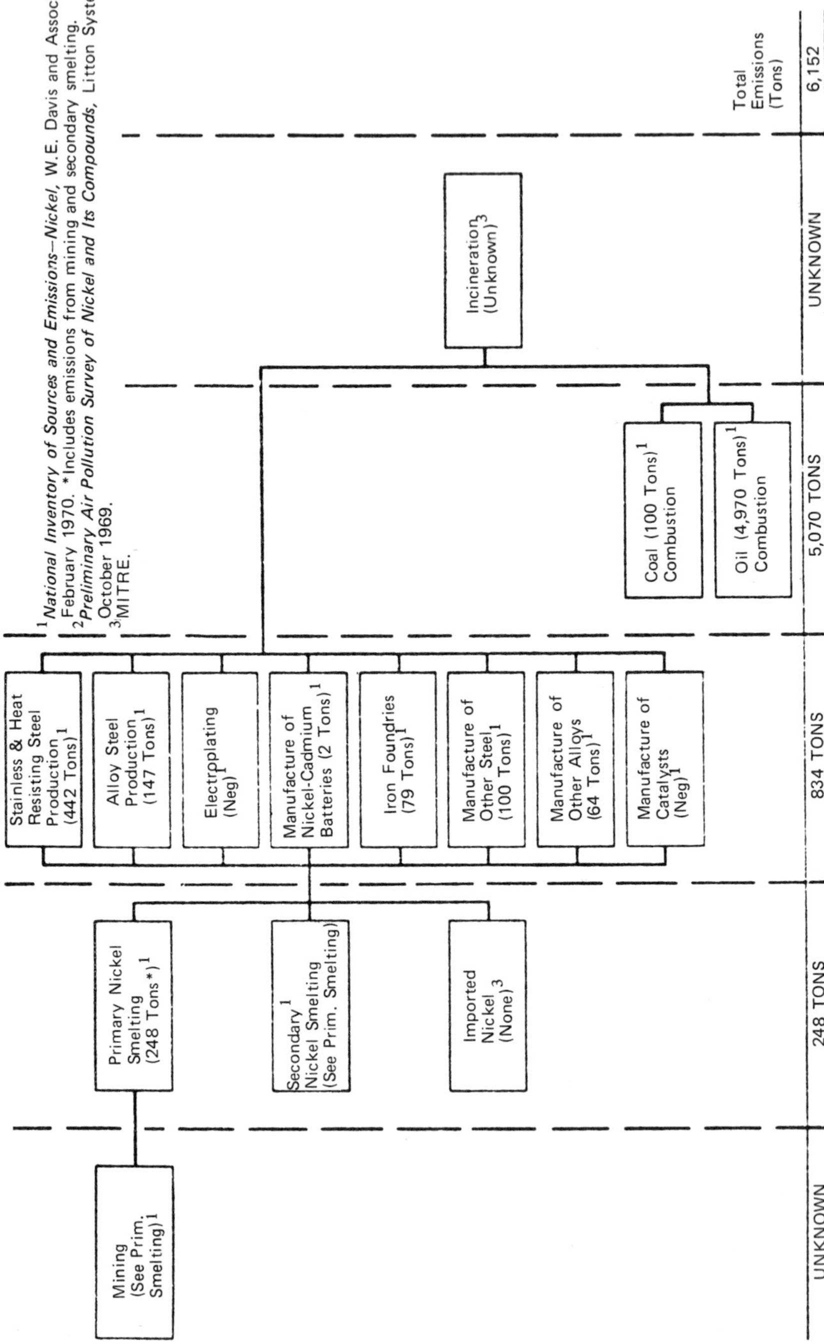

[1] *National Inventory of Sources and Emissions—Nickel*, W.E. Davis and Associates, February 1970. *Includes emissions from mining and secondary smelting.
[2] *Preliminary Air Pollution Survey of Nickel and Its Compounds*, Litton Systems, Inc., October 1969.
[3] MITRE.

Source: Reference (8)

and aircraft industries, and printing have been mentioned as potential sources of nickel (11). Golomb (12) states that generally, rinse waters may be considered to contain 1% of the plating bath concentration. Lowe (13) has reported that acid and sulfamate nickel plating baths generally contain nickel at 40,000 to 90,000 mg/kg of solution, while electrodeless process baths are of weaker concentration, containing 7,000 to 7,500 mg/kg of solution.

Industries concerned primarily with copper or brass plating or processing may also contain nickel in plant wastewaters, but generally at low levels, 1 mg/l or less, and are therefore of little concern. In addition, Hill (14) has reported nickel concentrations ranging from 0.46 to 3.4 mg/l in acid mine drainage.

Nickel exists in waste streams as a soluble ion. In the presence of complexing agents such as cyanide, nickel may exist in a soluble complexed form. The presence of nickel cyanide complexes interferes with both cyanide and nickel treatment, and the presence of this species may be responsible for increased levels of both cyanide and nickel in treated wastewater effluents. Nickel concentrations typically found in plating and metal processing wastes are summarized in Table 47.

TABLE 47: SUMMARY OF NICKEL CONCENTRATIONS IN METAL PROCESSING AND PLATING WASTEWATERS

Industry	Nickel Concentration, mg/l Range	Average
Tableware Plating		
Silver bearing waste	0-30	5
Acid waste	10-130	33
Alkaline waste	0.4-3.2	1.9
Metal Finishing		
Mixed wastes	17-51	
Acid wastes	12-48	
Alkaline wastes	2-21	
Business Machine Manufacture		
Plating wastes	5.0-35	11
Pickling wastes	6.0-32	17
Plating Plants		
(4 different plants)	2-205	
Rinse waters	2-900	
Large plants	up to 200	25

Source: Reference (15)

The use of nickel or its compounds in electroplating is a potential source for environmental contamination of the waters with soluble nickel salts. The plating solutions normally used contain nickel sulfate or nickel chloride. According to a survey conducted by Battelle in 1965, the estimated daily raw waste from nickel plating amounted to 54,430 kilograms per day of these nickel salts.

Also, only 17% of the total nickel salts used was consumed by plating. In addition to particles of the plating solution, raw waste may contain small concentrations of other heavy metals, depending on the kind of materials being processed.

Leakage from filters, pipes, and pumps is a secondary source of nickel wastes. The relatively high value of nickel has encouraged the adoption of process controls for minimizing losses into the rinse water after the plating. Also because of their relatively high value, nickel plating baths are rarely dumped. (16).

Nickel ions (+2,+3) may be present in significant amounts in the wastewater from the manufacture of nickel sulfate. This is of concern because of its toxicity to aquatic organisms at certain concentrations.

DETECTION

Dusts and fumes of nickel compounds may be collected by any method suitable for collection of other dusts and fumes; the impinger, electrostatic precipitator and filters are commonly used. The National Air Sampling Network uses a high volume filtration sampler (17). The sampling of $Ni(CO)_4$ requires a special sampling solution, and various sampling techniques are discussed with the quantitative methods.

Brief et al (18) mention five methods which have been used for the determination of nickel carbonyl. [1] An air sample can be bubbled through a saturated solution of sulfur in trifluoroethylene. The sulfur reacts with nickel to form a precipitate. Spectrographic examination is sensitive to 0.0003 ppm nickel carbonyl.

[2] An air sample may be drawn through a tube containing red mercuric oxide at 200°C and the liberated mercury may be determined spectrographically. A parallel stream of air is drawn through an oxidizing reagent to convert the CO to CO_2, and the stream is passed over mercuric oxide; the liberated mercury is again determined spectrographically. The difference in the amounts of mercury vapor measured corresponds to the nickel carbonyl content in the air. A sensitivity of 0.0014 ppm is reported.

[3] The nickel carbonyl may be absorbed in chloramine-B. The nickel determination is accomplished colorimetrically using dimethylglyoxime. For a 30 minute sample, at the suggested sampling rate of 0.5 liter per minute, a sensitivity of 0.01 ppm is obtained.

[4] Another colorimetric method uses iodine in carbon tetrachloride as the collection medium. The nickel is colored with dimethylglyoxime. A sensitivity of 0.1 ppm to nickel carbonyl is claimed.

[5] Nickel carbonyl may be collected in dilute sulfuric acid followed by spectrophotometry using sodium diethyldithiocarbamate as the coloring agent. Brief et al (18) collected a sample in dilute hydrochloric acid in a midget impinger (0.1 cfm for 30 minutes). The nickel was complexed with alpha-furildioxime and extracted with chloroform. The nickel content is determined spectrophotometrically, and the method is sensitive to 0.008 ppm.

Infrared spectrophotometry using multiple-reflection long-path cells has been used in the quantitative analysis of nickel carbonyl vapors. By using the very strong carbonyl stretching frequencies near 2,000 cm^{-1}, detectability limits on the order of 0.01 to 0.001 ppm were attained. The method is rapid, specific, and free from interference. Accuracy is estimated at ±10% (19). The American Industrial Hygiene Association indicates that a direct field instrument is commercially available for continuous monitoring of operations involving nickel carbonyl. The instrument can be used over a range of 10 ppb to 1,500 ppb (20).

Investigations to find a granulated chemical sorbent for the efficient absorption of nickel carbonyl at high rates of air passage were conducted. Of all the chemical sorbents investigated the best results were given by finely porous silica gel with grain size 0.25 to 0.5 mm impregnated with a solution of potassium iodate in sulfuric acid.

To prepare the sorbent the silica gel was purified by boiling with diluted hydrochloric acid (1:1), washed with hot water to a negative reaction to chloride, dried at 200°C and ignited for 30 minutes at 600°C. On cooling, the silica gel was impregnated with 5% solution of potassium iodate in 10% solution (by weight) sulfuric acid, using one volume of the iodate per three volumes powder.

The impregnated silica gel was distributed in a thin layer in a porcelain dish, heated to 180°C for 2 hours and then transferred, while still hot, to a flask with a tight fitting ground-glass stopper. The sorbent prepared was used to fill beaded glass tubes. It was found that with 3 ml sorbent, the air alloy should be sampled at the rate of 15 cubic meters per minute. The loss of nickel carbonyl did not exceed 1.5% even at comparatively high concentrations of the substance. The method makes possible the determination of thousandths of a milligram of nickel carbonyl in a cubic meter of air, by sampling 0.5 cubic meter of air in 30 minutes (21).

To detect nickel in particulate airborne contaminants collected in air filters, a small portion of the filter is rolled up into a cylinder and placed in a hollow graphite electrode. This portion of the air filter is directly excited by the condensed spark discharge in an oxygen atmosphere. The sample burns, exciting the spectra.

A photographic recording is used to interpret the spectra and, when improved precision is required, line intensities are measured with a microphotometer. The limits of detection are 0.1 to 1.0 micrograms. The results obtained from the rapid emission spectrographic method are compatible with results obtained with the more conventional technique of ashing the filter, mixing it with a spectrographic buffer, and exciting it in a DC arc (22).

Emission spectroscopy has been used by the National Air Pollution Control Administration for nickel analysis of samples from the National Air Sampling Network (23). The samples are ashed and extracted to eliminate interfering elements. The minimum detectable nickel concentration by emission spectroscopy is 0.0064 $\mu g/m^3$ for nonurban samples. The different sensitivities result from different extraction procedures required for urban samples (17).

Thompson et al (23) have reported that the National Air Pollution Control Administration uses atomic absorption to supplement analyses obtained by emissions spectroscopy. The method has a minimum detectable limit of 0.004 $\mu g/m^3$, based on a 2,000 m^3 air sample.

The ring-oven technique has been adapted to the measurement of nickel particulates.

Dimethylglyoxime is used as the coloring agent. The lower limit of identification is 0.08 μg and the normal range of measurement is 0.10 to 1.0 μg (24). In addition to the above test, a number of colorimetric procedures, all using dimethylglyoxime, have been described for the detection of nickel (25)-(30).

Nickel may be monitored in water by the following instrumental techniques (31): Polarographic, Spark Source Mass Spectrometry, and X-Ray Fluorescence. Nickel may be measured in water down to a detection limit of 6 ppb by atomic absorption.

ENVIRONMENTAL STANDARDS

For nickel carbonyl, the TLV (recommended) is 0.001 ppm in air (0.007 milligram per cubic meter of air). For nickel and insoluble nickel compounds, the TLV (recommended) is 1 milligram per cubic meter of air. For soluble nickel salts, the TLV (recommended) is 0.1 milligram per cubic meter of air.

The tolerance for nickel in irrigation waters is 0.05 mg/l on a continuous basis or 2.0 mg/l on a short-term basis in fine textured soil. The maximum allowable concentration of nickel in an effluent being discharged to a storm sewer or stream is 2.0 ppm.

HANDLING PROCEDURES

Care should be taken to avoid inhalation of airborne nickel dust. Skin contact with nickel compounds should be avoided. Any contact with nickel carbonyl should be minimized. Nickel powder or finely-divided nickel, also nickel catalysts, activated or spent, are classified as flammable solids (1). The Interstate Commerce Commission requires a yellow label on shipping containers and limits loads to 45 kilograms. These are not accepted for transport by air unless mixed with 40% water or some other suitable liquid.

Nickel carbonyl is classified as a flammable liquid by the ICC and is not accepted for transport by land or air. This compound is given the highest "toxic hazard rating" (1). No special transportation or other handling regulations exist for other nickel compounds with the exception of nickel cyanide (1).

REMOVAL FROM AIR

Nickel air pollution usually stems from particulate emissions from burning coal, oil, or incinerators or nickel-containing material and is normally an oxide. This can be controlled with the usual dust handling equipment such as bag filters, precipitators, and scrubbers (1).

As applied to reduce nickel emissions from coal-fired electric power plants, single-stage electrostatic precipitators have shown collection efficiencies averaging approximately 95% for total nickel removal (32). This compares favorably with efficiency ratios experienced for total particulate matter from incinerator waste gas (33), of which from 1 to 10% is particulate nickel.

Assuming the nickel oxide particle size to be similar to that for total particulate matter, the removal may be anticipated to range from 90 to 98% in such applications. Prefacing the precipitator with a spray chamber may increase total particulate collection efficiency by another 2 to 5%, depending upon design characteristics. Similar successes in particulate removal using the electrostatic precipitator have been noted in the iron and steel industry (34).

Bag collectors used in controlling emissions from ferroalloy furnaces reportedly have a 97% particulate metal oxide collection efficiency (35). In reducing particulate emissions from municipal incineration, scrubber efficiencies have been noted which range from 40% (jet-impingement type) to 98% (high-energy venturi) for micrometer size particles (36). For cleaning metal oxide fumes from carrier waste gas, scrubbers of the venturi type are recommended.

As noted, nickel carbonyl represents a unique air pollution control problem. Not only is it an extremely hazardous air pollutant from the standpoint of human health, but it can be formed where carbon monoxide in certain concentrations comes into contact with powdered nickel in a well-defined thermal environment.

Nickel carbonyl is an air pollution concern in its gaseous phase. In industrial processes where the compound is used or generated, control is normally effected by taking advantage of the fact that the compound decomposes at temperatures above 60°C (140°F). Gases containing nickel carbonyl normally are passed through a furnace, where the compound decomposes into its two components, nickel and carbon monoxide. The resulting nickel metal then can be removed from the decomposition furnace vent gas stream as a particulate (2), using any of the above-mentioned standard dust collection equipment, particularly the fabric filter.

A process developed by E.R. Breining et al (37) relates to the recovery and purification of metal carbonyl compounds and particularly to the recovery of such compounds from the exhaust gases of metallizing operations.

Nickel carbonyl, which at room temperature is a liquid, is commonly employed in metallizing by thermal decomposition of the carbonyl in contact with the heated substrate to be metallized. Frequently such decomposition is incomplete and undecomposed carbonyl, together with products of decomposition such as CO, and carrier gases such as CO_2 and N_2 form the exhaust gases of the operation.

Carbonyl recovery is important as many of the carbonyls are expensive and even in the case of nickel carbonyl efficient recovery is an important factor in commercial aspects of the operation. The exhaust gases of such procedure are passed in the process into a liquid, such as water at relatively low temperature to effect condensation of the carbonyl; being insoluble and heavier than water the condensed carbonyl is readily separable from the water, dried and passed to storage for further use in metallizing.

REMOVAL FROM WATER

Regardless of the type of treatment employed, the level of treatment required and associated costs can often be reduced by good housekeeping practices to reduce accidental loss, spillage, or leaks of plating or pickling solutions containing nickel. Process modifications to reduce the major contributor, dragout loss of plating bath solution into rinse waters, have been recommended by Domey and Steifel (38).

Treatment processes can be classed as destructive, based upon nickel precipitation and disposal, and nickel recovery, which can be accomplished by several means. Recovered nickel is normally recycled through the metal processing operations.

The formation and precipitation of nickel hydroxide is the basis for destructive treatment of nickel bearing wastes. Nickel forms insoluble nickel hydroxide upon addition of lime. The nickel hydroxide has a minimum theoretical solubility of 0.01 mg/l at pH 10 (39). The precipitation is most effective at a high pH, although little efficiency is gained above pH 10. Therefore, close pH control is not essential other than from considerations of the chemical costs associated with pH adjustment.

Kantawala and Tomlinson (40) have reported the reduction of 100 mg/l of nickel to 1.5 mg/l at pH 9.9, by addition of 250 mg/l lime and resultant formation of nickel hydroxide. The use of lime, followed by nickel hydroxide sludge conditioning with ferric chloride (to improve dewatering) and sand filtration reduced a combined waste effluent averaging 21 mg/l of nickel down to 0.09 to 1.9 mg/l. Total waste flow was 200,000 gallons per day.

Anderson and Iobst (41) reported 35 mg/l nickel could be reduced to 0.2 mg/l in the final effluent by use of lime. Since the final effluent included an equal volume of rinse water which acted to dilute the treated waste stream, treatment actually accomplished reduction to 0.4 mg/l for the nickel waste. Total flow was 0.45 mgd. Operating costs for the metal bearing waste stream were estimated at 80¢/1,000 gallons (1966 prices), and installation costs as $1,108/1,000 gallons per day capacity.

Hanson and Zabban (42) report that lime treatment for a waste which employed the coagulant aid, Separan NP-10, at a concentration of 1 to 2 mg/l was able to reduce nickel from 39 mg/l to 0.17 mg/l. No cost data were given. Stone (43) reported that nickel in the lime treated effluent of a copper, brass, cupro-nickel metal processing plant was 0.1 mg/l prior to sand filtration of the treated wastewater. Filtration acts as a polishing process to remove small nonsettleable metal hydroxide particles.

Tupps (44) reported a final concentration of 15 mg/l nickel in a treated plating waste in which chromium had been reduced to 0.05 mg/l and iron to 0.5 mg/l. The effluent pH was 8.1, which resulted in poor nickel hydroxide formation. Indications are that a pH greater than 8.5 is required for effective nickel hydroxide precipitation.

The Lancy process (45) is designed to treat plating bath dragout solution directly, instead of diluting it into a rinse water and treating the larger volume. The plated metal, with dragout adhering to it, is submerged in a vat containing a concentrated solution of the chemical (e.g., lime), employed to precipitate the nickel. This solution is continuously circulated from the vat to a reservoir, with sufficient detention time to allow settling of the metal hydroxide precipitate.

Sludge removal is provided in the reservoir. Makeup chemicals are added to the reservoir in proportion to the amount of precipitate formed. Subsequent rinses contain salts derived from the precipitation vat. Lancy (45) claims this process is the only means of guaranteeing nickel levels of less than 1 mg/l, although these levels were reported by other plants using direct lime precipitation (41)(42)(43).

A process developed by G. Hirs et al (46) involves removing nickel from plating rinse water or the like by adjusting upward the pH of an aqueous nickel solution to precipitate the nickel as a colloidal suspension, passing the precipitate suspension through a deepbed filter having a granular, synthetic organic medium therein to deposit the nickel precipitate on the filter medium, backwashing the filter medium to remove the deposited nickel precipitate thereform and then filtering the backwash liquid through a conventional filtration mechanism.

Recovery of nickel from process waste streams is more attractive than for some other metals, due to the high value of nickel. One report (47) estimates rinse water recovery values to be $0.80 to 7.00/1,000 gallons for nickel concentrations ranging from 150 to 900 mg/l. Recovery processes currently in use are ion exchange, evaporative recovery, electrolytic recovery, and direct reuse by countercurrent rinse techniques in systems where evaporative losses are high. In addition, reverse osmosis has been evaluated for recovery. Not all systems provide complete metal recovery, and final polishing treatment may be required to remove residual nickel.

Usually, total flow reduction is an integral part of recovery systems, in order to reduce capital costs. Barnes (48) used countercurrent flow multiple rinse tanks to allow direct recovery of nickel plating solution dragout. The first (and therefore most concentrated) rinse solution served as plating bath solution makeup to compensate for evaporation from the hot plating bath. Use of four sequential rinse tanks resulted in a nickel concentration of only 0.9 mg/l in the final rinse.

Recovery of nickel by ion exchange has been employed for some time. Reents and Stromquist (49) report on the use of cation-anion exchange to treat nickel sulfate plating bath rinse water containing 870 mg/l nickel. Complete nickel removal was reported, and the reclaimed water was acceptable for reuse in the rinse tank. The unit processed 1,300 gal/hour.

To be economical, evaporative recovery requires reduction in rinse water volume, thereby increasing nickel concentrations in the rinse waters to be processed by the recovery system. Countercurrent rinse serves this need.

Banerjee and Banerjee (50) have reported on nickel recovery by electrolysis from silver refinery waste. However, even at high nickel concentrations, recovery was only 65 to 70% effective. Thus, additional treatment was required to reach low nickel levels. No costs were given, but electrolytic recovery would not appear to be practical for plating applications.

Golomb (51) investigated the use of reverse osmosis as a bench scale process, for the recovery of nickel from plating bath rinse solutions. Treatment costs were quite high on both a volume and daily operating expense basis. For each level of operation, the evaporation rate was assumed to be ten times the dragout rate. With the reverse osmosis process, not all the permeate (i.e., filtered) solution returned to the process. The excess contained 22 mg/l of nickel, therefore requiring additional treatment.

Studies determined the nickel tolerance of the activated sludge process and its related capacity to remove nickel from sewage. Modest adverse effects of biological and chemical oxygen demands, suspended solids, and turbidity were noted at 2.5 ppm nickel. Digestion of primary and activated sludge was not impaired at influent concentrations up to 40 ppm. Treatment removed 30% of the influent nickel (52).

SOLID WASTE DISPOSAL

Nickel-containing hazardous wastes for land disposal in relatively small amounts are generated by the manufacture of nickel sulfate (53). They result from the treatment of wastewaters by raising the pH to precipitate metallic salts. Two plants in this segment directly sewer their wastes. One plant uses on-site landfill for the precipitates and another uses contractor landfill (general purpose). Perhaps the major source of nickel salts, however, is metal plating effluents.

Nearly all the small and medium size plating shops, and even some of the large ones, discharge their metal laden aqueous effluents to municipal sewers. Some plants have facilities to precipitate and clarify before discharge, but they are estimated to be about 15% of the total shops. In many areas, especially large metropolitan areas, this practice is allowed to continue as long as the toxicity of the nickel discharges has little or no effect on the continuous operation of any of the biological waste treatment processes of the sewage treatment plant.

Many of the metal cations, nickel included, precipitate with sulfides in the domestic sewer lines while in transit to the sewage treatment plant. Thus, a great proportion of them are removed in the primary treatment process and most of the remainder are bound up in the sludge resulting from the secondary treatment processes.

The method is considered adequate provided that: [1] the sewage treatment authorities can effectively regulate the discharge of toxic metal compounds into the sewer system at a rate which will not disturb the efficiency of the sewage treatment processes; and [2] the operation and design of the sewage treatment plant is such that effective treatment is accomplished and the effluent from the treatment plant complies with the local purity regulations for discharge to open waters. The secondary sludge must be landfilled such that the provisional limits are not exceeded or they must be placed in a California Class 1 type landfill (53).

When caustic or an alkaline stream in an industrial plant is combined with an acid plating shop effluent, the heavy metal hydroxides precipitate. These precipitates are normally gelatinous sludges with very high water content (70 to 80%). These sludges normally contain a wide mixture of materials and treatment after precipitation includes passage through a sand bed or similar clarifier. The sludge is removed for landfilling and the effluent is discharged to a municipal sewer or open waterway if permitted by the regulating authorities (53).

Some landfills are chartered to accept nickel antimonide, arsenide or selenide for ultimate encapsulation and landfill. They would most likely be combined with other acceptable materials in a steel drum, and the steel drum would be given a concrete collar before burial. Landfilling is adequate in California Class 1 type facilities.

The amounts of concentrated waste nickel antimonide, arsenide or selenide which could arise are expected to be very small. For this reason, careful packaging and shipment back to the vendor or manufacturer by prior arrangement is an easy and safe means of disposal. The manufacturer can either reprocess the material, if economically feasible, or could dispose of it in his own facilities. Return to the manufacturer or vendor should be considered before any other disposal option is considered.

ECONOMIC IMPACT OF CONTROLS

Lancy (45) indicates that waste treatment facilities can be expected to cost 10% or less of the installed cost of the associated industrial process equipment. Depending upon the industrial process, wastewater treatment method, and size of operation, capital costs totaling 3 to 5% appear to be mentioned most frequently.

Mattock (54) has reported that the extra cost of ion exchange equipment for recovery of nickel is rapidly offset by the value of the recovered material. He emphasizes the necessity for waste stream segregation, particularly in the presence of cyanide, in order that the nickel, once concentrated in the ion exchange regenerant, can be recovered by precipitation as the carbonate.

Heidorn and Keller (55) also employed a scheme of cation exchange for nickel removal, followed by weak base anion exchange. Reclaimed concentrations of 60,000 mg/l nickel and 24,000 mg/l sulfuric acid were achieved upon regenerating the exchange resins. Neutralization with nickel carbonate allowed reuse of the solution. Chemical costs were given as about 52¢/lb of nickel recovered.

With mixed metal wastes where no chemical recovery can be practiced, Ross (56) has demonstrated that for small waste volumes (35,000 gal/day), ion exchange is more than twice as costly as chemical precipitation (70¢ versus 34¢/1,000 gal), in terms of operating plus capital costs. Detailed comparative costs for a waste containing nickel, zinc and tin are presented in his discussion.

Culotta and Swanton (57)(58) report costs for evaporative recovery of nickel plating solutions ranging from $0.60 to $1.23/gallon, and provide formulas for comparison of destructive treatment costs versus evaporative recovery costs (57). Graphical analysis based upon their data was provided by the authors for zinc cyanide with a lower unit cost of zinc plating bath solution ($0.45/gal). The nickel recovery system would be expected to exhibit more favorable economics than zinc recovery, due to nickel's higher recovery value. Besselievre (59) reported that a commercially available evaporative recovery system that was capable of treating 400 gal/hr, would cost $45,000, and could recover $100,000 per year in chemicals.

Pinner (60) has reported on the use of the Lancy process for recovery of nickel by precipitation directly as the carbonate, and reuse after washing and redissolution of the precipitate. Both batch and continuous recovery systems have been developed. If recovered nickel carbonate amounts to less than 1 ton per year, recovery is accomplished once a week on a bath basis. Capital costs for simple bath systems vary from 800 to 1,200 English pounds and for continuous flow systems from 2,000 to 8,000 English pounds.

REFERENCES

(1) Office of Toxic Substances, *Preliminary Investigation of Effects on the Environment of Boron, Indium, Nickel, Selenium, Tin, Vanadium and Their Compounds, Vol. III, Nickel,* Report EPA 560/2-75-0056, Wash. D.C., Environmental Protection Agency (Aug. 1975).

(2) Sullivan, R.J., *Air Pollution Aspects of Nickel and Its Compounds,* Report PB 188,070, Springfield, Va., Nat. Tech. Information Service (Sept. 1969).

(3) Cothern, C.R., *Determination of Trace Metal Pollutants in Water Resources and Stream Sediments,* Report PB 213,369, Springfield, Va., Nat. Tech. Info. Service (1972).

(4) Schroeder, H.A., Ballassa, J.J. and Tipton, I.H., *J. Chron. Dis.* 15, 51 (1961).

(5) Luckey, T.D., Venugopal, B. and Hutcheson, D., *Heavy Metal Toxicity, Safety and Hormonology,* New York, Academic Press (1975).

(6) Booz-Allen Applied Research, Inc., *A Study of Hazardous Waste Materials, Hazardous Effects and Disposal Methods,* Report PB 221,466, Springfield, Va., Nat. Tech. Info. Service (July 1973).

(7) National Academy of Sciences, *Manganese,* Wash. D.C., (1975).

(8) Goldberg, A.J., *A Survey of Emissions and Controls for Hazardous and Other Pollutants,* Wash. D.C., Environmental Protection Agency (Feb. 1973).

(9) Anderson, D., *Emission Factors for Trace Substances,* Report PB 230,894, Springfield, Va., Nat. Tech. Info. Service (Dec. 1973).

(10) Banerjee, N.G. and Banerjee, T., "Recovery of Nickel and Zinc from Refinery Waste Liquors: Part I, Recovery of Nickel by Electrodeposition," *Jour. Sci. Industr. Res.* 11B:77-78 (1952).

(11) Dean, J.G., Bosqui, F.L. and Lanouette, K.H., "Removing Heavy Metals from Waste Water," *Env. Sci. Tech.,* 6:518-522 (1972).

(12) Golomb, A., "Application of Reverse Osmosis to Electroplating Waste Water Treatment. Part II, The Potential of Reverse Osmosis in the Treatment of Some Plating Wastes," *Plating,* 59:316-319 (1972).

(13) Lowe, W., "The Origin and Characteristics of Toxic Wastes with Particular Reference to the Metals Industries," *Wat. Poll. Control* (London), 1970:270-280.

(14) Hill, R.D., "Control and Prevention of Mine Drainage," presented at the Environmental Resources Conference on Cycling and Control of Metals; Battelle Columbus Laboratories, Columbus, Ohio (October 31 to November 2, 1972).

(15) Patterson, J.W. and Minear, R.A., *Wastewater Treatment Technology,* Report PB 216,162, Springfield, Va., Nat. Tech. Info. Service (February 1973).

(16) Environmental Protection Agency, *Development Document for Proposed Effluent Limitations Guidelines and New Source Performance Standards for the Copper, Nickel, Chromium and Zinc Segment of the Electroplating Point Source Category,* Report EPA 440/1-73-003, Wash. D.C., (Aug. 1973).

(17) *Air Pollution Measurements of the National Air Sampling Network, Analyses of Suspended Particulates, 1957-1961,* U.S. Public Health Service Publ. 978 (1962).

(18) Brief, R.S., Venable, F.S. and Ajemian, R.S., "Nickel Carbonyl: Its Detection and Potential for Formation," *Am. Ind. Hyg. Assoc. J.* 26:72 (1965).

(19) McDowell, R.S., "Metal Carbonyl Vapors: Rapid Analysis by Infrared Spectrometry," *Am. Ind. Hygiene Assoc. J.* 32, No. 9, 621-24, (Sept. 1971).

(20) "Nickel Carbonyl," Hygienic Guide Series, *Am. Ind. Hygiene Assoc. J.* 29, 304 (1968).

(21) Vol'berg, N.S. and Gerskhovich, E.E., "Determination of Small Amounts of Nickel Carbonyl in Air," *Hyg. Sant.* (Moscow), 33, No. 4-6, 226-29 (April-June 1968).

(22) Webb, R.L. and Webb, M.S.W., *A Rapid Emission Spectrographic Method for the Analysis of Air Filters,* Harwell, Atomic Energy Research Establishment (1971).

(23) Thompson, R.J., Morgan, G.B. and Purdue, L.J., "Analyses of Selected Elements in Atmospheric Particulate Matter by Atomic Absorption," Preprint. Presented at the Instrument Society of America Symposium, New Orleans, Louisiana (May 5 to 7, 1969).

(24) West, P.W., et al, "Transfer, Concentration and Analysis of Collected Airborne Particulates Based on Ring Oven Techniques," *Anal. Chem.* 32(8):943-6 (1960).

(25) Feigl, F. and Stern, R., "The Application of Spot Reactions in Qualitative Analysis," *Z. Anal. Chem.* 60:1 (1921).

(26) Kielczewski, W. and Supinski, J., "Determination of Microgram Quantities of Nickel by the Impregnated-Paper Method," *Chem. Anal.* (Warsaw), 10(4):677 (1965).

(27) Kobayashi, Y., "A Rapid Determination of a Small Amount of Nickel by Means of a Detector Tube," *J. Chem. Soc. Japan,* Ind. Chem. Sect. 59(10):728 (1955).

(28) Kutzelnigg, A., "New Reagent Papers for Porosity Testing and for the Determination of Metals," *Metalloberflaeche* 5:B 113 (1951).

(29) Marion, S.P. and Zlochower, I., "A Spot Test Analysis of the Group III Cations," *J. Chem. Educ.* 36:378 (1959).

(30) Tananaev, N.A., "Qualitative Analysis of the Elements of the First to Third Groups When Present Together, with Special Regard to Spot Tests," *Z. Anorg. Allgem. Chem.* 140:320 (1924).

(31) Lawrence Berkeley Laboratory, *Instrumentation for Environmental Monitoring - Water,* Publ. LBL-1, Volume 2, Berkeley, University of California (February 1, 1973).

(32) Cuffee, S., Gerstle, R., Orning, A., and Schwartz, C., "Air Pollutant Emissions from Coal-Fired Plants: Report No. 1," *Journal of the Air Pollution Control Association,* 14: September, 1964.

(33) Bump, R., "The Use of Electrostatic Precipitators on Municipal Incinerators," *Journal of the Air Pollution Control Association,* 18, 803-809 (December 1968).

(34) Smith, J., "Air Pollution Control in Oxygen Steelmaking," *Journal of Metals,* 13, 632-634, (September 1961).

(35) Person, R., "Control of Emissions from Ferroalloy Processing," *Journal of Metals,* 23, 17-29 (April 1971).

(36) Ellison, W., *Control of Air and Water Pollution from Municipal Incinerators Using the Wet-Approach Venturi Scrubber,* Chemical Construction Company, New York, 163, (1970).

(37) Breining, E.R. and Bolton, W.M., U.S. Patent 2,985,509, May 23, 1961, assigned to Union Carbide Corporation.

(38) Domey, W.H. and Steifel, H.C., "Wastewater Reduction in Metal Finishing Operations," presented at 43rd Ann. Conf. Water Poll. Control Fed., Boston, Mass. (Oct. 5, 1970).

(39) Jenkins, S.H., Keight, D.G. and Humphreys, R.E., "The Solubility of Heavy Metal Hydroxides in Water, Sewage and Sewage Sludge I, the Solubility of Some Metal Hydroxides," *Internat. Jour. Air Wat. Poll.* 8, 53-56 (1964).

(40) Kantawala, D. and Tomlinson, H.D., "Comparative Study of Recovery of Zinc and Nickel by Ion Exchange Media and Chemical Precipitation," *Wat. Sew. Works,* 111, R281-R286 (1964).

(41) Anderson, J.S. and Iobst, E.H., Jr., "Case History of Wastewater Treatment in a General Electric Appliance Plant," *Jour. Water Poll. Control Fed.,* 40: 1786-1795 (1968).

(42) Hansen, N.H. and Zabban, W., "Design and Operation Problems of a Continuous Automatic Plating Waste Treatment Plant at Data Processing Division, IBM, Rochester, Minn.," *Proc. 14th Purdue Industrial Waste Conf.,* pp. 227-249 (1959).

(43) Stone, E.F.F., "Treatment of Nonferrous Metal Process Waste at Kynoch Works, Birmingham, England," *Proc. 22nd Purdue Industrial Waste Conf.,* pp. 848-865 (1967).

(44) Tupps, C.C., "Treatment of Wastes for Automobile Bumper Finishing," *Industr. Wat. and Wastes,* 6, 111-114 (1961).

(45) Lancy, L.E., "An Economic Study of Metal Finishing Waste Treatment," *Plating* 54, 157-61 (1967).

(46) Hirs, G. and Kozar, R.S., U.S. Patent 3,630,892; December 28, 1971; assigned to Hydromation Filter Company.

(47) *State of the Art: Review on Product Recovery,* Water Pollution Control Research Services U.S. Department of Interior, Wash. D.C., November 1969.

(48) Barnes, G.E., "Disposal and Recovery of Electroplating Wastes," *Jour. Wat. Poll. Control Fed.,* 40, 1459-1470 (1968).

(49) Reents, A.C. and Stromquist, D.M., "Recovery of Chromate and Nickel Ions from Rinse Waters by Ion Exchange," *Proc. 7th Purdue Industrial Waste Conf.,* pp. 462-473 (1952).

(50) Banerjee, N.G. and Banerjee, T., "Recovery of Nickel and Zinc from Refinery Waste Liquors: Part I — Recovery of Nickel by Electro-deposition," *J. Sci. Industr. Res.* 11B: 77-78 (1952).

(51) Golomb, A., "Application of Reverse Osmosis to Electroplating Waste Treatment: Part I, Recovery of Nickel," *Plating,* 57, 1001-1005 (1970).

(52) McDermott, G.N., Post, M.A., Jackson, B.N., and Ettinger, M.B., "Nickel in Relation to Activated Sludge and Anaerobic Digestion Process," *J. Water Poll. Control Fed.* 37, No. 2, 163-77 (1965).

(53) Powers, P.W., *How to Dispose of Toxic Substances and Industrial Wastes,* Park Ridge, N.J., Noyes Data Corp. (1976).

(54) Mattock, G., "Modern Trends in Effluent Control," *Metal Finishing Jour.,* 14, 168-175 (1968).

(55) Heidorn, R.F., and Keller, H.W., "Methods for Disposal and Treatment of Plating Room Solutions," *Proc. 13th Purdue Industrial Waste Conf.,* pp. 418-426 (1958).

(56) Ross. R.D., *Industrial Waste Disposal,* New York, N.Y., Reinhold Book Corp. (1968).

(57) Culotta, J.W. and Swanton, W.F., "The Role of Evaporation in the Economics of Waste Treatment for Plating Operations," *Plating,* 55, 957-961 (1968).

(58) Culotta, J.W. and Swanton, W.F., "Case Histories of Plating Waste Recovery Systems," *Plating,* 57, 251-255 (1970).

(59) Besselievre, E.B., *The Treatment of Industrial Wastes,* New York, McGraw-Hill Book Co. (1969).

(60) Pinner, R., "Quantitive Recovery of Nickel Dragout Losses by the Integrated Method," *Metal Finishing Jour.,* 14, 272-273 (1968).

SELENIUM

Selenium is a paradoxical element, beneficial or essential in amounts varying from trace to parts per million concentrations for humans, animals and plants and quite toxic to animals at concentrations which may exist in the environment. Also, the sensitivity to selenium and its compounds is extremely variable in all classes (1).

Selenium production is a by-product of copper recovery both in the U.S. and elsewhere. The United States has slipped from its position as the world's leading selenium producer as a result of the 1967-1968 copper strikes. However, the production of selenium does not present any economic or strategic problems. At present, selenium imports amount to less than one-half of the production, and less than 60% of the selenium available to U.S. copper refiners is recovered. In practically all the applications for selenium, the chemically similar compounds of sulfur and tellurium can be substituted. However, only selenium is suitable in reusable photosensitive plates considering current technology.

Selenium is widely distributed in nature, but is found in greater concentrations associated with sulfide minerals of lead, iron, copper and other metals. The concentration in certain "seleniferous" soils is great enough that specific species of plants growing there pick up sufficient selenium to be toxic to animals. This natural source of selenium entering the environment is significant but less than that from the weathering of natural rock.

Selenium entering the environment from activities of man is estimated at 3,500 metric tons per year, of which most is attributed to the combustion of coal. These contributions appear to be small in comparison to contributions from natural sources.

TOXICITY

Although selenium had been reported to be toxic to animals as early as 1842, it was not until the early 1930s that its role as a poison received much attention

(2). At this time, the U.S. Department of Agriculture and the Wyoming and South Dakota agricultural experiment stations found this element to occur at toxic levels in plants growing on certain soils of the Great Plains and Rocky Mountain area and to cause the so-called alkali disease and certain heretofore unexplained losses of livestock. Several excellent reviews of this early work are presented elsewhere (3)-(6). Since then, there has been increased concern over the toxicity of selenium. The health effects of selenium have recently been reviewed by Science Applications, Inc. (7).

Selenium compounds, particularly the water-soluble compounds, are toxic to humans and animals (8). In humans, mild inhalation of selenium dusts, fumes, or vapors irritates the membranes of the eyes, nose, throat, and respiratory tract, causing lacrimation, sneezing, nasal congestion, coughing, etc. Prolonged exposure through inhalation can cause marked pallor, coated tongue, gastrointestinal disorders, nervousness, and a garlicky odor of breath and sweat. In animals, subacute selenium poisoning produces pneumonia and degeneration of the liver and kidneys. Furthermore, experiments with rats indicate that selenium may cause cancer of the liver.

The biochemical effects of elemental selenium and its compounds on humans is not as yet thoroughly understood. The selenium deficiency diseases found in animal species, as well as some of the frank selenium poisonings, have not been observed in man. Similarly, the carcinogenic hazard of selenium and the antagonistic effect of arsenic for selenium seen in animals are yet to be shown in humans. These are important factors that need clarification to properly evaluate the role of selenium and its compounds in air pollution.

Workers exposed to fine elemental selenium dust collect the dust in the upper nasal passages, producing catarrh, nose bleeding, and loss of smell. A few cases of selenium dermatitis have occurred. Exposure to fumes of elemental selenium produces frontal headache, intense irritation of the eyes and nasopharyngeal passages, slight difficulty in breathing, and uvular edema. Exposed workers recovered within three days and no ill effects persisted.

Most of the workers in a Japanese factory making selenium rectifiers are minors and children. Hypochromic anemia and drastic leucopenia were prevalent in workers who had been employed in the factory for a long time. Fingernails were damaged from regular contact with selenium, and increasing numbers of female workers reported irregular menses or menostasis (1).

Selenium dioxide readily dissolves in water producing selenious acid. Intense local irritation and inflammation of the skin and mucous membranes occur upon contact. Selenium dioxide, selenious acid, and selenite may be absorbed through the skin, resulting in internal accumulation. Allergic dermatitis to selenium dioxide may develop, especially in fairhaired people. Sodium thiosulfate solution is used to treat exposure.

Hydrogen selenide fumes, from an etching and imprinting operation on a steel strip using a selenious acid bath induced nausea, vomiting, metallic taste in the mouth, extreme lassitude, and fatigability. Olfactory fatigue and garlic breath also occurred.

Selenium oxychloride is a severe vesicant, capable of producing a third degree burn which is extremely painful and slow to heal (1). Dimethyl selenide produces acute sore throat and pneumonitis. A 13 year study of selenium intoxication in industrial workers found that the mortality rate was similar to that of the normal population. To date, no long-term systemic effects are known which can be attributed to selenium.

Acute dermatitis which proved to be selenium poisoning was observed in a rancher in South Dakota. Appreciable quantities of selenium were found in the water, meat, vegetables, and dairy products from his ranch. The patient was treated with bromobenzene to eliminate selenium from his system (1).

Many other cases of selenium dermatitis have been reported from North Dakota, South Dakota, Montana, Wyoming, and Nebraska. Symptoms included various degrees of skin rash, follicular rashes over the hairy parts of the body, extreme fatigue, and dizziness. The selenium sources were home-grown farm products in the diet. Recovery was achieved by removal of selenium from the diet.

Well water from the Wasatch geological formation in Utah contained 9 ppm selenium. Chronic selenosis occurred in the children, parents, and dog drinking the water. No selenium was found in the food. Lassitude, total or partial loss of hair, discoloration and loss of fingernails were symptoms of the condition. A halt in the use of the water brought regrowth of the hair and nails and increased mental alertness. People living in seleniferous regions should be conscious of the possible danger of selenium in foods and water (1).

Chronic selenosis has been known in an area of Colombia, South America since the 1600s. The black "peladero" soil in the area contains extremely high levels of selenium, which accumulate in the corn and other crops, and forage plants and are present in the streams in the area. Loss of hair and nails, and the birth of malformed babies have been symptoms of the selenosis in Indians living in the area. Animals grazing there lost hair and hoofs, showed malformations of lips and legs, a number of animals aborted, and eggs failed to hatch (1).

EXTENT OF EXPOSURE

As part of a national inventory of sources and emissions, W.E. Davis and Assoc. of Leawood, Kansas have reported on the nature, magnitude and extent of the emissions of selenium in the United States for the year 1969 (9).

Samples of snow, rain, and air taken in Boston, Mass., have shown that the selenium content of the air is averaging 0.001 $\mu g/m^3$. Based on the selenium-to-sulfur ratio in these samples, the atmospheric selenium was probably from terrestrial sources, including the fuels and ores used by industry. Another source may be the burning of trash containing paper products. Some papers when analyzed contain as much as 6 ppm selenium. Selenium in paper may come from accumulation by the original tree or plant, or possibly from the manufacturing of the paper (from the use of pyrites in the process). Any vegetation which is burned may be a possible source of atmospheric selenium. Another source could be the refining of sulfide ores, particularly copper and lead ores (8).

A rough estimate as to the magnitude of selenium in the atmosphere might be made from the concentration of sulfur in the atmosphere. This method would be valid if the sources of these two pollutants are sulfide ores, fossil fuels, or igneous and sedimentary rocks, since in these materials the average weight ratio of selenium to sulfur is 1×10^{-4}.

Selenium, or in some cases sodium selenate, sodium selenite, or barium selenite is added to glass in small amounts to decolor it or in larger amounts to color it. Material may be lost to the environment when the raw materials are handled. Salts of selenium, which are soluble, serve as an example of this. Mixing ingredients for glass making, especially in batch processing, is done with dry materials causing some loss by dusting. A large amount of the elemental selenium used is volatilized during melting as was confirmed by stack analysis.

The glass industry estimates that 184 metric tons of selenium were emitted into the atmosphere in 1969 (1). Except for possible local concern, this should not pose an environmental problem. Data is not available on selenium loss to the environment via other channels. Considering the nature of the processes involved, however, this figure should be insignificant.

For the most part the photocopy processes of charging, exposure, image development, image transfer, image fixing, and photoceptor cleaning take place in an enclosed environment, presenting little chance for contamination. The exception is in the manufacture of photocopy equipment, where selenium is lost to the atmosphere in the vacuum plating operation. The industry estimates that about one kilogram per metric ton of selenium processed or around 135 kilograms in 1972 escaped into the atmosphere (1).

The levels of airborne selenium in xerography rooms has been the subject of a recent publication (10) and response in a "Letter to the Editor" (11). The original publication reported concentrations well above ambient but the letter cited disputes the published results and questions the analytical technique used.

Selenium-containing pigments are dispersed in liquid media, applied by painting, spraying, screen printing or dipping, and fired at 480° to 870°C. Bag filters are used to control emission to the atmosphre, and losses to the environment via other routes are very small. Among other uses of selenium, its use as an insecticide, particularly for carnations and crysanthemums, allows small amounts to get into the environment.

Selenium recovery in the United States is principally a by-product of copper refining. The environmental contamination from selenium is closely related to the mining and milling of copper-bearing ores during the recovery of selenium. Unfortunately, analyses of the wastes are not available.

A major producer of selenium and precious metals by recovery from the copper-bearing ore has recently started analyzing waste streams. The analyses cover a long list of elements including selenium and are complicated by the extremely small concentration levels. Some qualitative information is provided on losses of selenium to the environment. Selenium was found in a number of waste streams within the operation. The precious metals recovery is a probable source of loss of selenium. In the selenium recovery process there are several points where selenium vapor may be lost. Besides the melting operation and drying oven, which represent small potential losses of selenium, the bisulfate fusion is of great concern. In the original installation a wet scrubber was installed on the exhaust side followed by an electrostatic precipitator. In 1971 new hooding and wet scrubbers followed by electrostatic precipitators were installed at all vapor discharge points. This eliminated visual emissions to the atmosphere, but no analyses of the exhausts are available.

Selenium is also present in the very hot exhaust gases from the reverberatory furnaces and converters in the copper smelter. This is revealed by chemical analysis of particulate matter collected in the balloon flues and electrostatic precipitators through which the exhaust gases flow. Selenium is also found in waste water streams from the selenium recovery plant and from the wet scrubbers ahead of the sulfuric acid plant. Selenium may appear in the water used in leaching metal values from waste dumps (1).

Figure 32 and Table 47 give some detailed breakdowns of airborne emissions of selenium (12).

TABLE 47: SELENIUM EMISSIONS

Source	Amount in Tons	% This Pollutant
Primary Copper		
Roasting	17	1.99
Reverberatory Furnace	8	0.94
Converters	22	2.57
Material Handling	5	0.59
Primary Zinc		
Roasting	27	3.16
Primary Lead		
Sintering	5	0.59
Blast Furnace	2	0.23
Secondary Copper, Zinc, Lead	1	0.12
Glass Manufacture	203	23.74
Paint, Varnish, Lacquer Manufacture	1	0.12
Incineration	1	0.12
Power Plant Boilers		
Pulverized Coal	360	42.11
Stoker Coal	41	4.80
Cyclone Coal	12	1.40
All oil	19	2.22
Industrial Boilers		
Pulverized Coal	16	1.87
Stoker Coal	56	6.55
Cyclone Coal	8	0.94
All Oil	14	1.64
Residential/Commercial Boilers		
Coal	5	0.59
Oil	32	3.74
Total	855	100.03

Source: Reference (12)

FIGURE 32: MATERIAL FLOW THROUGH THE ECONOMY SHOWING PRIMARY EMISSION SOURCES—SELENIUM

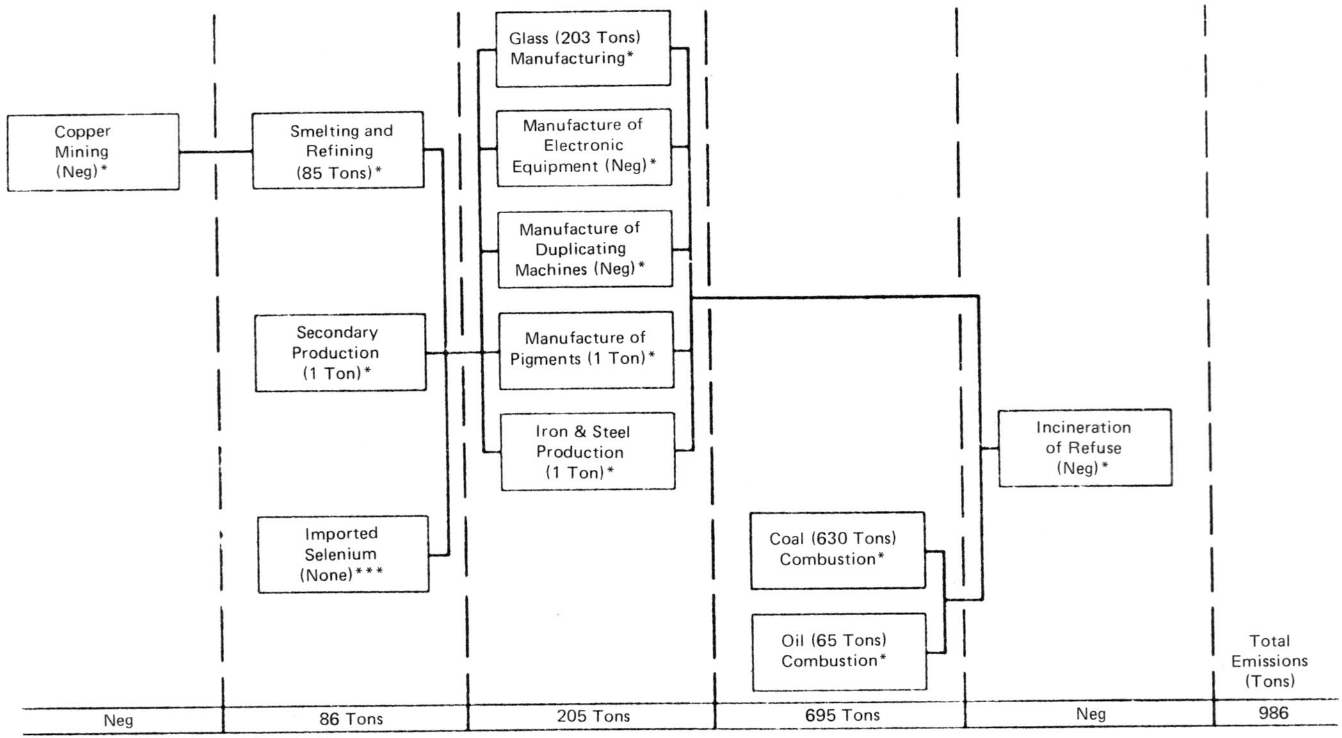

National Inventory of Sources and Emissions—Selenium, W.E. Davis and Associates, April 1972.

***Preliminary Air Pollution Survey of Selenium and Its Compounds,* Litton Systems, Inc., October 1969. ***MITRE.

Source: Reference (12)

Selenium is not found in significant quantities as wastes (13). One source of selenium waste that has been identified is in the manufacture and reconditioning of Xerox drums. Selenium wastes generated from these operations each year include: [1] about 700,000 lb of solid wastes containing cotton linter, steel, aluminum, and 300 to 400 lb of selenium and arsenic; and [2] 50,000 lb of liquid waste consisting primarily of caustic solutions and 0.3% selenium and arsenic.

Selenium as cadmium sulfoselenide is also found in paint wastes. However, it is estimated that only 370 lb of selenium are lost in the waste solvent-based paint sludges from paint manufacture each year, and an additional 2,600 lb as paint residue left in used paint containers.

DETECTION

Methods of analysis are available that can measure selenium quantitatively in the parts per billion or submicrogram region (8). However, none of the methods is simple, rapid, or applicable to continuous monitoring, and many of the methods are not specific for selenium. An EPA report (1) states that "there are no data available on the monitoring for selenium of air or water."

Air samples for selenium determination have been collected by means of electrostatic precipitators (14), filters (15), and liquid impingers (14)(15)(16). All of these methods can be used for selenium dusts and fumes, while only the last can be used for collection of vapors or gaseous compounds.

The use of filters and liquid impingers are the most commonly employed sampling methods today. Some air samples were taken by passing 100 m^3 of air at the rate of 1 m^3/hr through a 1 micron pore diameter Millipore filter. Because selenium is found in many filter papers, Millipore filters or other selenium-free collection media (e.g., glass fibers) should be used. However, some filter papers can be washed with sodium sulfide and water to remove traces of selenium. The liquid used in the impingers has been water (15)(16), and a solution of 40 to 48% hydrobromic acid with 5 to 10% bromine. The water can be used with nonvolatile water-soluble compounds such as selenium dioxide (14).

The advantage of the hydrobromic acid-bromine solution is that it converts selenium and most of its compounds to the soluble selenium tetrabromide. This solution, after collection, can be analyzed directly after decolorization and neutralization (16) or the selenium may be separated by distillation. It has also been suggested that soda lime or silica gel could be used to collect vapors of selenium compounds (14).

Little study has actually been given to air sampling for selenium analysis however (2). Most of the work reported to date has dealt only with the particulate matter that can be removed by filters. Tabor et al (17) published a tentative method for determining atmospheric selenium. However, Pillay and Thomas (18) found the element to be divided into gaseous and suspended forms, and improvements in collecting the gaseous form are needed.

Water containing selenium should be sampled by commonly accepted methods.

If it contains suspended material, this should be removed by filtration within a few hours, preferably immediately after the sample is collected. The selenium content of the sediment can be determined separately if it is essential that this be known. The filtered sample should be stored at 0° to 5°C to reduce the possibility of microbial action that could result in the precipitation or removal of selenium. Usually, an appropriate amount of the sample is made alkaline to phenolphthalein and then evaporated to a small volume or to dryness for analysis. This assumes that no forms of the element are present that would be volatile under these conditions. Normally, the assumption can be made, but additional studies on ways of reducing this volume seem needed (2).

Detailed discussions of methods of selenium analysis have been presented elsewhere, (19)(20) and the subject will be only briefly reviewed here. Many procedures are available for use, and these will be discussed in general terms without attempting to completely cover the literature. They may be divided into two general classes: those that do not require the destruction of the sample (nondestructive) and those that require getting rid of the organic matter before the selenium is measured (destructive).

In the nondestructive category, one can use x-ray fluorescence analysis, but this method lacks sensitivity and has not been widely used. On the other hand, neutron activation analysis (NAA) can be used either with or without destruction of the sample, and it has become increasingly popular in the latter use. When used with destruction of the sample, the selenium is usually separated from interfering elements following activation and prior to its measurement.

The chief disadvantage of NAA for selenium analysis is the cost of the equipment. This is so great that it can hardly be justified for selenium analysis only. In survey types of work where multielement analysis is conducted on large numbers of samples, it may be the method of choice, since it can greatly speed the work (when used nondestructively) and since it has excellent sensitivity for a number of elements. There is evidence, however, that at submicrogram levels of selenium the analysis of natural materials by this method can yield inaccurate results (19), so the method should probably be tested in each laboratory on a variety of samples of known selenium content before it is used routinely.

A quick qualitative estimation of selenium can be made by taking advantage of the reaction of aromatic 1,2-diamines with selenious acid [or SE(IV) cation] to form colored piaselenoles. Elemental selenium and selenides readily oxidize in nitric acid or hydrogen peroxide to selenious acid. Selenate salts, or Se(VI) cations, give only weak colors.

The most common diamine employed is 3,3'-diaminobenzidine, which forms a yellow dipiaselenole in acid and is sensitive to 10 ppm. The 4-dimethylamino and 4-methylthio derivatives of 1,2-phenylenediamine (21) will also react with selenious acid to give stable bright red and blue-purple colors, respectively. Reaction with 2,3-diaminonaphthalene has been used as a spot test (22). Other ions, particularly copper, iron, tellurium, chromium, nickel, and cobalt, may cause interference in high concentration, but some may be masked by complexing agents such as EDTA.

Other methods of detection of selenious acid that have been reported are reaction with [a] iodides (23), [b] thiourea (24), [c] diphenylhydrazine (25), and

[d] pyrrole (26), as well as the catalytic effect on the reaction of methylene blue with alkali sulfide (27).

Several quantitative methods of analyzing selenium are based on the reaction of selenious ion with 3,3'-diaminobenzidine. West and Cimerman (28) used this reagent in connection with the ring-oven technique. This colorimetric method is basically free from interferences when masking reagents are used and is applicable at concentrations of 0.1 to 0.5 μg; limit of identification is 0.08 μg (28)(29). Kawamura and Matsumoto (16) spectrophotometrically determined the selenium in air samples collected in impingers by measuring the absorbance of the solutions treated with this reagent. The benzidine reagent has also been used in fluorometric determination of selenium in biological materials (30)(31). Detection in these methods was on the order of 10 to 50 ppb selenium.

Walkinson (32), as well as Allaway and Cary (33), found that the product from 2,3-diaminonaphthalene and selenium showed a fluorescence sensitivity in the neighborhood of 0.5 to 5 ppb selenium, a sensitivity greater than that with the benzidine product.

Neutron activation methods have been used to determine selenium from air samples (via filters and aqueous bubbler) and from samples of snow and rain (15). A sensitivity of 0.01 μg of selenium was obtained when gamma radiation was measured with a scintillation spectrometer. This technique has also been used to determine traces of selenium in biological materials (34) and in metals (35).

West and Ramakrishna (36) have developed a method for determining trace amounts of selenium based on its catalytic effect in the reduction of methylene blue by sodium sulfite. Color comparisons are made in the range of 0.1 to 1.0 μg of selenium. There is serious interference from copper if present in an excess of 10 μg. Kawashima and Tanaka (37) also developed a catalytic method based on the reduction of 1,4,6,11-tetraazanaphthacene; this method is subject to interference from several ions.

Walkinson (38) has discussed some of the newer methods that are used in the analysis of selenium in biological material, including polarography, x-ray fluorescence, and atomic absorption spectrophotometry.

In most methods used for the analysis of selenium, the selenium must be in solution as the quadrivalent ion. It is then either reduced and precipitated as the element or measured volumetrically by oxidation to the hexivalent state, or by iodimetry. Selenium can also be determined polarographically. The only specific spot test for selenium depends on its catalytic action on the reduction and decoloration of methylene blue by alkali sulfides. A drop of the solution to be tested is added to a few crystals of sodium sulfite, followed by a little sodium sulfide solution, and then the methylene blue. The sensitivity is 0.08 micrograms (1).

A spectrophotometric method for the determination of submicron quantities of selenious acid was developed in 1971. The procedure is based on the oxidation of hydroxylamine hydrochloride to nitrous acid by selenous acid followed by the subsequent coupling of the diazonium salt with N(1-naphthyl)-ethylenediamine dihydrochloride. Optimum conditions have been established. The range of determination is 0.01 to 0.20 mg Se^{+4} per liter. The method is simple, sensitive, and reproducible with no common interferences which cannot be easily obviated (1).

The current level of selenium tolerated in public waters is 10 ppb (0.01 mg/l) which is beyond the capability of classical analytical methods due to limitations of volume of sample, time of analysis, or removal of interfering substances. A fluorometric method was developed to analyze the selenium by the Analytical Research Department of Abbott Laboratories, North Chicago, Ill.

Elemental selenium is converted to selenious acid by the action of a bromine-bromide redox buffer. Selenious acid reacts with 2,3-diaminonaphthalene in acid solution to form the strongly fluorescent naphtho-(2,3-d)-2-seleno-1,3-diazole. Selenates are not reduced to selenites under these conditions, hence the method is specific for elemental selenium and selenium in the four valance state. Plots of fluorescent intensity vs concentration in the region of 0.005 to 0.2 μg selenium^{+4} are linear and practically free from reagent interference at excitation frequency of 366 nm and a fluorescent emission minimum of 522 nm (1).

Selenium may be measured in water by flame atomic absorption or by neutron activation down to detection limits of 400 to 600 ppb. Flameless atomic absorption or polarography may be used to measure selenium down to detection limits in the 10 to 20 ppb range.

ENVIRONMENTAL STANDARDS

Threshold Limit Values (TLV)

Elemental selenium	0.2 mg/m^3
U.S.S.R.: (1959)	0.1 mg/m^3
Hydrogen selenide (H_2Se)	0.05 ppm (equivalent to ~0.2 mg/m^3)
Selenium hexafluoride (SeF_6)	0.05 ppm or ~0.4 mg/m^3

From public water supplies the permissible criterion for selenium (+2, +4, +6) is 0.01 mg/l (10 ppb) (1).

The State of Illinois has established the regulation for public water that the "maximum allowable twelve-month average selenium concentration" is 0.01 mg/l. No single analysis shall show a concentration at any time of more than 1.5 times this value. The justification given for this criterion is that "to maintain the rather narrow margin of safety imposed by the background ingestion of selenium from food, drinking water must not contain more than 0.01 mg/l of selenium" (1).

A tentative allowable concentration of 1 mg/l of selenium in urine was recommended for rural populations living on seleniferous soil and for workers exposed to selenium or its compounds.

American Feed Manufacturers Association, Inc., sent a petition to the Food and Drug Administration to amend the Food Additive Regulations to permit the safe use of selenium as a nutrient in the feed of chickens, turkeys, and swine. The proposal provides that selenium as sodium selenite or selenate may be added at a level not to exceed 0.1 ppm for chickens and swine and 0.2 ppm for turkeys. The status of the petition is dependent on the comments arising from the publication of the proposed changes (1).

For continuous water use in irrigation, the selenium tolerance is 0.05 mg/l and the permissible concentration for short-term use in irrigating fine textured soil is also 0.05 mg/l. The maximum allowable concentration of selenium in effluents being discharged to storm sewers and streams is 0.01 ppm.

HANDLING PROCEDURES

The chief industrial control methods are the use of a good ventilation system and personal protective equipment, including safety goggles and a respirator when working in a low airborne concentration and a supplied air respirator in heavily contaminated areas (39). Selenium-containing liquid wastes are disposed of by washing them down the drain, while solid wastes are buried.

Commercial elemental selenium is relatively inert and may be handled without special precaution (1). This is also true for stable metal selenides, such as copper and lead. All other selenium compounds should be treated with caution since some of these compounds can enter the body through the lungs or skin and may affect the body organs. This includes the reactive selenides, the volatile and soluble compounds, and particularly hydrogen selenide and organoselenium compounds.

Selenic acid, with a melting point of 58°C is classified as a corrosive liquid, and transport containers must carry a white label. Cargo aircraft cannot accept more than 25 liters of selenic acid and carriage by passenger aircraft is prohibited. When handled as a solid, selenic acid requires a poison label. Quantities of solid selenic acid up to 25 kg can be shipped in passenger aircraft and up to 85 kg can be shipped in cargo aircraft.

REMOVAL FROM AIR

The air pollution aspects of selenium and its compounds have been reviewed in some detail by Stahl (40). To the extent to which selenium emissions are associated with the particulate phase, conventional particulate control devices such as bag filters and wet scrubbers appear to be applicable as available control technology (41).

However, several selenium compounds, including the oxides, exist as vapors at temperatures common for stack gases. Thus, cooling devices may be needed before selenium removal by particulate control technology is feasible.

Furthermore, an electrostatic precipitator was comparatively ineffectual for removing particulate selenium in one study of a coal fired power plant; precipitator efficiency for selenium ranged from less than 29 up to 60%. To predict the removal efficiency of conventional particulate devices, a size distribution of the emissions is needed. Most selenium pollutants are generally water soluble, except for selenium metal.

A process developed by A. Yomiyama and S. Yonekawa (42) involves the recovery of selenium or selenium compounds used in the production of unsaturated aldehydes or nitriles from unsaturated hydrocarbons.

A process developed by H. Kakuta et al (43) is a method for the recovery of selenium from combustion waste gas, which comprises:

> contacting the combustion waste gas with a first absorbing solution which contains at least one absorbent selected from the group consisting of an alkali metal sulfite and an alkali metal bisulfite, whereby the major portion of the selenium component of the waste gas is absorbed therein, and wherein the waste gas is cooled so that the humidity of the gas is increased;
> passing the gas through a moistened glass fiber filter of glass wool having a bulk density of 150 to 250 kg/m^3, so that the residual selenium in the gas is collected thereon;
> combining the selenium component collected on the filter with the first absorbing solution; and
> acidifying the combined solution to reduce and precipitate the selenium component to metallic selenium.

Reduction of industrial emissions of selenium may occur with implementation of other pollution control actions. The close association of selenium with sulfur in its occurrence and their similar chemical properties would suggest a corresponding reduction in the emission of selenium as sulfur control and associated removal of particulates become effective. Copper, lead and zinc smelters were estimated to recover 26% of the sulfur contained in the smelter feed as sulfuric acid in 1970, an increase from the estimated 20% in 1960.

Most promulgated emission-control regulations for nonferrous smelters specify achievement of about 90% sulfur control. Similar sulfur-emisson control programs are being instituted for the burning of coal and fuel oil. Improved particulate-control programs at plants processing sludges containing selenium and at plants consuming selenium would reduce emissions of selenium along with other pollutants.

REMOVAL FROM WATER

Essentially no information is available on levels of selenium in industrial wastewaters, treatment methods for selenium wastes or costs associated with removal of selenium from industrial wastewaters in water pollution or industrial waste literature as well as the technical literature of the industries using selenium. Industries using selenium include paint, pigment and dye producers, electronics, glass manufacturers, and insecticide industries. The subject of selenium wastewaters and potential methods of treatment is discussed by J.S. Patterson and R.A. Minear (44).

Johnson (45) has reported that selenium is present in almost all conceivable types of paper, and it might be concluded from this information that pulp and paper mill wastes could contain selenium. Johnson's primary concern was with the release of selenium as an air pollutant, upon incineration of paper products. However, he did report selenium wastewater levels as follows.

For incinerator fly ash quench water, soluble selenium was measured at 5 to 23 μg/l, while incinerator residue quench water contained 3 μg/l of soluble selenium. The incinerator handled a solid waste containing approximately 55 to 69% paper.

Hutchinson (46) reports that selenium occurs as an impurity in the form of selenide, $Se^{=}$, in metallic sulfide ores. When the sulfide ores are roasted in air (to convert the sulfide and drive off SO_2) the selenide is oxidized to selenium dioxide and is released in the flue gas. If the flue gas is quenched, selenium dioxide reacts with the quench water to form selenious acid, H_2SeO_3. The selenious ion appears to be the most common form of selenium in wastewater except for pigment and dye wastes, which contain the selenide (e.g., yellow cadmium selenide). Other forms decompose to yield the $SeO_3^{=}$ ion.

Hutchinson also reports that selenium dioxide is readily reduced, to precipitate finely divided elemental selenium. It is probable that this same reaction might occur for selenious ion at acidic pH. Selenide may be precipitated as the metallic salt, which is highly insoluble.

Secondary municipal sewage treatment plant effluents containing 2 to 9 μg/l of selenium has been reported (47). A tertiary sequence of treatment, which included lime treatment to pH 11, sedimentation, mixed media filtration, activated carbon adsorption and chlorination yielded selenium removal of 0 to 89%. Best removal was obtained at higher initial selenium concentrations.

Raible (48) has noted the presence of selenium in a pharmaceutical wastewater, but did not report concentrations. Linstedt, et al (49), measured 2.3 μg/l of selenium in the effluent from a secondary sewage treatment plant. Efficiencies of removal of selenium, as $SeO_3^{=}$, by several advanced wastewater treatment processes have been investigated and reported (44)(50). The results are shown in Table 48.

TABLE 48: REMOVAL OF SELENIUM BY ADVANCED WASTEWATER TREATMENT PROCESSES (BENCH-SCALE)

Process	Removal (%)
Lime coagulation-settling	16.2
Cation exchange	0.9
Cation plus anion exchange	99.7
Process Sequence*	
(1) Sand filtration	9.5
(2) Activated carbon	43.2
(3) Cation exchange	44.7
(4) Anion exchange	99.9

*Cumulative removal after indicated process

Source: Reference (44)

The authors concluded that efficient removal of selenium could be achieved with a strong acid-weak base ion exchange system. R.M. Ahlgren has presented capital and operating costs for such a weak base system, as a function of dissolved solids content of the wastewater (50). The nonselective nature of ion exchange resins is a drawback when attempting to remove a low level contaminant in the presence of significant quantities of other ions. The simultaneous presence of other ions rapidly increases ion exchange costs.

According to the work of Linstedt and his coworkers, however, other commonly employed industrial metal treatment processes, e.g., lime coagulation, settling, and sand filtration were ineffective in recovering selenium, at least as the negatively charged anion. According to J.D. Hem (51), selenium is most likely to occur in aqueous solution in anionic form, and presumably would not be treated for removal in that form. In that event, ion exchange appears to be the most effective technique reported.

A process developed by W.N. Marchant (52) is one whereby selenium is removed from solution in acidic waste water by treatment of the water with a metallic reducing agent. The process is particularly effective for removal of selenium from zinc smelter effluent by reaction of the effluent with powdered zinc.

SOLID WASTE DISPOSAL

The selenium-containing solid wastes from the manufacture and reconditioning of Xerox drums are disposed of by burial in a segregated section of a landfill site for industrial wastes. Waste sludges containing selenium and toxic heavy metals from the manufacture of paint and pigments are also disposed of in approved sanitary landfill sites.

It is felt that sanitary landfill is an acceptable and adequate method for the disposal of selenium wastes, provided that the sites used are located over nonwater-bearing sediments or have only unusable groundwater underlying them and are completely protected from flooding, surface runoff or drainage, so that all waste materials and internal drainage are restricted to the site. In California any sanitary landfill sites approved for the disposal of selenium wastes must meet the criteria for a Class 1 disposal site (53).

ECONOMIC IMPACT OF CONTROLS

The general economics of particulate removal and their application to selenium have been discussed by McFarland (41). R.M. Ahlgren (50) has presented some data on costs of selenium removal from wastewaters by ion exchange.

REFERENCES

(1) Office of Toxic Substances, *Preliminary Investigation of Effects on the Environment of Boron, Indium, Nickel, Selenium, Tin and Vanadium—Vol. IV—Selenium,* Washington, D.C., Environmental Protection Agency (Aug. 1975).

(2) National Research Council, *Selenium,* Report EPA-600/1-76-014, Research Triangle Park, N.C., Environmental Protection Agency (Jan. 1976).

(3) M.S. Anderson, H.W. Lakin, K.C. Beeson, F.F. Smith and E. Thacker, *Selenium in Agriculture,* U.S. Dept. of Agriculture Handbook 200, Washington, D.C., U.S. Government Printing Office (1961).

(4) J.R. Harr and O.H. Muth, "Selenium Poisoning in Domestic Animals and Its Relationship to Man," *Clin. Toxicology* 5, 175-86 (1972).

(5) A.C. Moxon, *Alkali Disease of Selenium Poisoning,* Bull. No. 311, Brookings, S.D., South Dakota Agricultural Experiment Station (1937).

(6) I. Rosenfeld and O.A. Beath, "Chemistry of Selenium" in *Selenium: Geobotany, Biochemistry, Toxicity and Nutrition,* New York, Academic Press (1964).

(7) Science Applications, Inc., *Health Effects of Selenium,* Report PB 250 568, Springfield, Va., Nat. Tech. Information Service (Jan 1976).
(8) Booz-Allen Applied Research, Inc., *A Study of Hazardous Waste Materials, Hazardous Effects and Disposal Methods* Report PB 221 466, Springfield, Va., Nat. Tech. Information Service (July 1973).
(9) W.E. Davis and Assoc., *National Inventory of Sources and Emissions: Barium, Boron, Copper, Selenium and Zinc; Section IV–Selenium,* Report PB 210 679, Springfield, Va., Nat. Tech. Information Service (May 1972).
(10) J.M. Harkin, A. Dong and G. Chesters, *Nature,* 259, 204 (1976).
(11) R.A. Parent, Letter to the Editor in *Chem. and Eng. News,* pp 3, 27 (May 31, 1976).
(12) A.J. Goldberg, *A Survey of Emissions and Controls for Hazardous and Other Pollutants,* Washington, D.C., Environmental Protection Agency (Feb 1973).
(13) R.S. Ottinger, J.L. Blumenthal, D.F. Dal Porto, G.I. Gruber, M.J. Santy and C.C. Shih, *Recommended Methods of Reduction, Neutralization, Recovery or Disposal of Hazardous Waste,* Report PB 224 580, Springfield, Va., Nat. Tech. Info. Serv. (Aug. 1973).
(14) F.A. Patty, (Ed.), *Industrial Hygiene and Toxicology,* vol. II, 2nd ed. (New York: Interscience, p. 887, 1963).
(15) Y. Hashimoto and J.W. Winchester, "Selenium in the Atmosphere," *Environ Sci. Technol.* 1:338 (1967)
(16) M. Kawamura and K. Matsumoto, "Determination of Small Amounts of Hydrogen Selenide in Air," *Japan Analyst* (Tokyo) 14(9):789 (1965).
(17) E.C. Tabor, M.M. Braverman, H.E. Bumstead, A. Carotti, H.M. Donaldson, L. Dubois, and R.E. Kupel, "Tentative Method of Analysis for Selenium Content of Atmospheric Particulate Matter," *Health Lab. Sci. Suppl.* 7, No. 1, 96-101 (1970).
(18) K.K.S. Pillay and C.C. Thomas, Jr., "Determination of the Trace Element Levels in Atmospheric Pollutants by Neutron Activation Analysis," *J. Radioanal. Chem.* 7, 107-118 (1971).
(19) O.E. Olson, I.S. Palmer and E.I. Whitehead, "Determination of Selenium in Biological Materials," in *Methods of Biochemical Analysis,* D. Glick, Ed., New York, John Wiley and Sons (1973).
(20) J.H. Watkinson "Analytical Methods for Selenium in Biological Material" in *Symposium: Selenium in Biomedicine,* O.H. Muth, Ed., Westport, Conn, AVI Publishing Co. (1967).
(21) E. Sawicki, "New Color Test for Selenium," *Anal. Chem.,* 29:1376 (1957).
(22) F. Feigl, *Qualitative Analysis by Spot Tests,* 3rd ed. New York, Elsevier, p. 266 (1946).
(23) N.S. Poluektov, "Detection of Selenium and Tellurium in the Presence of One Another," *Mikrochemie* 15:32 (1934).
(24) G. Deniges, "Detection of Selenium, Application to Natural Waters," *Bull. Soc. Pharm. Bordeaux* 75:197 (1937).
(25) F. Feigl and V. Demant, "Microchemical Detection of Selenium," *Mikrochim. Acta,* 1:322 (1937).
(26) M. Suzuki, "Colorimetric Method for Determining Selenium in Crude Copper," *J. Chem. Soc. Japan* 56:323 (1953).
(27) F. Feigl and P.W. West, "Test for Selenium, Based on a Catalytic Effect," *Anal. Chem.* 19:351 (1947)
(28) P.W. West and C. Cimerman, "Microdetermination of Selenium with 3,3'-Diaminobenzidine by the Ring Oven Technique and Its Application to Air Pollution Studies," *Anal. Chem.* 36:2013 (1964).
(29) P.W. West, "Chemical Analysis of Inorganic Pollutants," *Air Pollution* II:163 (1968).
(30) K.L. Cheng, "Determination of Traces of Selenium; 3,3'-Diaminobenzidine as Selenium(IV) Organic Reagent," *Anal. Chem.* 28:1738 (1956).
(31) L.M. Cummins et al, "A Rapid Method for Determination of Selenium in Biological Material," *Anal. Chem.* 36:382 (1964).
(32) J.H. Walkinson, "Fluorometric Determination of Selenium in Biological Material with 2,3-Diaminonaphthalene," *Anal. Chem.* 38:92 (1966).
(33) W.H. Allaway and E.E. Cary, "Determination of Submicrogram Amounts of Selenium in Biological Materials," *Anal. Chem.* 36:1359 (1964).

(34) H.J.M. Bowen and P.A. Cawse, "The Determination of Selenium in Biological Material by Radioactivation," *Analyst* 88:721 (1963).
(35) F.J. Conrad and B.T. Kenna, "Determination of Selenium by Activation Analysis and Dry Volatilization," *Anal. Chem.* 39:1001 (1967).
(36) P.W. West and T.V. Ramakrishna, "A Catalytic Method for Determining Traces of Selenium," *Anal. Chem.* 40:966 (1968).
(37) T. Kawashima and M. Tanaka, "Determination of Submicrogram Amounts of Selenium (IV) by Means of Catalytic Reduction of 1,4,6,11-Tetraazanaphthalene," *Anal. Chim. Acta.* 40:137 (1968).
(38) J.H. Walkinson, "Analytical Methods for Selenium in Biological Material," in *Symposium: Selenium in Biomedicine,* O.H. Muth, Ed. Westport, Conn: AVI Pub. Co. (1967).
(39) "Selenium and Its Compounds," *Nat. Safety News* 93, 5 (1966).
(40) Q.R. Stahl, *Air Pollution Aspects of Selenium and Its Compounds,* Report PB 188 077, Springfield, Va., Nat. Tech. Information Service (Sept. 1969).
(41) A.R. McFarland, Ed., *Control Technology for Toxic and Hazardous Air Pollutants,* Chicago, Ill. Institute for Environmental Quality (Dec. 1974).
(42) A. Yomiyama and S. Yonekawa; U.S. Patent 3,052,522; September 4, 1962; assigned to Asahi Kasei Kogyo KK, Japan.
(43) H. Kakuta, S. Matoba, K. Shimizu and S. Yamashita; U.S. Patent 3,966,889; June 29, 1976; assigned to Asahi Glass Co., Ltd.
(44) J.W. Patterson and R.A. Minear, *Wastewater Treatment Technology,* Report PB 204 521; Springfield, Va., Nat. Tech. Information Service (Aug. 1971).
(45) H. Johnson, "Determination of Selenium in Solid Waste," *Envir. Sci. Tech.,* 4, 850-853 (1970).
(46) E. Hutchinson, *Chemistry: The Elements and Their Reactions,* Philadelphia, W.B. Saunders Co. (1959).
(47) D.G. Argo and G.L. Culp, "Heavy Metals Removal in Wastewater Treatment Processes: Part 2–Pilot Plant Operation," *Water Sew. Wks.,* 119, 28-132 (1972).
(48) J.A. Raible, "Fluorometric Determination of Selenium in Effluent Streams with 2,3-Diaminonaphthalene," *Envir. Sci. Tech.,* 6, 621-622 (1972).
(49) K.D. Linstedt, C.P. Houck and J.T. O'Connor, "Trace Element Removals in Advanced Wastewater Treatment Processes," *Jour. Water Pollution Control Fed.,* 43, 1507-1513 (1971).
(50) R.M. Ahlgren, "Membrane vs Resinous Ion Exchange Demineralization," *Indust. Water Engrs.,* 8, 12-14 (1969).
(51) J.D. Hem, "Study and Interpretation of the Chemical Characteristics of Natural Water," *Geological Survey Water - Supply Paper 1473,* U.S. Dept. Interior, Washington, D.C., (1959).
(52) W.N. Marchant; U.S. Patent 3,933,635; January 20, 1976; assigned to the U.S. Secretary of the Interior.
(53) P.W. Powers, *How to Dispose of Toxic Substances and Industrial Wastes,* Park Ridge, N.J., Noyes Data Corp. (1976).

TIN

Pure tin was used in Egypt as early as 600 BC and as an alloy in bronze implements many centuries earlier, possibly 3200 to 3500 BC. While tin touches many facets of our daily lives, it is used to improve the properties and characteristics of other materials and its functional applications are usually hidden. It is used in coating cans, joining pipes or electrical conductors, and in bearings of all types. Tin in metal form is used in collapsible tubes, pewter ware and pipes in the food processing industry.

TOXICITY

A Versar, Inc. report (1) has this to say about occupational exposure to tin.

Inorganic Tin: Exposure to the dust or fumes of tin causes a benign pneumoconiosis-stannosis. Workmen involved in bagging tin oxide, tending furnaces in which metallic tin is burnt to tin oxide, or processing tin ores may receive substantial exposures. Small dense shadows appear in lung x-rays. The cases of stannosis were not disabled, and lung function tests were within normal limits, as were the blood and urine tests.

Organic Tin: Brief contact with organic tin compounds may cause acute skin burns. Tributyl and dibutyl compounds are especially caustic. Lesions consisted of reddening and swelling of the hair follicles two to three hours after exposure. The reaction became more intense over the next eight hours, and minute pustules over the follicular openings formed during the second day. The skin between the hair follicles was little affected. Only faint erythema and scaling remained after a week.

Subacute lesions are characterized by itching and may affect the skin of the lower abdomen, thighs, and groin of workmen handling these compounds through clothing contamination. A faint erythematous eruption develops, but disappears rapidly upon removal from contact. In five young persons employed in a chemical laboratory, acute organotin intoxications occurred. In all cases, headaches and nausea, with visual defects were reported. Electroencephalogram showed general disturbances and dysrhythmia, as well as occipital foci of short waves.

As far as epidemological exposure to tin, the following can be said.

Inorganic Tin: Colloidal metallic tin and tin arsenate were tested as treatments for typhoid and staphylococcal infections. Little therapeutic effect was observed, but no toxic reactions were reported either.

An orange juice and apple juice containing 250 to 385 ppm tin were under suspicion as the causes of an outbreak of food poisoning in Kuwait in 1967. Experiments with animals indicated that tin levels, as tin(IV) citrate, must be at least 1,400 ppm before toxic signs appear, and that toxicity is due to the local irritation of the mucous membrane of the digestive tract. The major toxic effect was emesis.

Organotin: Attempted therapy of bacterial infections by organic tin compounds has led to disastrous results. In France, a preparation called Stalinon was used for the treatment of various infections. The active ingredient was diethyltin diiodide, but high amounts of impurities as monoethyl- and triethyltin produced 217 cases of poisoning, 100 of which were fatal. Triethyltin is ten times as toxic to rats as diethyltin and contributed up to 10% of the theoretical weight of diethyltin.

Poisoning symptoms appeared after a latent period of about four days, beginning with a severe persistent headache. Vomiting, vertigo, abdominal pain, and urine retention followed. Paralysis occurred in severe cases; death, preceded by convulsions or coma, was due to cardiac or respiratory failure. Edema of the white matter of the brain was the most significant lesion at necropsy.

As pointed out by Schroeder (2), tetraethyltin is highly toxic, whereas inorganic tin is not. However, as pointed out by Versar, Inc. (1), tin salts in the blood stream are highly toxic, producing paralysis and other neurologic damage. The toxicity of tin and its compounds has been reviewed by Luckey et al (3).

EXTENT OF EXPOSURE

The major source of tin entering the environment is tin plate used to coat steel cans for containment of food and other materials. Since little of this tin is reclaimed, essentially all of the roughly 20,000 metric tons of tin per year utilized in this manner may be assumed to enter the environment. Some wastes also arise from the use of tin in solder (1).

Contribution from tin-bearing rocks and soils may be considered much smaller than the above value. Because of the relatively high value of tin, smelting and other processing wastes are minimal, and much of the tin used for purposes other than tin plate and solder is reclaimed. Waste hazards from manufacture and use of the toxic organotin compounds do not appear generally significant because the waste quantities are small and the compounds are minimally soluble in water. Perhaps the most significant aspect of tin plate solid waste is the availability of the tin, much of which would be recoverable under economic conditions more favorable than at present.

Most tin containers are used to store food. A study of the tin content of condensed milk samples was made (1). Where the product was not over four weeks old, the tin content was at most 40 ppm, and rose inappreciably for an additional five months. In storage experiments lasting more than two years, however, the tin content rose to 160 ppm in exceptional cases. While there are no reports of poisoning by the tin present in condensed milk, it has been suggested that the date of filling be stated on the cans so that only new cans be used for infant feeding.

The only tin smelter-refiner in operation in the United States uses a process in which dust losses occur during the crushing of the feed ore, the roasting step, leaching, drying, and smelting in the reverberatory furnaces. The gas and particulates from these processes go through electrostatic precipitators and then are discharged through an 80-meter-high stack. The recovered particulates are returned to the process.

The waste slag, which is inert and insoluble, is used for landfill on the property. All liquid wastes are neutralized, clarified, and discharged through approved effluent points as provided in state regulations (1). Some modifications have been made recently to improve the operations, but details are not available.

The Environmental Protection Agency, state and local agencies conduct air sampling activities to obtain information about the quality of the air in the United States. Tin was one of twelve metals checked in suspended particulate matter. The last published detailed survey, based on sampling during 1968 to 1969, showed tin concentrations in urban and nonurban regions during the various seasons. The detectable concentrations of tin were 0.001 microgram per cubic meter for urban samples and 0.0003 for nonurban areas. The tin concentration was decidedly higher in urban areas with the highest concentrations occurring during the colder months. The highest maximum values reported were 0.120 microgram per cubic meter for Jersey City and 0.100 for Philadelphia.

Figure 33 and Table 49 give some details of airborne tin emissions (4).

TABLE 49: TIN EMISSIONS

Source	Amount in Tons	Percent this Pollutant
Iron and steel		
open hearth	260	73.03
Power plant boilers		
Pulverized coal	68	19.10
Stoker coal	8	2.25
Cyclone coal	2	0.56
Industrial boilers		
Pulverized coal	3	0.84
Stoker coal	11	3.09
Cyclone coal	2	0.56
Residential/commercial		
coal	1	0.28
All boilers, oil	1	0.28
Total	356	99.99

Source: Reference (4)

FIGURE 33: MATERIAL FLOW THROUGH THE ECONOMY SHOWING PRIMARY EMISSION SOURCES–TIN

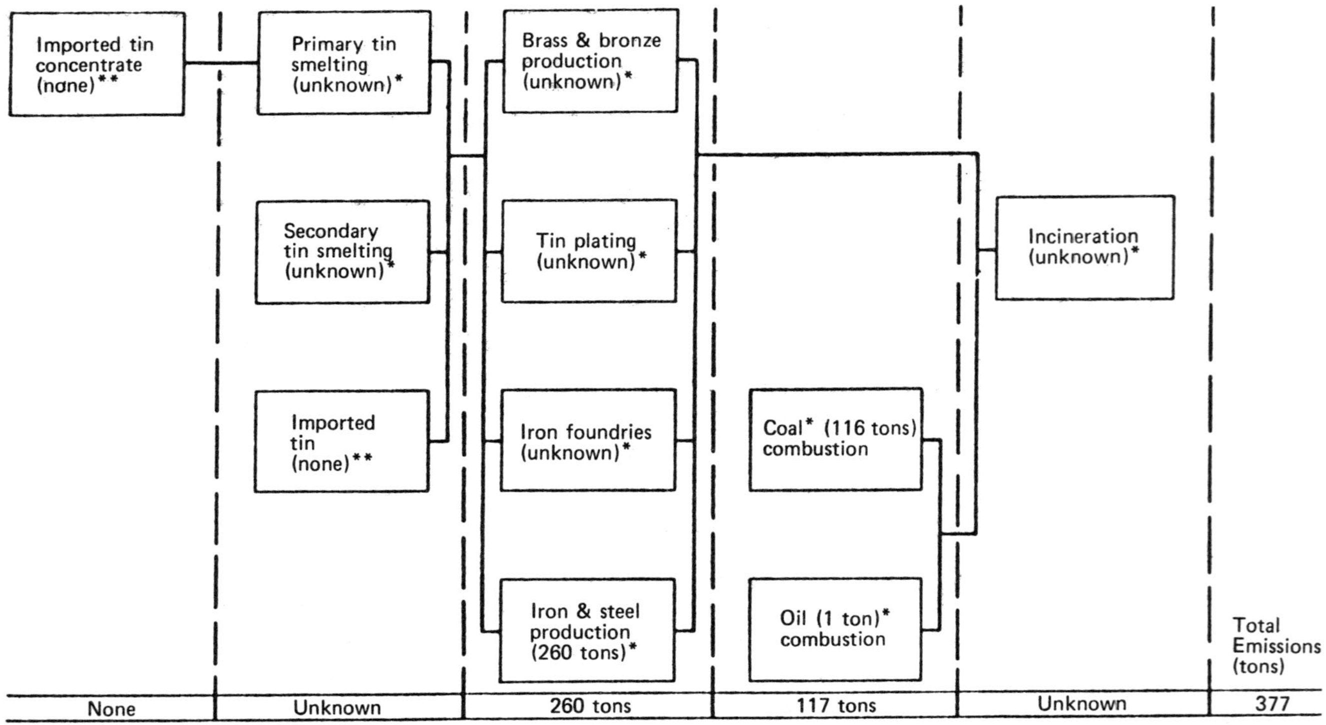

*Mitre

***Mineral Facts and Problems*, BOM

Source: Reference (4)

DETECTION

Tin may be monitored in air by emission spectroscopy (5). Organotin compounds are highly toxic, so a method of detection in the air was developed. The measured air is passed through a mixture of 30 to 35% hydrogen peroxide and concentrated sulfuric acid. This is followed by colorimetric determination of the stannic ion in the form of a yellow-orange compound by the standard series method or with a photoelectrocolorimeter. The sensitivity is 0.1 microgram per sample. Interference of ferric ion or antimony (3+) is eliminated by the addition of ascorbic acid (6).

Tin may be measured in water by: flame absorption spectroscopy, flameless absorption spectroscopy, neutron activation and polarography.

ENVIRONMENTAL STANDARDS

The TLV's set forth by the American Conference of Industrial Hygienists (7) are as follows:

	Milligrams per Cubic Meter
Tin (inorganic compounds except SnH_4 and SnO_2) as Sn	2.0
Tin (organic compounds) as Sn–cutaneous exposure	0.1
Tin oxide	10.0

The maximum allowable concentration of tin in effluents being discharged to a storm sewer or stream is 2.0 ppm according to L.E. Lancy (8). The "agreed" limit on tin content in foods is 300 mg/kg in the United States and 286 ppm (mg/kg) in the United Kingdom (1).

HANDLING PROCEDURES

The only tin compound in the "List of Explosives and Other Dangerous Articles" is tin tetrachloride, anhydrous. Tin tetrachloride is classed as a corrosive liquid and requires a white label. The maximum quantity which can be shipped in one outside container by rail express is 1 liter (1). Other tin compounds except for tin oxide are considered toxic and must be so labeled.

REMOVAL FROM AIR

No problem with airborne tin is known and no techniques have been developed to deal with such emissions.

REMOVAL FROM WATER

A process developed by R.C. Williamson (9) relates to the removal of tin and fluoride from aqueous solutions of the same. Certain industrial operations generate aqueous waste that contains both tin and fluoride in solution. One of these is electrolytic tinplating from halogen tin lines. In such operations, wash solutions and other aqueous wastes are produced from rinsing sprays, the scrubbing of effluent fumes, the liquid effluent from tin recovery operations, and from other sources.

Both tin and the fluoride must be removed from or substantially reduced in such waste solutions, for two reasons. In the first place the loss of large quantities of recoverable tin is economically unacceptable and in the second place the fluoride is a pollutant that cannot be pumped into streams and rivers in large quantity.

In the past, it has been conventional practice to remove and recover tin from aqueous solutions of this type by treating the solutions with sodium carbonate so as to precipitate the tetravalent tin solute as stannic hydroxide. However, such treatment was ineffective to precipitate or otherwise remove the fluoride from the solution.

A separate treatment had to be provided for fluoride removal. Even if a single reagent could have been found that would precipitate both the tin and the fluoride, the recoverable stannic hydroxide sludge would be greatly diluted by precipitated fluoride and would be of correspondingly decreased market value. Accordingly, it is an object of the process to provide methods for the removal both of tin and of fluoride from aqueous solutions of the same, while at the same time recovering a tin concentrate of desirably high tin content.

The process is based on the discovery that both tin and fluoride can be removed from aqueous solution by precipitation with calcium hydroxide and that, moreover, the precipitation can be carried out stepwise in such a way that in at least one early precipitation stage a greater percentage of the original tin solute can be precipitated at fairly low pH compared to that of the original fluoride solute, while during a later precipitation stage at fairly high pH, a greater proportion of the original fluoride solute can be precipitated than of the original tin solute.

In other words, the process can be used to produce first a tin-rich and second a fluoride-rich precipitate. The tin-rich precipitate is valuable for recovery of its tin values. The fluoride precipitate assures that the liquid can be discharged without giving rise to objectionably high levels of fluoride contamination. The process is best adapted for the treatment of large volumes of water where the tin and fluoride are in relatively low concentration. For small volumes of highly concentrated tin solute, it is preferable to rely on tin recovery by precipitation with sodium carbonate if fluoride is present only at such a low level as not to give rise to undesirable fluoride contamination. Wastewater from rinsing detinned steel may contain small amounts of caustic, sodium nitrate, and tin.

SOLID WASTE DISPOSAL

Detinning operations generate tin-containing muds as a waste material. Tin oxides can be landfilled.

ECONOMIC IMPACT OF CONTROLS

No information is available on the economics of pollution control methods for reducing tin contents of air or water.

REFERENCES

(1) Versar, Inc., *Preliminary Investigation of Effects on the Environment of Boron, Indium, Nickel, Selenium, Tin, Vanadium and Their Compounds, Vol V–Tin,* Washington, D.C. Environmental Protection Agency (August 1975).

(2) Schroeder, H.A., *The Poisons Around Us,* Bloomington, Indiana University Press (1974).

(3) Luckey, T.D., Venugopal, B. and Hutcheson, D., *Heavy Metal Toxicity and Hormonology,* New York, Academic Press (1975).

(4) Goldberg, A.J., *A Survey of Emissions and Controls for Hazardous and Other Pollutants,* Washington, D.C., Environmental Protection Agency (February 1973).

(5) Lawrence Berkeley Laboratory, *Instrumentation for Environmental Monitoring Air,* Publ. LBL-1, Vol 1, Berkeley, University of California (May 1, 1972).

(6) Selnorklin, P.I., "Determination of Tin Tetrabutyl and Tin Tetraethyl in Factory Air," *Hyg. Sanit.* 31, No. 8, 270-72 (August 1966).

(7) American Conference of Govermental Industrial Hygienists, *Threshold Limit Values for Chemical Substances in Workroom Air,* Cincinnati, Ohio (1974).

(8) Lancy, L.E. in *Industrial Pollution Control Handbook,* H.F. Lund, Ed., New York, McGraw-Hill Book Co. (1971).

(9) Williamson, R.C.; U.S. Patent 3,284,350; November 8, 1966; assigned to National Steel Corp.

VANADIUM

Vanadium, a relatively abundant element in the earth's crust, is recovered principally as a coproduct of other metals. As an alloying element, vanadium is used in the manufacture of specialty steels and superalloys; as the metal, it has important potential nuclear applications; in the form of compounds, vanadium is used as a catalyst for important industrial processes (1).

TOXICITY

Vanadium is toxic to humans and animals; through inhalation of relatively low concentrations (less than 1,000 $\mu g/m^3$), it has been found to result in inhibition of the synthesis of cholesterol and other liquids, cysteine, and other amino acids and hemoglobin. Low concentrations also act as strong catalysts on serotinin and adrenaline (2).

Chronic exposure to environmental air concentrations of vanadium has been statistically associated with incidences of cardiovascular diseases and certain cancers. Human exposure to high concentrations of vanadium (greater than 1,000 micrograms per cubic meter) results in a variety of clinically observable adverse effects whose severity increases with increasing concentrations.

These effects include irritation of the gastrointestinal and respiratory tracts, anorexia, coughing (from slight to paroxysmal), hemoptysis, destruction of epithelium in the lungs and kidneys, pneumonia, bronchitis and bronochopneumonia, tuberculosis, and effects on the nervous system ranging from melancholia to hysteria. The medical and biologic effects of vanadium as an environmental pollutant have been reviewed by the National Academy of Sciences (35).

Vanadium has been recognized as a health hazard since the turn of the century as a result of occupational exposure (3). Since then, most attention to occupational vanadium exposure has focused primarily on respiratory effects. Inhaled vanadium oxide is associated with a wide variety of respiratory problems such as bronchitis, rhonchi, dyspnea, bronchospasm, rhinorrhea, both dry and productive cough, wheezing, rales, infected pharynx, soreness behind the sternum, and mucous membrane irritation. Inhaled vanadium oxide is also linked with nausea, vomiting, conjunctivitis, dermatitis, fatigue, depression, tremor of the hands, and green tongue. Sensitization, an increased severity of symptoms with repeated exposures of lesser time and intensity, is reported to occur. Ingested vanadium results in gastrointestinal disorders and green tongue.

Several groups of people may be especially susceptible to the toxic effects of vanadium; these include persons with: respiratory diseases; a variety of chronic diseases (tuberculosis, arthritis, cancer) which partially result in depressed sulfur metabolism; Wilson's disease; and an $alpha_1$ antitrypsin deficiency (3).

Little data are available concerning the relationship of vanadium air pollution to diseases in the general population. Two studies, however, have statistically associated vanadium concentrations in the atmosphere with several categories of common respiratory diseases, heart diseases, and cancers (4)(5).

Other community studies indicate that vanadium may play a role in causing respiratory disease. A synergistic effect on the respiratory system has been noted to occur between vanadium and sulfur dioxide. Research has also been conducted on the subclinical actions of vanadium in interference with metabolic systems on a cellular and molecular level in both human beings and test animals. Of special significance are studies on vanadium-induced obstruction of sulfur metabolism.

The earliest description (6) of vanadium poisoning in industry referred to chronic systemic intoxication, primarily in the lungs, kidneys and intestinal tract. Death was reported to occasionally result, often from pulmonary tuberculosis, to which vanadium workers were believed to be especially susceptible. Research conducted in the 1950s and 1960s focused primarily on the local respiratory tract effects of vanadium exposure. Data were drawn from two types of occupational exposure: acute and chronic.

Workers annually cleaning vanadium-containing residue from the insides of oil-burning boilers were found to contract a variety of respiratory problems within a few hours or days of the exposure (3). The problems lasted from several days to a few weeks. Sensitization to vanadium was recorded to occur after the initial exposure causing a more severe reaction with repeated exposures of lesser time and intensity.

The results of chronic exposure to vanadium pentoxide have been studied in workers involved in the mining, processing, and industrial use of vanadium (3). Comparison of these exposed workers to control groups showed disease manifestations similar to those caused by acute exposure. One experimental study of inhalation toxicity was performed in 1967. Vanadium may also play a role in the etiology of hypertension (3).

Injection of vanadium in humans (20 mg V_2O_5 as sodium tetravanadate) while used as a possible cure for syphilis resulted in the disappearance of pulse, cessation of respiration, lacrimation, salivation, diarrhea, and an increase of up to 3°F in temperature. An intravenous dose of 30 mg V_2O_5 is thought to be fatal to humans. Ingestion of 4.5 to 20 mg for several weeks has produced black, loosened stools, increased intestinal activity, cramps, and green tongue (3). The toxicity of vanadium and its compounds has been reviewed by Luckey et al (7).

EXTENT OF EXPOSURE

The production and use of vanadium in the United States has been traced and charted for the year 1968 by W.E. Davis and Associates (8).

Many occupations involve exposure to aerosols and dusts of vanadium compounds (9). Mining and milling of vanadium-containing ores such as patronite involve health risks from handling of the inactive ore before roasting, the active ore after the roast, and in the final processing. Atmospheres in a Peruvian operation contained 0.18 to 58.82 mg vanadium per cubic meter of air in the active processing, the highest concentrations of 12.77 mg/m^3 occurring in the final stages of processing (9). Dust concentrations in the production of vanadium pentoxide and vanadates at one time might have amounted to 6.5 mg vanadium pentoxide per cubic meter.

The production of ferrovanadium and other vanadium alloys, vanadium catalysts, and pure vanadium compounds such as vanadium trioxide, vanadium chloride, and vanadium carbide, also involve high vanadium atmospheres (9).

Occupational exposure to vanadium is not limited to production of vanadium compounds. Since most crude oils contain up to 0.072% vanadium by weight, occupations involving exposure to ash and fumes derived from these fuels may involve hazardous vanadium levels. The ash of these oils is about 65% vanadium, with some ship's boiler ash containing 82% vanadium. Men who clean such oil-fired burners are likely to be exposed to rather high concentrations of vanadium dust containing all valences of vanadium compounds (9).

The major sources of vanadium emissions are the metallurgical processes producing vanadium metal and concentrates; the alloy industry; the chemical industry; power plants and utilities using vanadium-rich residual oils; and, to a lesser extent, the coal and oil refining industries. Vanadium production is concentrated in the states of Colorado, Utah, Idaho, and New Mexico, while the highest concentration of industries producing vanadium chemicals is found in New Jersey and New York (2). Figure 34 and Table 50 give details of airborne vanadium emissions estimates by EPA (10).

Vanadium in the atmosphere is usually found as vanadium pentoxide, but is believed to also be found as other oxides and in chemical association with one or more of the other metals contained in the fossil fuel burned (3). Vanadium is widely dispersed and predominantly of respirable size (59% of particles $<1\ \mu$). It has been hypothesized that combustion creates vanadium compounds in a vapor state, which cooling condenses to a fine particulate fume. It has been further suggested that the larger vanadium pentoxide particles form the core of large sulfuric acid aerosols which settle close to the industrial sources. Vanadium oxide may also catalyze the oxidation of sulfur dioxide to sulfuric acid, in the presence of water vapor:

$$SO_2\ (g) \xrightarrow{H_2O} H_2SO_3\ (l) \xrightarrow[V_2O_5]{O_2} H_2SO_4\ (l)$$

Atmospheric vanadium concentrations tend to be higher during the heating season than in the summer, which supports the contention that fuel oil and other combustion sources are the major sources of emissions.

FIGURE 34: MATERIAL FLOW THROUGH THE ECONOMY SHOWING PRIMARY EMISSION SOURCES–VANADIUM

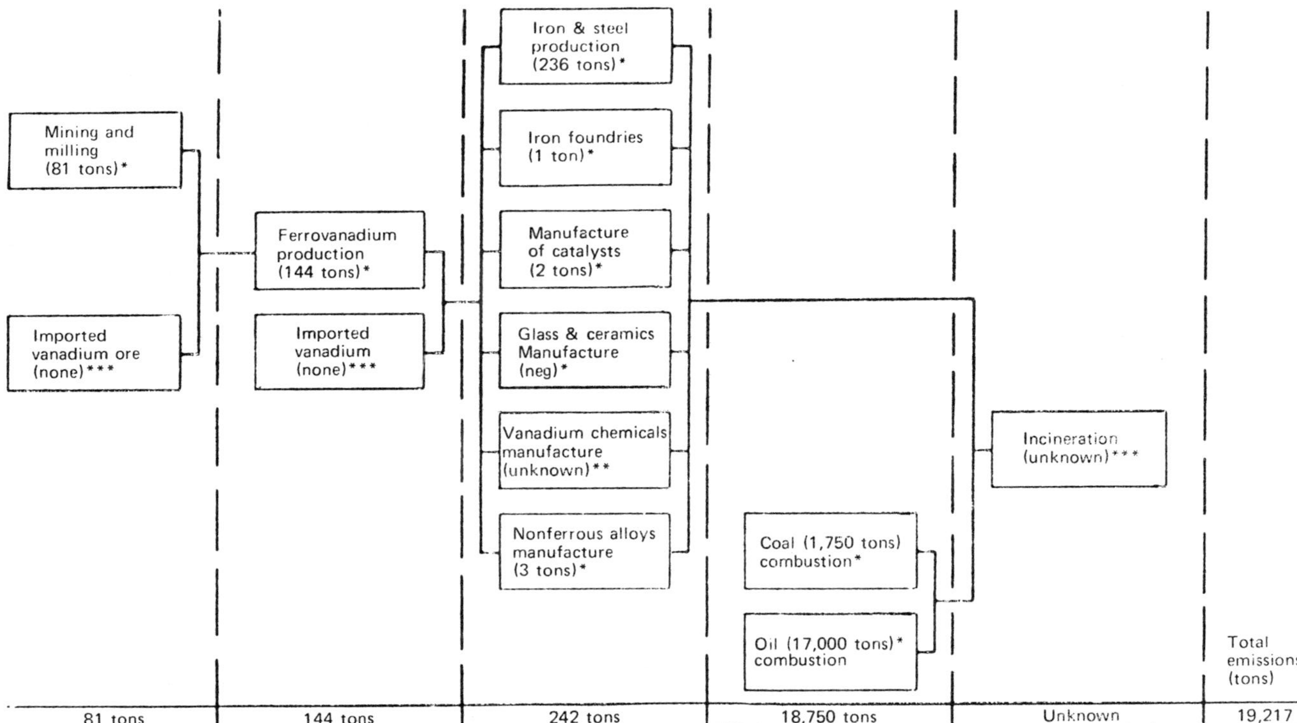

**National Inventory of Sources and Emissions–Vanadium,* W.E. Davis and Associates (June 1971).
***Preliminary Air Pollution Survey of Vanadium and its Compounds,* Litton Systems, Inc. (October 1969).
***MITRE

Source: Reference (10)

TABLE 50: VANADIUM EMISSIONS

Source	Amount in Tons	% This Pollutant
Vanadium refining	81	0.45
Iron and steel:		
Blast furnace	63	0.35
Open hearth furnace	166	0.93
Basic oxygen furnace	7	0.04
Gray iron foundry:		
Cupola	1	0.01
Ferro alloys:		
Electric furnace	115	0.64
Material handling	29	0.16
Nonferrous alloys:		
Furnaces	3	0.02
Vanadium chemical preparation	4	0.02
Ceramic coating preparation	Neg	Neg
Power plant boilers:		
Pulverized coal	1,013	5.67
Stoker coal	116	0.65
Cyclone coal	35	0.20
All oil	4,930	27.58
Industrial boilers:		
Pulverized coal	45	0.25
Stoker coal	158	0.88
Cyclone coal	23	0.13
All oil	2,740	15.33
Residential/commercial boilers:		
Coal	14	0.08
Oil	8,330	46.61
Total	17,873	100.00

Source: Reference (10)

Total production of vanadium pentoxide in the United States has risen from 4,000 short tons in 1950 to 12,000 tons by 1970. Production has been projected to increase to 16,000 short tons by 1980. Approximately 80% of the vanadium consumed in the United States is used in the production of high strength steel alloys (4,500 short tons consumed in 1972). Another 10% of the annual consumption is used in nonferrous alloys.

About 3% of the annual vanadium consumption is by the chemical industry, primarily as catalysts. Catalytic use of vanadium in oil refineries may represent an important emission source. Vanadium may also be used in the future as an active ingredient in automobile exhaust catalytic systems (3).

Little data are available quantifying emissions from these sources. Table 51 shows vanadium particulate emission factors from a variety of sources, such as (11). The projected increase in industrial vanadium consumption may result in higher emissions from these sources.

The Quality Assurance and Environmental Monitoring Laboratory of the National Environmental Research Center, state, and local organizations sample air in a systematic manner to obtain information about air quality in the United States.

TABLE 51: VANADIUM PARTICULATE EMISSION FACTORS (UNCONTROLLED EMISSIONS)

Source	Vanadium Emission Factor
Steel production	
Blast furnace	0.03 lb V/ton of pig iron produced
Open hearth furnace	
No oxygen lance	0.006 lb V/ton of steel produced
Oxygen lance	0.01 lb V/ton of steel produced
Basic oxygen furnace	0.009 lb V/ton of steel produced
Electric furnace	Not determined
Cast iron production	0.2 lb V/10^3 ton of charge
Catalysts	20 lb V/ton of vanadium processed
Miscellaneous (steel alloying, magnetic alloys, paint oils, color film)	10 lb V/ton of vanadium processed
Coal*	2×10^{-4} to 1.7×10^{-3} x (% ash) lb V/ton coal
Cement plants*	
Kilns:	2.28×10^{-3} to 2.45×10^{-2} lb V/ton cement
Driers, grinders, etc.:	3.2×10^{-4} to 9.6×10^{-3} lb V/ton cement
Municipal incinerators*	1.4×10^{-4} to 3×10^{-3} lb V/ton refuse burned
Zn-Cu smelters*	3×10^{-6} to 1.2×10^{-3} lb V/ton ore produced
Brass-bronze smelters*	3×10^{-6} to 7.3×10^{-3} lb V/ton metal charged

*The vanadium particulate emission factor (lb V/ton material) was obtained by multiplying the range of concentrations of vanadium in a total particulate sample (parts per million by weight) times the range of total particulate emission factors (pound particulate/ton material) from various sources:

$$\frac{\text{lb V}}{\text{ton material}} = \frac{\text{lb V}}{10^6 \text{ lb particulate}} \times \frac{\text{lb particulate}}{\text{ton material}}$$

Source: Reference (11)

The last comprehensive survey was carried out in 1968-1969. Vanadium present in suspended particulate matter was one of the twelve metals whose concentration was checked. The analysis emphasized urban and rural differences. Spot checks of the air quality have also been conducted in such extreme regions as in highly industrialized areas and very remote regions such as the Artic or Antartic. Also, many state and local government air pollution control organizations maintain some surveillance of air quality.

Analysis of the air over relatively unpopulated areas of the earth, such as the Pacific Ocean area between San Diego and Honolulu, show some vanadium, but only about one-thousandth of the amount found over populated areas.

The vanadium found over a remote area of northwestern Canada is almost ten times that cited above. This suggests that in addition to vanadium particulates circulated through the air by the wind, some is injected into the atmosphere by wind erosion of rocks and soil (9).

Monitoring of rivers and other waters for vanadium has not been reported.

DETECTION

Methods of quantitative analysis of vanadium in the environmental air include colorimetry, atomic absorption spectroscopy, emission spectrography, and recently polarography. The trend is toward more use of spectrographic and spectrophotometric methods, some of which are more sensitive than the other methods and easily automated. Sensitivities on the order of 0.001 $\mu g/m^3$ are reported (2).

At low concentrations of vanadium in air, high volume samplers are used which operate during the 24 hour period sampling 50 cubic feet of air per minute, or 2,200 cubic meters of air. Preweighed glass-fiber filters (about 10 inches in diameter) are used. The filters should be equilibrated at a standard temperature (75°F or less) with relative humidity of 50% or less, and then weighed to determine the concentration of particulate matter. Afterwards, an aliquot of the sample is ashed (at 100°C) and then extracted with nitric and hydrochloric acids. For nonurban samples, extracts are made that are up to five times more concentrated than these for urban samples (12).

Kuz'micheva (13) described the following sampling method for the colorimetric determination of aerosols of vanadium and its compounds in metallurgical plants. Air samples were passed through filters of polyvinyl chloride fabric and then placed in porcelain dishes, treated with 2 ml of a 50% HNO_3 solution, and evaporated to dryness. Ashing was done in a muffle furnace at 500°C, and the residue was treated with 2 ml of a 10% NaOH solution which dissolved the vanadium, leaving iron in the residue.

Membrane ultrafilters (having a pore width of 0.6 to 0.9 μ) were used in a study by von Jerman (14), who used a polarographic method to determine vanadium concentrations in the air of alloy and chemical plants.

The following techniques are applicable to the analysis of vanadium in airborne materials, according to Versar, Inc. (9).

A section of an air filter is rolled up into a cylinder and placed in a hollow graphited electrode, then enclosed in an atmosphere of oxygen. A condensed spark discharges to burn the filter sample and excite the spectra. An internal standard compensates for the small pieces of paper and ash which may be ejected during the process. The spectra are interpreted by using a photographic recording and the line intensities are measured with a microphotometer to provide greater precision. The limits of detection varied from 0.1 to 1 microgram; the results agreed favorably with the more conventional technique of ashing the filter, mixing with a spectrographic buffer, and exciting in a DC arc.

Vanadium in compounds can be measured by irradiating the samples with neutrons and counting the gamma rays emitted by the neutron capture product, 3.8-minute vanadium-52 (with lithium-drifted germanium gamma ray detectors), without prior chemical separation. In some cases, particularly in marine environments, gamma rays from sodium and chlorine may obscure those of vanadium-52, making it necessary to add a chemical step before counting sensitivity of the method. Values as low as 7×10^{-7} $\mu g/m^3$ have been observed in Antarctica. Based on the sensitivity of this method of analysis, vanadium can serve as an indicator of wide-scale movement of particulates.

A very simple, inexpensive, and specific method for the determination of vanadium in air is the ring-oven technique. The relative error is said to be within the range from 5 to 10%, which at the microgram level compares well with other more sophisticated methods. The limit of identification is 0.01 μg, and its range 0.01 to 3.0 μg. The wavelength used is 3184 A, and the concentration giving 1% absorption is 1.5 μg of vanadium per milliliter (14).

Two other colorimetric techniques have been recently described. The first is based on the oxidation of vanadium and its compounds to vanadium pentoxide and its further reaction with hydrogen peroxide in acid medium. This method has a sensitivity of 17.8 μg of vanadium pentoxide or 10 μg of vanadium.

The second method is based on the development of a greenish-yellow color when pentavalent vanadium reacts with sodium tungstate in neutral medium. The method is not specific and there is interference from alkalies and mineral acids. Kuz'micheva (13) used a method based on the formation of yellow phosphotungstovanadic acid when vanadium or vanadium compounds react with phosphoric and sodium tungstate.

The determination was made on a 5 ml aliquot placed in colorimetric test tubes. The sensitivity was 5 μg of vanadium pentoxide in 5 ml. No interference was observed with aluminum, calcium, silicon dioxide, or iron. Colored chromium compounds were found to interfere when present in amounts greater than 40 μg. In general, colorimetric methods are being replaced by more sophisticated and sensitive methods.

Vanadium in the range of 500 to 1,000 μg/l can be determined by atomic absorption spectroscopy in an oxyacetylene or nitrous oxide-acetylene flame. For use with oxyacetylene flames, vanadium is extracted as vanadium cupferrate into a mixture of ketone and acid, and the resultant product is aspirated by the flame. For use with the nitrous oxide-acetylene flame, an aqueous solution of vanadium is aspirated directly (15).

A polarographic method has been outlined for the determination of vanadium in air. The method described was designed to be used for determining vanadium in the air of the working environment in the alloy industry, where vanadium is used as input, and in chemical manufacture, where vanadium is used as a catalyst. Dusts in such environments contain iron, aluminum, and magnesium, but these elements are not expected to interfere with the test.

Following mineralization of the sample in 45% nitric acid, the polarographic levels of vanadium were recorded from a conductive solution of borax, ammonia, and Chelaton III. The method is said to be sensitive to 1.5 μg/ml of vanadium pentoxide (14).

The method used by the National Air Sampling Network (16) for the determination of vanadium concentration in aliquots obtained from 24 hour samples after ashing and extraction is emission spectrography. In 1966, improvements in sensitivity made this method accurate enough for the determination of vanadium in many nonurban air samples. The minimum detectable concentration of vanadium by this method is 0.003 $\mu g/m^3$ for urban samples and 0.0005 $\mu g/m^3$ for nonurban samples.

Other analytical methods, used mostly for the determination of vanadium in biomaterial, are paper chromatography (17)(18), neutron radioactivation (19)(20), electrophoresis (21), low energy x-ray mass absorption (22), and autoradiography (23). Vanadium may be monitored in air by the following instrumental techniques (24): colorimetric, emission spectroscopy and neutron activation.

In solution, atomic absorption spectroscopy in oxy-acetylene or nitric oxide-acetylene flames can be used for the analysis of vanadium in the range of 0.5 to 100 mg/l. With oxy-acetylene flames, vanadium is extracted as the cupferrate into a mixture of methyl isobutyl ketone and oleic acid and the organic phase aspirated to the flame. The sensitivity is 0.7 mg/l of vanadium in the organic phase.

An extraction procedure eliminates potential interferences from a large number of both cations and anions. Excess cupferron must be added if Fe^{+3}, Sb^{+3} or Zn^{+4} are present since they are preferentially complexed. When nitrous oxide-acetylene flames are used, an aqueous solution of vanadium salts is aspirated directly and the sensitivity is improved by the use of methyl isobutyl ketone, Al^{+3} and diethylene glycol or diethyl ether (9).

Vanadium may be monitored in water by the following instrumental techniques (25): neutron activation and spark source mass spectrometry. Vanadium may be measured in water by flame atomic absorption down to a detection limit of 2 ppb.

A process developed by D.G. Biechler and D.E. Jordan (26) is one in which the vanadium-containing sample is dissolved in a suitable acid or mixture of acids and the vanadium is then oxidized to the pentavalent state by the use of suitable oxidizing agents. Either prior to or after the oxidation of the vanadium to the pentavalent state, phosphoric acid and a suitable tungstate salt are added to the solution to form the phosphotungstate complex of vanadium.

The complex thus formed is extracted from the acid solution using aliphatic alcohols containing from 6 to 12 carbons. The vanadium in the alcoholic extract is then determined colorimetrically by use of a photoelectric colorimeter or spectrophotometer.

Several analytical methods have been utilized for determining trace or microgram amounts of vanadium in the presence of other metallic ions. Many of these procedures employ the step of forming a pentavalent vanadium complex compound which can be at least partially extracted with a suitable solvent, or can be separated by precipitation from the solution in which the complex is formed. The previous separatory and analytical methods each have disadvantages which distract from the universality of their usefulness.

For example, many of the solvents which have been proposed for extraction of pentavalent vanadium compounds do not completely or consistently extract the vanadium from the solution in which it is contained. Moreover, the form in which the vanadium is extracted frequently does not permit the extract to be subjected to simple and rapid analysis, or, in some instances, extraneous undesirable materials are extracted with the vanadium which interfere with subsequent analysis or render the analysis inaccurate. In some of the analytical procedures proposed, the pentavalent vanadium is complexed with other materials, and the vanadium determined by colorimetric methods while the vanadium is still present in the original solution. One of the most successful of these methods, insofar as the determination of trace amounts of vanadium is concerned, is based upon the formation of a phosphotungstovanadate complex in an acid solution by the addition to an acidic solution of pentavalent vanadium, of phosphoric acid and a suitable tungstate salt.

The complex, the precise chemical nature of which is obscure, is a stable yellow color and the vanadium content of the solution from which it is derived is generally determined by spectrophotometric procedures while the complex remains in the original sample solution.

The latter aspect of the analytical procedure does not, of course, lend itself to the separation or removal of the vanadium from the other materials in the sample, and while the analysis is recognized to be characterized by excellent accuracy and repeatability, other or additional procedures must be adopted if the vanadium is to be recovered from the sample. Thus, in summary, in many instances, the analytical technique adopted does not ultimately result in the separation of the vanadium from the material in which it occurs, and conversely, the separatory technique does not place the vanadium in a state suitable for analysis, or else concurrently with the extraction of the vanadium, extracts other materials which interfere with the vanadium analysis. Moreover, the analytical techniques heretofore proposed generally are not both rapid and accurate.

ENVIRONMENTAL STANDARDS

The recommended threshold limit values (TLV) are as follows (27):

Vanadium pentoxide dust: 0.5 mg/m^3 as V_2O_5
Vanadium pentoxide fume: 0.1 mg/m^3 as V_2O_5

German studies (8) recommend maximum work area concentrations of:

V dust (probably as V_2O_5): 0.5 mg/m^3
V smoke (probably as V_2O_5): 0.1 mg/m^3
Ferrovanadium dust: 1 mg/m^3

In the U.S.S.R., the maximum allowable concentrations adopted for occupational exposure to industrial hazards are (28):

V_2O_5 condensation aerosol: 0.1 mg/m^3
V_2O_5 comminution aerosol: 0.5 mg/m^3
Vanadates and vanadium chlorides: 0.5 mg/m^3
Ferrovanadium and vanadium-aluminum alloys: 1.0 mg/m^3
Vanadium carbide: 4.0 mg/m^3

There is some difficulty, especially in foreign articles, in distinguishing between regulation and standard. The foreign regulations presented below must therefore be interpreted with reservations.

One German article (29) states that "...the maximum allowable working place concentration of vanadium in air is uniform for all industrial nations." Maximum allowable emission concentrations, however, differ. In the U.S.S.R. the maximum is 0.002 mg/m^3; in the Federal Republic of Germany a maximum allowable emission concentration of 0.001 mg vanadium/m^3 is under discussion.

Only one article, a 1960 Russian piece, expressed concern with the vanadium content of water. The writer suggested a vanadium concentration of 0.1 mg/l as a maximum permissible limit for water basins. The allowable concentration of vanadium in irrigation water is 10.0 mg/l on a continuous use basis and is also 10.0 mg/l on a short-term use basis in fine textured soil.

HANDLING PROCEDURES

Vanadium pentoxide and ammonium-metavanadate are packaged in standard 50 kilogram fiber drums or shipped in bulk. They should be stored in a cool, dry location. A nose or mouth respirator should be used when handling large quantities (9).

Vanadium oxytrichloride is shipped in tank cars, tank trucks and cylinders. It must be stored in a cool, dry location. Skin contact must be avoided and the fumes should be regarded as an acid gas and respiratory system irritant. Vanadium trioxychloride is classified as a corrosive liquid by the Interstate Commerce Commission. This is the only commercially significant vanadium product which requires special handling. In transport, the chemical is identified with a white label, is not accepted on passenger planes and no more than one liter is permitted on cargo planes.

REMOVAL FROM AIR

The air pollution aspects of vanadium and its compounds have been reviewed in some detail by Athanassiadis (30).

The portion of total particulates in the less than 10 μ range may be reduced from 60 to 40% in the case of high vanadium content oils. This is significant, since particulate size distribution is an important parameter of abatement efficiency. Centrifugal collectors are preferred over electrostatic precipitators because they reduce the difficulties associated with the acid character of the ash, especially where bag filters and scrubbers are used. However, for any given particle size distribution, the efficiency of various centrifuges may vary as much as from 50 to 65% and from 70 to 85%, around the particle size range of 5 and 10 μ, respectively (31).

The use of efficient fly ash control equipment in modern coal fired power plants may considerably reduce the emission of particulates containing vanadium. The control equipment most used are cyclones and electrostatic precipitators. When additives are used, which result in the formation of larger ash particles, cyclones are more efficient and economical to use than electrostatic precipitators (32).

In general vanadium and its compounds can be controlled by standard particle removal devices. Cyclones and electrostatic precipitators are commonly used to control fly ash emissions. Cyclones become much more effective in controlling particulate emissions when larger ash particles are formed due to the use of additives such as magnesium oxide. The portion of total particulates in the size range less than 10 microns may be reduced from 60 to 40% in the case of high vanadium content oils. Cyclones also minimize the difficulties associated with the acid character of the ash. Studies have been made on coal fired plants using a cyclone collector and a cyclone and electrostatic precipitator in series. In one study, vanadium was analyzed by emissions spectroscopy with an estimated accuracy of ± 50%. An electrostatic precipitator operating on a coal fired steam plant was extremely efficient (99.1%) in minimizing vanadium emissions (33).

REMOVAL FROM WATER

There seems to be little or no information on the removal of vanadium ion from wastewaters; most of the attention seems to have been given to airborne vanadium.

SOLID WASTE DISPOSAL

Vanadium pentoxide may be disposed of as described by P.W. Powers (34).

ECONOMIC IMPACT OF CONTROLS

Little information is available on the economic losses due to vanadium air pollution or on the costs of abatement. One report indicated that measures taken to reduce the loss of vanadium to the atmosphere from an oil-fired steam generator resulted in recovery of commercially valuable vanadium pentoxide, thereby producing a profit from air pollution abatement (2).

REFERENCES

(1) Griffith, R.F., "Vanadium" in *Mineral Facts and Problems,* Washington, D.C., U.S. Bureau of Mines (1970).

(2) Booz-Allen Applied Research, Inc., *A Study of Hazardous Waste Materials, Hazardous Effects and Disposal Methods,* Report PB 221 466, Springfield, Virginia, National Technical Information Service (July 1973).

(3) Environmental Health Resource Center, *Air-Borne Vanadium Health Effects and Recommended Standards,* Report PB 241 608, Springfield, Virginia, National Technical Information Service (October 1974).

(4) Stocks, P., "On the Relations Between Atmospheric Pollution in Urban and Rural Localities and Mortality from Cancer, Bronchitis and Pneumonia, with Particular Reference to 3,4-Benzopyrene, Beryllium, Molybdenum, Vanadium and Arsenic," *British Journal Cancer* 14, 397 (1960).

(5) U.S. Department of Health, Education and Welfare, "Analyses of Suspended Particulates, 1953-57" in *Air Pollution Measurements of the National Air Sampling Network,* Cincinnati, Ohio (1958).

(6) Dutton, W.F., "Vanadiumism," *J. Amer. Med. Assoc.* 56, 1648 (1911).

(7) Luckey, T.D., Venugopal, B and Hutcheson, D., *Heavy Metal Toxicity, Safety and Hormonology,* New York, Academic Press (1975).

(8) Davis, W.E. and Associates, *National Inventory of Sources and Emissions: Vanadium—1968,* Report PB 221 655, Springfield, Virginia, National Technical Information Service (June 1971).

(9) Versar, Inc., *Preliminary Investigation of Effects on the Environment of Boron, Indium, Nickel, Selenium, Tin, Vanadium and Their Compounds, Vol. VI—Vanadium,* Report EPA-560/2-75-005F, Washington, D.C. Environmental Protection Agency (Aug. 1975).

(10) Goldberg, A.J., *A Survey of Emissions and Controls for Hazardous and Other Pollutants,* Washington, D.C., Environmental Protection Agency (February 1973).

(11) Anderson, D., *Emission Factors for Trace Substances,* Report PB 230 894, Springfield, Virginia, National Technical Information Service (December 1973).

(12) Stern, A.C. (Ed.), *Air Pollution,* Vol. II, New York, Academic Press (1968).

(13) Kuz'micheva, M.N., "The Determination of Vanadium in Air," text in Russian, *Gigiena i Sanit.* 31, 229 (1966).

(14) von Jerman, L., et al, "Polarographische Bestimmung von Vanadin in der Luft von Arbeitsraumen," *Zeit. Hyg.* (Berlin), 14(1):12 (1968).

(15) Sachdev, S.L., et al, "Determination of Vanadium by Atomic Absorption Spectrophotometry," *Anal. Chim. Acta* 37, 12 (1967).

(16) *Air Quality Data, 1966,* National Air Sampling Network, U.S. Department of Health, Education, and Welfare, Public Health Service, Cincinnati, Ohio (1968).

(17) Miketukova, V., "Detection of Metals on Paper Chromatograms with Rhodamine B," *J. Chromatog.* 24, 302 (1966).

(18) Yamakawa, K., et al, "Organometallic Compounds. II. Gas Chromatography of Metal Acetylacetonates," *Chem. Pharm. Bull.* (Tokyo) 11, 1405 (1963).

(19) Comar, D., et al, "Concentration of Vanadium in the Rat and Its Influence on Cholesterol Synthesis. Studies by the Technic of Neutron Radioactivation and the Method of Isotopic Equilibrium," text in French, *Bull. Soc. Chim. Biol.* 49, 1357 (1967).

(20) Livingston, H.D., et al, "Estimation of Vanadium in Biological Material by Neutron Activation Analysis," *Anal. Chem.* 37, 1285 (1965).

(21) Kaser, M.M., et al, "The Separation and Identification of Vanilmandelic Acid and Related Compounds by Electrophoresis on Cellulose Acetate," *J. Chromatog.* 29, 378 (1967).

(22) Carter, R.W., et al, "Low Energy X-Ray Mass Absorption Coefficients from 1.49 to 15.77 kev for Scandium, Titanium, Vanadium, Iron, Cobalt, Nickel and Zinc," *Health Phys.* 13, 593 (1967).

(23) Soremark, R., et al, "Autoradiographic Localization of V-48-Labeled Vanadium Pentoxide (V_2O_5) in Developing Teeth and Bones in Rats," *Acta Odontol. Scand.* 20, 225 (1962).

(24) Lawrence Berkeley Laboratory, *Instrumentation for Environmental Monitoring-Air,* Publ. LBL-1, Vol. 1, Berkeley, University of California (May 1, 1972).

(25) Lawrence Berkeley Laboratory, *Instrumentation for Environmental Monitoring-Water,* Publ. LBL-1, Vol. 2, Berkeley, University of California (February 1, 1973).

(26) Biechler, D.G. and Jordan, D.E., U.S. Patent 3,345,126; October 3, 1967; assigned to Continental Oil Co.

(27) American Conference of Governmental Industrial Hygienists, *Threshold Limit Values for Chemical Substances in Workroom Air,* Cincinnati, Ohio (1974).

(28) Roskchin, I.V., "Toxicology of Vanadium Compounds Used in Modern Industry," *Gig. Sanit.* (Moscow) 32, 4-6 (1967).

(29) Michaels, S., "The Biological Effect of Vanadium with Respect to a Limitation of Its Emission," *Staub Reinhaltung Luft* 33, No. 7, 276-79 (July 1973).

(30) Athanassiadis, Y.C., *Air Pollution Aspects of Vanadium and Its Compounds,* Report PB 188 093, Springfield, Virginia, National Technical Information Service (Sept. 1969).

(31) Burdock, J.L., "Fly Ash Collection from Oil-Fired Boilers," Paper presented at the 10th meeting of the New England Air Pollution Control Association, Hartford, Conn. (1966).

(32) Gerstle, R.W., et al, "Air Pollutant Emissions from Coal-Fired Power Plants: Report No. 2, *J. Air Pollution Control Assoc.* 15 (2), 59 (1965).

(33) McFarland, A.R., Ed., *Control Technology for Toxic and Hazardous Air Pollutants,* Chicago, Illinois Institute for Environmental Quality (December 1974).

(34) Powers, P.W., *How to Dispose of Toxic Substances and Industrial Wastes,* Park Ridge, New Jersey, Noyes Data Corp. (1976).

(35) National Academy of Sciences, *Vanadium—Medical and Biologic Effects of Environmental Pollutants,* Washington, D.C. (1974).

ZINC

Major quantities of zinc are consumed in four widely divergent applications. Zinc is used as the metal in the form of castings and rolled zinc, as a major alloying ingredient in brass, as a protective coating for steel, and as a chemical compound in rubber and paints.

TOXICITY

The most common effects of zinc poisoning in humans are nonfatal metal-fume fever, caused by inhalation of zinc oxide fumes, and illnesses arising from the ingestion of acidic foods prepared in zinc-galvanized containers. Zinc chloride fumes, though only moderately toxic, have produced fatalities in one instance of highly concentrated inhalation. Zinc stearate has been mentioned as a possible cause of pneumonitis. Zinc salts, particularly zinc chloride produce dermatitis upon contact with the skin (1).

Accidental poisoning of cattle and horses has occurred from inhalation of a combination of lead- and zinc-contaminated air. Zinc oxide concentrations of 400 to 600 $\mu g/m^3$ are toxic to rats, producing damage to lung and liver, with death resulting in approximately 10% of the cases. Although dogs and cats tolerate high concentrations (up to 1,000,000 μg/day) of zinc oxide for long periods, glycosuria and damage to the pancreas may result. Concentrations of 40,000 to 50,000 $\mu g/m^3$ of zinc ammonium sulfate produce no appreciable effects on cats (1). Despite the fact that specific effects attributed to certain compounds of zinc have been noted, the common association of zinc with other metals, and the frequent presence of toxic contaminants (such as cadmium) in zinc materials, raise questions which have yet to be answered concerning the synergistic effects of these metals.

EXTENT OF EXPOSURE

As part of a national inventory of sources and emissions, W.E. Davis and Assoc.

FIGURE 35: MATERIAL FLOW THROUGH THE ECONOMY SHOWING PRIMARY EMISSION SOURCES—ZINC

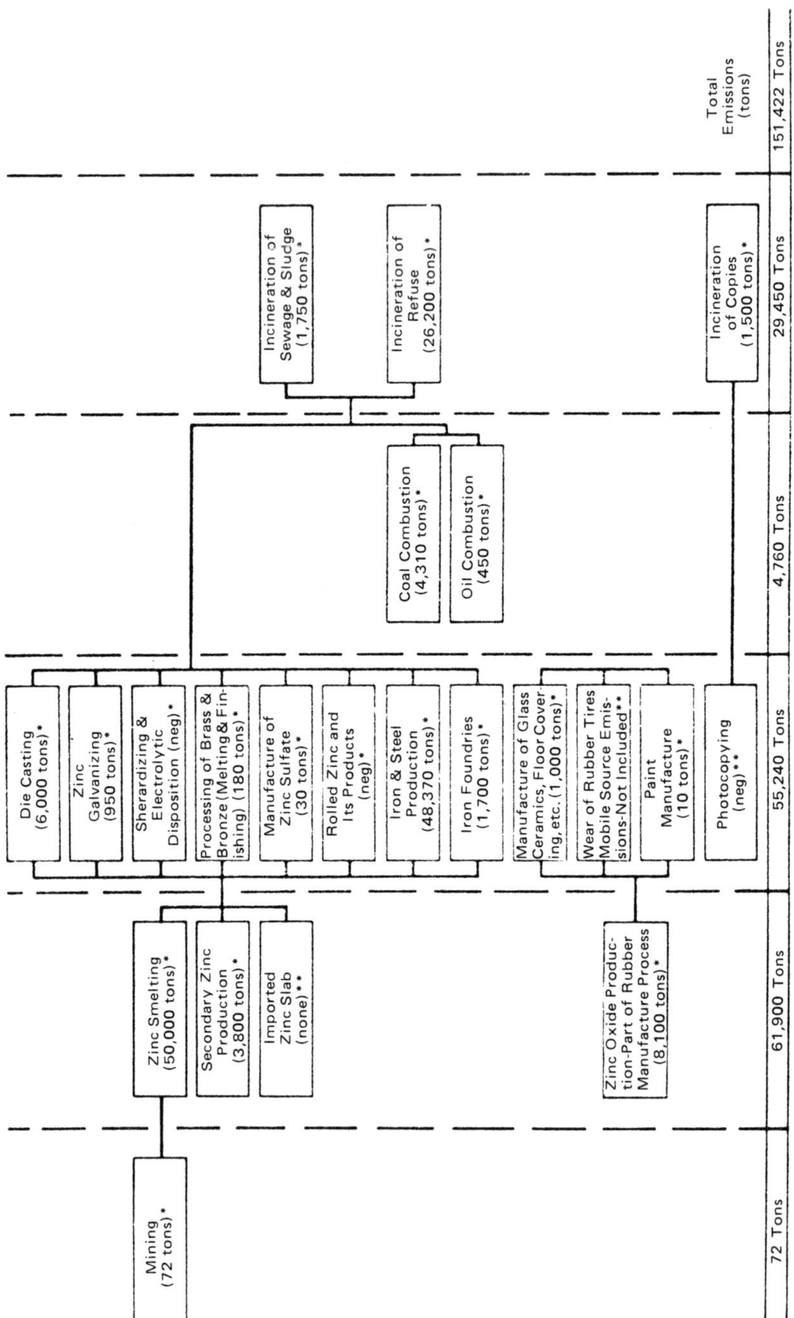

National Inventory of Sources and Emissions—Zinc, W.E. Davis and Associates, May 1972 **MITRE

Source: Reference (3)

of Leawood, Kansas have provided information regarding the nature, magnitude and extent of the atmospheric emissions of zinc in the United States for the year 1969 (2). The primary sources of zinc compounds in the atmosphere are: the zinc-, lead-, and copper-smelting industries; secondary-processing operations which recover zinc from scrap, namely, brass-alloy manufacturing and reclaiming; and galvanizing processing. Average annual production and consumption of zinc in the U.S. have increased steadily during this century, and it is predicted that this trend will continue.

As the emission of zinc into the atmosphere, in most of these operations, represents an economic loss of the zinc material, control procedures are normally employed to prevent emission to the atmosphere. In those industries where zinc is a by-product, control procedures for zinc are not as effective, and greater quantities of zinc therefore escape into the environment. Figure 35 and Table 52 give details of airborne emissions of zinc (3).

TABLE 52: ZINC EMISSIONS

Source	Amount, tons	Pollutant, %
Zinc mining	72	0.05
Primary zinc		
Roasting	31,818	21.13
Sintering	13,637	9.05
Distillation	4,545	3.02
Zinc oxide production	8,100	5.38
Secondary copper		
Wire burning	135	0.09
Sweating furnace	14	0.01
Blast furnace	14	0.01
Smelting, reverberatory		
Furnace	14	0.01
Smelting, rotary furnace	3	neg
Secondary zinc		
Sweating furnaces	2,890	1.92
Distillation furnaces	950	0.63
Iron and steel		
Blast furnace	1,070	0.71
Open hearth furnace	39,000	25.89
Basic oxygen furnace	900	0.60
Electric arc furnace	7,400	4.91
Gray iron foundry		
Cupola	1,700	1.13
Ferro-alloys		
Blast furnace	2,500	1.66
Electric furnace	500	0.33
Material handling	neg	neg
Zinc galvanizing	950	0.63
Zinc chemical preparation	1,130	0.75
Paint, varnish, lacquer		
Manufacture	10	0.01
Incineration	29,450	19.55
Power plant boilers		
Pulverized coal	2,457	1.63

(continued)

TABLE 52: (continued)

Source	Amount, tons	Pollutant, %
Stoker coal	282	0.19
Cyclone coal	85	0.06
All oil	130	0.09
Industrial boilers		
Pulverized coal	109	0.07
Stoker coal	382	0.25
Cyclone coal	55	0.04
All oil	99	0.07
Residential/commercial boilers		
Coal	34	0.02
Oil	221	0.15
Total	150,656	100.04

Source: Reference (3)

Industries discharging wastewater streams which carry significant quantities of zinc, according to J.W. Patterson et al (4) include zinc and brass metal works; zinc and brass plating; silver and stainless steel tableware manufacturing; viscose rayon yarn and fiber production; groundwood pulp production and newsprint paper production. In addition, recirculating cooling water systems employing cathanodic treatment contain zinc which is discharged during blowdown (5). Rock (5) estimates that 10,000 tons of waste zinc per year are discharged from viscose rayon production, and that 5 tons of zinc may be used per day in a large groundwood pulp mill. Similar quantities were estimated for a typical brass company (6).

The primary source of zinc in wastewaters from plating and metal processing industries is the solution adhering to the metal product after removal from pickling or plating baths. The metal is washed free of this solution, referred to as dragout and the contaminants are thus transferred to the rinse water. The pickling process consists of immersing the metal (zinc or brass) in a strong acid bath to remove oxides from the metal surface. Finished metals are brightened by submergence in a bright dip bath containing strong chromate concentrations in addition to acid.

Plating solutions typically contain 5,000 to 34,000 mg/l of zinc. The concentration of zinc in the rinse water will be a function of the bath zinc concentration, drainage time over the bath, and the volume of rinse water used.

Zinc and brass plating solutions generally contain cyanide as zinc and copper-cyanide complexes. Waste concentrations of zinc range from less than 1 to more than 1,000 milligrams per liter in various waste streams reported in the literature. Average values, however, seem to be between 10 and 100 milligrams per liter.

Table 53 shown on the following page summarizes values reported for various zinc-bearing wastewaters.

TABLE 53: CONCENTRATIONS OF ZINC IN PROCESS WASTEWATERS

Process	- - -Zinc Concentration, mg/l- - - Range	Average
Metal processing		
Pickle baths	4.3-41.4	
Bright dip wastes	0.2-37.0	
Brass mill wastes	40-1,463	
Brass mill wastes	8-10	
Pickle bath	0.5-37	
Plating		
General	2.4-13.8	8.2
General	55-120	
General	15-20	15
Zinc	70-150	
Brass	11-55	
Zinc	70-350	
Brass	10-60	
General	7.0-215	46.3
Silver plating		
Silver-bearing wastes	0-25	9
Acid waste	5-220	65
Alkaline	0.5-5.1	2.2
Rayon wastes		
General	250-1,000	
General	20*	

*After process recovery of zinc by ion exchange

Source: Reference (4)

DETECTION

Limited means are available for the determination of concentrations of zinc in the ambient air. These methods of analyses, however, are not considered adequate for air pollution monitoring purposes since they do not effectively discriminate between the zinc and other metals and they lack sensitivity (1).

Dusts and fumes of zinc compounds may be collected by any method suitable for collection of other dusts and fumes; the impinger, electrostatic precipitator, and filters are commonly used. The National Air Sampling Network uses a high volume filtration sampler. The ring-oven technique of analysis has been adapted to the determination of zinc in the range 0.05 to 1 $\mu g/m^3$. The reagent, o-mercaptothenalaniline is used as coloring agent. The method is specific for zinc and may be used in air pollution studies. The limit of detection is 0.04 μg, although previous studies (7) had suggested 0.1 μg as the lower limit.

A spectrophotometric method has been developed for the determination of zinc; concentrations down to 0.1 μg can be measured using 1-(2-thiazolylazo)-2-naphthol (8). An instrument capable of monitoring and recording air concentration of zinc sulfate pigment (a phosphorescing tracer) on a real-time basis has been developed. The instrument has a limit of detection of about 0.25 $\mu g/m^3$ (9). Emission spectroscopy has been used by the National Air Pollution Control Administration for zinc analysis of samples from the National Air Sampling Network (10). The samples are ashed and extracted to eliminate interfering elements. The minimum detectable zinc concentration by emission spectroscopy is 0.24 $\mu g/m^3$ for urban samples and 0.08 $\mu g/m^3$ for nonurban samples. The difference in sensitivities results from the different extraction procedures required for urban samples (11). Thompson et al (11) have reported that the National Air Pollution Control Administration uses atomic absorption to supplement analyses obtained by emission spectroscopy. The method has a minimum detectable limit of 0.0002 $\mu g/m^3$ based on a 2,000 m^3 air sample. Zinc may be monitored in air by the following instrumental techniques (12): emission spectroscopy, atomic absorption and x-ray fluorescence.

Zinc may be monitored in water by the following instrumental techniques (13): atomic fluorescence, electrochemical, spark source mass spectrometry, and x-ray fluorescence. Zinc may be measured down to a detection limit of 2 to 5 ppb by atomic (flame) absorption or by polarography. Neutron activation is effective only to a detection limit of 400 ppb but flameless atomic absorption permits the detection of zinc down to about 0.02 ppb.

A technique developed by C.J. Overbeck and J.J. Hickey (14) is one whereby analysis for small amounts of zinc in waters which also contain high hardness levels may be made by using as a titrant diethylenetriaminopentaacetic acid and a dithizone indicator. A method developed by C.A. Noll (15) involves determining the zinc concentration in an aqueous medium. More specifically, the method is directed for use to aqueous media which contain contaminated and perhaps interfering metal ions, such as the cations of aluminum, iron and copper.

According to the method, the sample of the aqueous medium is acidified, treated with a buffered complexing agent to complex any aluminum and iron ions present, treated with an organo-sulfur compound which will complex with any copper ions present and discriminate against any zinc ions present. An indicator compound which will react with the zinc to produce a color has been added to the aqueous medium and the color intensity of the resulting solution is ascertained. The color intensity is then compared with the color intensity of known quantities of zinc and the indicator compound, and the concentration of the zinc is ascertained accordingly.

ENVIRONMENTAL STANDARDS

The threshold limit values set by the American Conference of Governmental Industrial Hygienists for zinc compounds are as follows: 1 mg/m^3 zinc chloride fume; 5 mg/m^3 zinc oxide fume; and 10 mg/m^3 zinc stearate. The permissible concentration of zinc in domestic water supplies is 5 mg/l. The desirable situation is that zinc be virtually absent. The tolerance limit for zinc in irrigation water is 5.0 mg/l on a continuous use basis and is 10.0 mg/l on a short-term use basis in fine-textured soil. The maximum allowable concentration of zinc in effluent being discharged to a storm sewer or stream is 1.0 ppm.

HANDLING PROCEDURES

The obvious precautions to be taken with respect to zinc is to avoid inhalation of zinc chloride in view of its toxicity. Similarly, inhalation of zinc fume or dusts of zinc compounds is to be avoided. Skin contact with zinc chloride is to be avoided as well.

REMOVAL FROM AIR

Extensive air pollution abatement methods are in general use by the zinc industry. Control devices include precipitators, scrubbers, baghouses, and collectors. The efficiency of the various control methods varies widely. However, in many instances air pollution control devices are not used in the general metals industries. Thus, relatively large quantities of zinc or zinc compounds are still being emitted into the atmosphere by industrial plants processing zinc or other compounds containing zinc (1). The air pollution aspects of zinc and its compounds have been reviewed in some detail by Athanassiadis (16).

The problem of reducing zinc emissions to the atmosphere from the industrial sources is essentially one of controlling small particles. As noted, in nearly all cases zinc effluxes in the form of particulate zinc oxide fume, with particle size ranging from about 0.1 μm to well over 100 μm. To control small particle emissions, the most efficacious apparatus includes electrostatic precipitators, wet scrubbers, and filters (17).

REMOVAL FROM WATER

Treatment processes employed for wastewater zinc removal may involve either destruction or recovery. The destruction process is essentially chemical precipitation, with disposal of the resultant sludge. Recovery processes include ion exchange and evaporative recovery. Barnes (18) has pointed out that recovery of plating wastes frequently proves to be more economical on an overall basis than conventional destructive treatment.

The precipitation process most frequently involves adjustment of pH to achieve alkaline conditions and precipitation of zinc hydroxide. Lime addition has been the most widely accepted method for pH adjustment, in spite of the concurrent precipitation of calcium sulfate in the presence of high sulfate levels in pickling bath wastes. The precipitation of calcium sulfate along with the zinc hydroxide increases the total amount of sludge to be disposed of. Banerjee and Banerjee (19) report the use of hydrogen sulfide at pH 2 for complete removal of zinc from an electrolytic nickel recovery system. However, this approach is not mentioned to any extent in the literature encountered.

In the case of cyanide and/or chromate in the zinc waste stream (both are frequently encountered in zinc and brass plating wastes), the waste must undergo oxidative treatment for cyanide destruction, followed by reductive treatment of hexavalent chromium before precipitation. Zinc is then removed by precipitation along with other metal species present. For a metal-bearing waste containing 55 to 120 mg/l of zinc, Nyquist and Carroll (20) reported treated effluent concentrations ranging up to 5.4 mg/l over one month of monitoring, but with an average value less than 1 mg/l. The waste flow varied from 2,750 to 3,000 gal/hr and required a lime dose of 56 lb/1,000 gal. The theoretical minimum dosage, based upon the stoichiometry of the chemical reaction involved in treating the waste, was 52 lb/1,000 gal. The use of a Dorr Clariflocculator reduced zinc levels to 0.5 to 2.5 mg/l in a 1.5 to 2.0 MGD waste stream. After installation of sand filters, the effluent zinc was reduced to 0.1 to 0.5 mg/l. Process pH was 8.5 after lime addition (21). Sharda and Namwannan (22) have reported that lime treatment reduces 20 mg/l zinc to less than 1 mg/l by precipitation of zinc hydroxide. Rock (5) reported 95% or greater removal of zinc from rayon acid wastes upon addition of sodium hydroxide. This treatment yielded a residual zinc concentration of 3.5 mg/l for the initial 70 mg/l zinc contained in the waste stream.

For a tableware plant with three waste streams, one of 5,000 gal/day with an average zinc concentration of 9 mg/l; one of 46,000 gal/day with zinc concentration of 65 mg/l; and one of 160 gal/day with a zinc concentration of 2.2 mg/l, lime treatment (after cyanide oxidation and hexavalent chromium reduction) followed by sand filters produced a combined effluent zinc content of 0.02 to 0.23 mg/l. This represents a net reduction of zinc exceeding 99%.

Lancy (23) claims that his integrated (and patented) treatment system is the only way to guarantee effluent zinc levels less than 1.0 mg/l in plating and pickling systems, due to difficulty in settling metal hydroxide precipitates. Others have also reported that the gelatinous zinc hydroxide sludges formed in the precipitation process are difficult to settle or filter. Stone (21) reported, however, that a 10% solids sludge was obtained prior to vacuum dewatering. No mention was made of polyelectrolyte use. Rock (5) reported that a viscose rayon manufacturing zinc waste sludge initially capable of settling only to 0.5% solids after 18 hours could be thickened to 5% solids by constantly recycling the solids. This was accomplished without settling, dewatering or the use of polymers.

A process reported by DuPont (24) was developed primarily for small zinc and cadmium cyanide plating operations. A proprietary formulation called Kastone (composed of a catalyst, hydrogen peroxide and stabilizers) oxidizes the cyanide and promotes the formation of metal oxides instead of hydroxides. The former are readily removed by filtration apparatus which most small plating plants already have in operation. No removal levels are given, but the company claims the method offers a great advantage in metal removal.

Unless other process modifications are necessary, or extremely high levels of zinc are present and minimal purification of recovered material is required, zinc recovery is generally not economical. Precipitation can be considered as a recovery process in situations when zinc is essentially the only ion precipitated and the precipitant is regenerated by acid dissolution. This situation has been reported in viscose rayon manufacturing (25). Recovery processes include the following: (a) ion exchange (which may also function as a polishing unit for final effluent); (b) evaporative recovery; or (c) simple process modification such as countercurrent rinsing. Kantawala and Tomlinson (26) reported on laboratory studies which indicated that at flow rates of 2.5 gal/hr/ft^2 ion exchange was not as effective as lime precipitation for zinc removal.

Zinc recovery by ion exchange is practiced in viscose rayon manufacturing (22), where the acid regenerant can be reused in the manufacturing process directly without purification. A residual effluent zinc concentration of 20 mg/l was reported in one case after ion exchange recovery, necessitating further waste treatment. The major source of zinc-bearing wastewaters is from the metal processing and plating operations and specifically from pickling or plating dragout into rinse water, although periodic dumping of the concentrated plating bath solution may occur as a result of contaminant buildup. A major problem in treating dilute wastewater is the large volumes involved and attendant capital and operation costs. Direct recycle or evaporative recycle of pickling and plating bath rinse water is made feasible by countercurrent rinse flow.

Lancy (23) claims rinse water volumes can be reduced to one-third to one-fourth of normal volumes by the use of cascade or countercurrent rinse practices. This reduction in volume yields higher concentrations in the rinse water. Barnes (18) was able to reduce zinc cyanide plating rinse requirements from 310 to 12 gal/hr for two units by employing countercurrent rinse. The remaining rinse waters were of sufficient zinc concentration to allow evaporative concentration and recovery of the plating dragout. Final rinse dragout required no further treatment, due to the low zinc levels present. Lancy (23) points out however, that complete elimination of dragout by evaporative recovery does result in impurity buildup in the plating baths, and subsequent need for periodic dumping of the bath solutions.

SOLID WASTE DISPOSAL

The ultimate disposal of zinc compounds has been discussed by P.W. Powers (27).

ECONOMIC IMPACT OF CONTROLS

Costs for zinc removal systems cannot usually be isolated from the cost of treatment for other constituents contained in the particular waste. Neutralization must be accomplished even in waste streams containing no zinc; cyanide must be oxidized and chromium reduced. Costs directly related to zinc treatment result from the lime requirements, clarifier costs in the absence of other metal hydroxides and sludge dewatering and disposal costs per ton of zinc hydroxide produced. Culotta and Swanton (29) point out that in the past precipitation of metal hydroxides was relatively inexpensive, but that costs have increased considerably in achieving more effective treatment.

Nyquist and Carroll (20) reported waste treatment plant installation and treatment costs for the Conneaut, Ohio General Electric plant. Capital costs (1957 prices) for the 30,000 gallons per day plant were \$276,000, or \$9,200/1,000 gallons per day treatment capacity. These costs are itemized below.

Treatment building	\$121,300
Meter building	3,400
Utilities	11,600
Installed equipment*	107,200
Design costs	28,000
Miscellaneous	4,500
	\$276,000

*Largest single equipment expenditure was for a vacuum filter unit at \$30,000

The operating costs were itemized as shown on the following page.

	Per 1,000 Gallons
Mixing	0.060
Neutralization	0.360
Filtration	0.913
Sludge removal	0.280
Manpower	2.135
Miscellaneous	0.045
	3.793

Thus, total operating costs were \$3.79/1,000 gallons for that waste stream, with almost two-thirds of this cost being due to manpower requirements. Sludge disposal costs, also included in the total, represented \$0.28/1,000 gallons of wastewater treated. Zinc was reduced to an average effluent of less than 1 mg/l from a raw waste level of 55 to 120 mg/l (20). The waste stream also contained copper at 204 to 385 mg/l, reduced concurrently with zinc to an average of 0.5 mg/l. Lancy (23) estimates that in general, chemical costs for zinc waste treatment are \$0.05 to \$0.07/1,000 ft^2 of zinc plated surface, or \$0.70 to \$1.00/1,000 pounds of zinc plated small parts.

Ross (28) estimated operating costs for lime treatment of 35,000 gal/day of a zinc-nickel waste, containing 5.5 mg/l of nickel and 3.5 mg/l of zinc, to be \$0.34/1,000 gal. Lime quantity was estimated at 185 lb/day and sludge disposal amounted to approximately \$0.03/1,000 gal of wastewater treated.

Culotta and Swanton (29) reported on treatment of a zinc and brass plating operation by evaporative recovery. Plant modifications reduced process water from 3,000 to 50 gal/hr and resulted in process chemical savings of \$18,000/yr plus equal savings in chlorine and other treatment chemicals. Capital and operating costs associated with such a closed loop system were reported to be approximately the same as for destructive recovery.

Ross (28) compared the treatment of zinc and nickel wastes by ion exchange versus lime treatment, and found ion exchange to be more than twice as costly as lime treatment for a waste flow of 35,000 gallons per day. Precipitation costs were \$0.34/1,000 gal. Recovery value from the ion exchange process was credited for water reuse only.

REFERENCES

(1) Booz Allen Applied Research, Inc. *A Study of Hazardous Waste Materials, Hazardous Effects and Disposal Methods,* Report PB-221,466, Springfield, Va., Nat. Tech. Information Service (July 1973).

(2) W.E. Davis and Assoc., *National Inventory of Sources and Emissions: Barium, Boron, Copper, Selenium and Zinc, Section V, Zinc,* Report PB-210,680, Springfield, Va., Nat. Tech. Information Service (May 1972).

(3) A.J. Goldberg, *A Survey of Emissions and Controls for Hazardous and Other Pollutants,* Washington, D.C., Environmental Protection Agency (February 1973).

(4) J.W. Patterson and R.A. Minear, *Wastewater Treatment Technology,* 2nd Ed., Chicago, Illinois Institute for Environmental Quality (February 1973).

(5) D.M. Rock, "Hydroxide Precipitation and Recovery of Certain Metallic Ions from Wastewaters," presented at Annual Meeting, American Institute for Chemical Engineers, Chicago (November to December 1970).

(6) F.X. McGarvey, R.E. Tenhoor and R.P. Nevers, "Brass and Copper Industry: Cation Exchangers for Metals Concentration from Pickle Rinse Waters," *Ind. Engr. Chem.,* 44, 534-541 (1952).

(7) P.W. West et al, "Microdetermination of Zinc by Means of Reagent Crayons and the Ring-Oven Technique," *Anal. Chim. Acta* 37:246 (1967).
(8) A. Kawase, "2-(2-Thiazolylazo)phenol Derivatives as Analytical Reagents. VII. Spectrophotometric Determination of Zinc with 1-(2-thiazolylazo)naphthol," *Talanta* 12:195 (1965).
(9) P.W. Nickola, *A Sampler for Recording the Concentration of Airborne Zinc Sulfide on a Real Time Scale,* Pacific Northwest Laboratory, Richland, Washington (1965).
(10) *Air Pollution Measurements of the National Air Sampling Network-Analyses of Suspended Particulate, 1957-1961,* U.S. Dept. of Health, Education, and Welfare, Public Health Service Publication No. 978, U.S. Government Printing Office, Washington, D.C. (1962).
(11) *Air Quality Data from the National Air Sampling Networks and Contributing State and Local Networks,* 1966 ed., U.S. Dept. of Health, Education, and Welfare, National Air Pollution Control Administration Publication No. APTD 68-9, U.S. Government Printing Office, Washington, D.C. (1968).
(12) Lawrence Berkeley Laboratory, *Instrumentation for Environmental Monitoring-Air,* Publ. LBL-1, Vol. 1, Berkeley, Univ. of California (May 1, 1972).
(13) Lawrence Berkeley Laboratory, *Instrumentation for Environmental Monitoring-Water,* Publ. LBL-1, Vol. 2, Berkeley, Univ. of California (February 1, 1973).
(14) C.J. Overbeck and J.J. Hickey; U.S. Patent 3,479,152; November 18, 1969; assigned to Nalco Chemical Co.
(15) C.S. Noll; U.S. Patent 3,706,532; December 19, 1972; assigned to Betz Laboratories, Inc.
(16) Y.C. Athanassiadis, *Air Pollution Aspects of Zinc and Its Compounds,* Report PB-188,072, Springfield, Va., Nat. Tech. Information Service (September 1969).
(17) A.R. McFarland, *Control Technology for Toxic and Hazardous Air Pollutants,* Chicago, Illinois Institute for Environmental Quality (December 1974).
(18) G.E. Barnes, "Disposal and Recovery of Electroplating Wastes," *Jour. Wat. Poll. Control* Fed., 40, 1459-1470 (1968).
(19) N.G. Banerjee and T. Banerjee, "Recovery of Nickel and Zinc from Refinery Waste Liquors: Part 1 - Recovery of Nickel by Electrodeposition," *J. Sci. Industr. Res.* 11B, 77-78 (1952).
(20) O.W. Nyquist and H.R. Carroll, "Design and Treatment of Metal Processing Wastewaters," *Sew. Ind. Wastes,* 31, 941-948 (1959).
(21) E.H.F. Stone, "Treatment of Nonferrous Metal Process Waste at Kynoch Works, Birmingham, England," *Proc. 25th Purdue Industrial Waste Conf.,* pp. 848-865 (1967).
(22) C.P. Sharda and K. Namwannan, "Viscose Rayon Factory Wastes and Their Treatment," *Technology, Sindri,* 3, 58-70 (1966); *Water Poll. Abstr.* 41:No. 1698 (1968).
(23) L.E. Lancy, "An Economic Study of Metal Finishing Waste Treatment," *Plating,* 54, 157-161 (1967).
(24) "New Process Detoxifies Cyanide Wastes," *Env. Sci. Technol.,* 5, 496-497 (1971).
(25) K.L. Saxena and R.N. Chakraborty, "Viscose Rayon Wastes and Recovery of Zinc Therefrom," *Technology, Sindri,* 3, 29-33, (1964); *Wat. Poll. Abstr.,* 41:No. 1699 (1968).
(26) D. Kantawala and H.D. Tomlinson, "Comparative Study of Recovery of Zinc and Nickel by Ion Exchange Media and Chemical Precipitation," *Water Sew. Works,* III, R281-R286 (1964).
(27) P.W. Powers, *How to Dispose of Toxic Substances and Industrial Wastes,* Park Ridge, N.J., Noyes Data Corp. (1976).
(28) R.D. Ross, *Industrial Waste Disposal,* New York, Reinhold Book Company (1968).
(29) J.W. Culotta and W.F. Swanton, *The Role of Evaporation in the Economics of Waste Treatment for Plating Operations,* Plating, 55, 957-961 (1968).

NOTICE

HOW TO DISPOSE OF TOXIC SUBSTANCES AND INDUSTRIAL WASTES 1976

by Philip W. Powers

Environmental Technology Handbook No. 4

This book discusses all recognized ultimate disposal methods in detail and contains a long list of specific recommendations for specific substances plus alternative disposal or recovery methods.

Ultimate waste disposal implies the final disposition of non-degradable, persistent, harmful, and cumulative wastes that may be solid, liquid, or gaseous. Workable solutions to ultimate disposal problems as described in this book include conversion to harmless end products, subsurface storage in ponds or landfills, and disposal in the ocean. This condensed information will enable anyone to establish a sound background for action towards disposal of toxic and hazardous materials with safety.

The five general categories of hazardous wastes are:

1. Toxic Chemical
2. Radioactive
3. Flammable
4. Explosive
5. Toxic Biological

Many toxic or hazardous wastes contain valuable materials. Whenever this is the case, recovery and reuse is one of the most desirable methods of hazardous waste avoidance.

The book is based on government-sponsored reports (often elaborated by highly qualified personnel from industrial companies), U.S. patents, and on pertinent articles in authoritative journals. A partial and condensed table of contents follows here. Chapter headings are given in full whenever possible.

ISBN 0-8155-0615-5 **500 pages**

ENVIRONMENTAL SOURCES AND EMISSIONS HANDBOOK 1975

by Marshall Sittig

Environmental Technology Handbook No. 2

Environmental pollution with its far-reaching effects is only now beginning to be understood. In this practical handbook the various modifications a given pollutant can assume, are traced back to their origins, and the intermedia transfers are discussed fully. Intermedia transfers include direct transfer (removal of a pollutant from one medium and its disposal in another) or indirect (pollution created in another medium and usually in another form by a basic change in a process or industry).

Also, an example of insidious conversion is mercury. Metallic mercury is relatively innocuous, but when organic molecules or dead organisms are present in rivers and lakes, mercury reacts with these molecules to form toxic methylmercury compounds which are excreted very slowly by fish and man.

The implications are formidable: Chemical, biochemical, technological, statistical, mineralogical, zoological, pharmacological, medical, legal, legislative and public health involvements are all too obvious.

This volume therefore surveys the origins of both air and water pollution. Significant emphasis is placed on altered pollution which can result when an air pollutant is intentionally or accidentally transferred to an aqueous stream or vice versa.

Sometimes pollutants react with each other or initiate deleterious chain reactions. Yet this same reactivity may be the key to efficient removal: by adding certain chemicals or flocculants or microorganisms, toxic substances may be carried off physically or converted not only to nontoxic substances, but also into useful products.

The why and where is in this book, commensurate with present-day technology, without becoming too sophisticated or losing sight of the all important economic considerations.

Operators or potential operators of processes which produce pollutants will find this volume quite useful. Besides discussing all sorts of pollution sources, it should help to define industry-wide emission practices and magnitudes.

A partial and condensed table of contents follows here. The book contains 200 subject entries and 382 tables with figures and graphs. Descriptions are mostly based on studies conducted by industrial and engineering firms or university research teams under the auspices of various government agencies. As is the case with other handbooks in this series the entries are arranged in an alphabetical and encyclopedic fashion:

Aluminum Production, Primary
Aluminum Industry, Secondary
Aluminum Sulfate Manufacture
Ammonia Manufacture
Ammonium Nitrate Manufacture
Ammonium Sulfate Production
Asbestos Products Industry
Asphalt Roofing Manufacture
Asphaltic Concrete Plants (Asphalt Batching)
Automobile Body Incineration
Automotive Vehicle Operation
Bauxite Refining
Beet Sugar Manufacture
Boat & Ship Operation
Brass & Bronze Ingot Prod. (Secondary Copper Industry)
Brick & Related Clay Products Manufacture
Builders Paper Manufacture
Calcium Carbide Manufacture
Calcium Chloride Manufacture
Carbon Black Manufacture
Castable Refractory Manufacture
Cement Manufacture
Ceramic Clay Manufacture
Charcoal Manufacture
Chlor-Alkali Manufacture
Clay & Fly Ash Sintering
Coal Cleaning
Coal Combustion, Anthracite
Coal Combustion, Bituminous
Coffee Roasting
Coke (Metallurgical) Manufacture
Combustion Sources
Concrete Batching
Conical Burners
Copper (Primary) Smelting
Cotton Ginning
Dairy Industry
Dry Cleaning
Electroplating
Explosives Manufacture
Feedlots
Fermentation Industry
Ferroalloy Production
Fertilizer Industry
Fiberglass Manufacture
Fish Processing
Frit Manufacture
Fruit & Vegetable Industry
Fuel Oil Combustion
Glass Manufacture
Gold & Silver Mining & Production
Grain & Feed Mills & Elevators
Gypsum Manufacture
Hydrochloric Acid Manufacture
Hydrofluoric Acid Manufacture
Hydrogen Peroxide Manufacture
Incineration, Municipal Refuse
Inorganic Chemical Industry
Iron Foundries
Iron & Steel Mills
Lead Smelting, Primary
Lead Smelting, Secondary
Leather Industry
Lime Manufacture
Liquefied Petroleum Gas Combustion
Magnesium Smelting, Secondary
Meat Industry
Mercury Mining & Production
Mineral Wool Manufacturing
Natural Gas Combustion
Nitric Acid Manufacture
Nitrogen Fertilizer Manufacture
Nonferrous Metals Industry
Open Burning
Orchard Heating
Organic Chemicals Manufacture
Paint & Varnish Manufacture
Paperboard Manufacture
Perlite Manufacture
Pesticide Manufacture
Petroleum Marketing & Transportation
Petroleum Refining
Petroleum Storage
Phosphate Chemicals Manufacture
Phosphate Fertilizer Manufacture
Phosphate Rock Processing
Phosphoric Acid Manufacture
Phosphorus Manufacture
Phosphorus Oxychloride Manufacture
Phosphorus Pentasulfide Manufacture
Phosphorus Pentoxide Manufacture
Phosphorus Trichloride Manufacture
Phthalic Anhydride Manufacture
Plastics Industry
Potassium Chloride Production
Potassium Dichromate Manufacture
Potassium Sulfate Manufacture
Poultry Processing
Power Plants
Printing Ink Manufacture
Pulp & Paper Industry
Railway Operation
Refractory Metal (Mo, W) Production
Rubber Industry, Synthetic
Sand & Gravel Processing
Seafood Processing Industry
Sewage Sludge Incineration
Soap & Detergent Manufacture
Sodium Bicarbonate Manufacture
Sodium Carbonate Manufacture
Sodium Chloride Manufacture
Sodium Dichromate Manufacture
Sodium Metal Manufacture
Sodium Silicate Manufacture
Sodium Sulfite Manufacture
Starch Manufacture
Stationary Engine Operation
Steel Foundries
Stone Quarrying & Processing
Sugar Cane Processing
Sulfur Production
Sulfuric Acid Manufacture
Surface Coating Application
Synthetic Fiber Manufacture
Terephthalic Acid Manufacture
Textile Industry
Timber Industry
Tire & Inner Tube Manufacture
Titanium Dioxide Manufacture
Urea Manufacture
Wood Veneer & Plywood Products Industry
Wood Waste Combustion
Zinc Processing, Secondary
Zinc Smelting, Primary

ISBN 0-8155-0568-X **521 pages**

CORROSION RESISTANT MATERIALS HANDBOOK 1976

Third Edition

by Ibert Mellan

This famous book, first published in 1966 and now already in its third, greatly enlarged and completely revised edition, will help you cut losses due to corrosion by enabling you to choose the proper *commercially available* corrosion resistant materials for your particular purpose.

The great value of this outstanding reference work lies in the extensive cross-indexing of thousands of substances. The 151 tables are arranged by types of **corrosion resistant materials.** The **Corrosive Materials Index** is organized by *corrosive chemicals* and other *corrosive substances:* it refers you to specific recommendations in the tables. New in this edition are a separate **Trade Names Index** and a listing of **Company Names and Addresses.**

For the first time there appears also a group of 13 tables comparing the respective anticorrosive merits of commercial engineering and construction materials essential to industry.

The tables in this book represent selections from manufacturers' literature made by the author at no cost to, nor influence from, the makers or distributors of these materials.

Contents:

ISBN 0-8155-0628-7 **665 pages**

POLLUTION CONTROL IN METAL FINISHING 1973

by Michael R. Watson

Pollution Technology Review No. 5

Electroplating and other metal finishing waste streams are significant contributors to water pollution, either directly by their content of toxic and corrosive materials, such as cyanide, acids, and metals, or indirectly through the deleterious effect these effluents exert on sewage treatment systems.

Federal, state and municipal regulations fixing the allowable concentrations have been established and enforcement may be expected to become increasingly strict.

There is ample technology available for treating chromium and cyanide rinse waters to any required degree of detoxification. The problems facing the plater are:

1. Volume and composition of the rinse waters.
2. EPA requirements for an acceptable waste effluent.
3. Capital outlay for treatment facilities.
4. Operating costs for the treatment processes.
5. Disposal costs of unusable recovered materials.
6. Credits from the recovery of metals and decrease in sewage fees.

Solutions to these problems are afforded by this book which is based on various government surveys and reports with practical examples detailed from late U.S. patents. A partial and condensed table of contents follows. Chapter headings are given here, followed by examples of important subtitles.

ISBN 0-8155-0505-1 **295 pages**